D1256222

Applied Mathematical Sciences

Volume 71

Editors

F. John J.E. Marsden L. Sirovich

Advisors

M. Ghil J.K. Hale J. Keller

K. Kirchgässner B. Matkowsky

J.T. Stuart A. Weinstein

Applied Mathematical Sciences

1. *John:* Partial Differential Equations, 4th ed.
2. *Sirovich:* Techniques of Asymptotic Analysis.
3. *Hale:* Theory of Functional Differential Equations, 2nd ed.
4. *Percus:* Combinatorial Methods.
5. *von Mises/Friedrichs:* Fluid Dynamics.
6. *Freiberger/Grenander:* A Short Course in Computational Probability and Statistics.
7. *Pipkin:* Lectures on Viscoelasticity Theory.
9. *Friedrichs:* Spectral Theory of Operators in Hilbert Space.
11. *Wolovich:* Linear Multivariable Systems.
12. *Berkovitz:* Optimal Control Theory.
13. *Bluman/Cole:* Similarity Methods for Differential Equations.
14. *Yoshizawa:* Stability Theory and the Existence of Periodic Solution and Almost Periodic Solutions.
15. *Braun:* Differential Equations and Their Applications, 3rd ed.
16. *Lefschetz:* Applications of Algebraic Topology.
17. *Collatz/Wetterling:* Optimization Problems.
18. *Grenander:* Pattern Synthesis: Lectures in Pattern Theory, Vol I.
20. *Driver:* Ordinary and Delay Differential Equations.
21. *Courant/Friedrichs:* Supersonic Flow and Shock Waves.
22. *Rouche/Habets/Laloy:* Stability Theory by Liapunov's Direct Method.
23. *Lamperti:* Stochastic Processes: A Survey of the Mathematical Theory.
24. *Grenander:* Pattern Analysis: Lectures in Pattern Theory, Vol. II.
25. *Davies:* Integral Transforms and Their Applications, 2nd ed.
26. *Kushner/Clark:* Stochastic Approximation Methods for Constrained and Unconstrained Systems
27. *de Boor:* A Practical Guide to Splines.
28. *Keilson:* Markov Chain Models—Rarity and Exponentiality.
29. *de Veubeke:* A Course in Elasticity.
30. *Sniatycki:* Geometric Quantization and Quantum Mechanics.
31. *Reid:* Sturmian Theory for Ordinary Differential Equations.
32. *Meis/Markowitz:* Numerical Solution of Partial Differential Equations.
33. *Grenander:* Regular Structures: Lectures in Pattern Theory, Vol. III.
34. *Kevorkian/Cole:* Perturbation methods in Applied Mathematics.
35. *Carr:* Applications of Centre Manifold Theory.
36. *Bengtsson/Ghil/Källén:* Dynamic Meteorology: Data Assimilation Methods.
37. *Saperstone:* Semidynamical Systems in Infinite Dimensional Spaces.
38. *Lichtenberg/Lieberman:* Regular and Stochastic Motion.
39. *Piccini/Stampacchia/Vidossich:* Ordinary Differential Equations in R^n.
40. *Naylor/Sell:* Linear Operator Theory in Engineering and Science.
41. *Sparrow:* The Lorenz Equations: Bifurcations, Chaos, and Strange Attractors.
42. *Guckenheimer/Holmes:* Nonlinear Oscillations, Dynamical Systems and Bifurcations of Vector Fields.
43. *Ockendon/Tayler:* Inviscid Fluid Flows.
44. *Pazy:* Semigroups of Linear Operators and Applications to Partial Differential Equations.
45. *Glashoff/Gustafson:* Linear Optimization and Approximation: An Introduction to the Theoretical Analysis and Numerical Treatment of Semi-Infinite Programs.
46. *Wilcox:* Scattering Theory for Diffraction Gratings.
47. *Hale et al.:* An Introduction to Infinite Dimensional Dynamical Systems—Geometric Theory.
48. *Murray:* Asymptotic Analysis.
49. *Ladyzhenskaya:* The Boundary-Value Problems of Mathematical Physics.
50. *Wilcox:* Sound Propagation in Stratified Fluids.
51. *Golubitsky/Schaeffer:* Bifurcation and Groups in Bifurcation Theory, Vol. I.
52. *Chipot:* Variational Inequalities and Flow in Porous Media.
53. *Majda:* Compressible Fluid Flow and Systems of Conservation Laws in Several Space Variables.
54. *Wasow:* Linear Turning Point Theory.

Donald E. Catlin

Estimation, Control, and the Discrete Kalman Filter

With 13 Illustrations

Randall Library UNC-W

Springer-Verlag
New York Berlin Heidelberg
London Paris Tokyo

Donald E. Catlin
Department of Mathematics and Statistics
University of Massachusetts
Amherst, MA 01003
USA

Editors

F. John
Courant Institute of
Mathematical Sciences
New York University
New York, NY 10012
USA

J.E. Marsden
Department of
Mathematics
University of California
Berkeley, CA 94720
USA

L. Sirovich
Division of
Applied Mathematics
Brown University
Providence, RI 02912
USA

Mathematics Subject Classification (1980): 93C55, 93E11, 93E14, 93E20

Library of Congress Cataloging-in-Publication Data
Catlin, Donald E.
Estimation, control, and the discrete Kalman filter / Donald E. Catlin.
p. cm. — (Applied mathematical sciences ; v. 71)
Bibliography: p.
Includes index.
1. Kalman filtering. 2. Estimation theory. 3. Control theory.
I. Title. II. Series: Applied mathematical sciences (Springer
-Verlag New York Inc.) ; v. 71.
QA1.A647 vol. 71
[QA402.3]
510 s—dc19
[629.8′312] 88-20031

Printed on acid-free paper.

© 1989 by Springer-Verlag New York Inc.
All rights reserved. This work may not be translated or copied in whole or in part without the written permission of the publisher (Springer-Verlag, 175 Fifth Avenue, New York, NY 10010, USA), except for brief excerpts in connection with reviews or scholarly analysis. Use in connection with any form of information storage and retrieval, electronic adaptation, computer software, or by similar or dissimilar methodology now known or hereafter developed is forbidden.
The use of general descriptive names, trade names, trademarks, etc. in this publication, even if the former are not especially identified, is not to be taken as a sign that such names, as understood by the Trade Marks and Merchandise Marks Act, may accordingly be used freely by anyone.

Printed and bound by R.R. Donnelley & Sons, Harrisonburg, Virginia.
Printed in the United States of America.

9 8 7 6 5 4 3 2 1

ISBN 0-387-96777-X Springer-Verlag New York Berlin Heidelberg
ISBN 3-540-96777-X Springer-Verlag Berlin Heidelberg New York

QA
402.3
.C37
1989

To my mother,
Marian L. Catlin,
and the memory of my father,
F.H. Catlin

Preface

In 1960, R.E. Kalman published his celebrated paper on recursive minimum variance estimation in dynamical systems [14]. This paper, which introduced an algorithm that has since been known as the discrete Kalman filter, produced a virtual revolution in the field of systems engineering. Today, Kalman filters are used in such diverse areas as navigation, guidance, oil drilling, water and air quality, and geodetic surveys. In addition, Kalman's work led to a multitude of books and papers on minimum variance estimation in dynamical systems, including one by Kalman and Bucy on continuous time systems [15]. Most of this work was done outside of the mathematics and statistics communities and, in the spirit of true academic parochialism, was, with a few notable exceptions, ignored by them. This text is my effort toward closing that chasm. For mathematics students, the Kalman filtering theorem is a beautiful illustration of functional analysis in action; Hilbert spaces being used to solve an extremely important problem in applied mathematics. For statistics students, the Kalman filter is a vivid example of Bayesian statistics in action.

The present text grew out of a series of graduate courses given by me in the past decade. Most of these courses were given at the University of Massachusetts at Amherst. During the latter part of 1984, however, I had the privilege of presenting this material in a nonacademic setting, namely, to the Advanced Systems Department of The Analytic Sciences Corporation (TASC) of Reading, Massachusetts. I have, therefore, had "student feedback" from graduate students in mathematics, statistics, and engineering, as well as from practicing industrial scientists. This information has not only been interesting and fun to collect but has been of inestimable value to me.

This text is specifically designed as a one-semester text for students in mathematics, statistics, or engineering who have completed at least a one-semester course in real analysis, meaning measure, and integration. This having been said, I know from past experience that there will be readers who do not fit this mold. For this reason, I have provided a rather extensive series of appendices on various topics in algebra and analysis. The reader lacking in sufficient analysis could learn enough from the appendices to read the main text, although not without considerable effort. In addition, it is convenient for me (and presumably for the student) to be able to reference a particular result I need without sending the student on interminable trips to the library.

The text is done in four parts. The first part is Chapter 1. Here, I develop the basic theoretical probability that I will need. The next part, Chapters 2 and 3, explains how rational beliefs are used to render a difficult practical problem into a tractable mathematical problem and how the theory developed in Chapter 1 is useful in this regard. The third, and main part of the text, is the development of the discrete Kalman filter. This constitutes Chapters 4–7 and, with the first three chapters, may well be enough material for a one-semester graduate course. The last two chapters are on stochastic control and smoothing, two very important applications of Kalman filtering techniques. I should like to say that the choice of these two topics was made by virtue of the importance and elegance of the results and *not* by the aesthetic beauty of the development. Both of these topics involve tedious and involved matrix algebra, and might well be left for the student to read at his or her leisure.

If one designs a one-semester text, then naturally there will be certain omissions—choices must be made. My text is no exception. For example, although I do mention controllability, there is no mention of the dual concept, observability. In the stochastic world, neither controllability nor observability have the dramatic interpretation that they do in deterministic systems. In developing the quadratic tracking problem, I begin with a deterministic system having perfectly observable states. Here, observability is a triviality but controllability is meaningful. By the time in the development I consider imperfect measurements, the discussion is already stochastic and I really do not need the notion of observability, hence I omit it. Of course, had I decided to discuss stability of Kalman filters, then both observability and controllability would have had to be introduced. The inclusion of such a chapter was tempting and may well find its way into some future revision of this work; but, I am well aware that in spite of all the theory involving stability, unstable Kalman filters are sometimes implemented in practice, and they seem to work just fine. So, I took the easy way out and chose to not discuss a topic that I do not fully comprehend. So it went, with these and other topics.

The most glaring omission is, of course, that I do not treat continuous time systems. In fact, I do have a set of notes on continuous systems, based on the generalized processes of Gelfand and Shilov, and these may comprise a sequel to the present text. However, I chose to do discrete time systems because they were historically the first in which Kalman filtering was done, they are the filters that are actually implemented in practice (I have never heard of anyone actually implementing a continuous Kalman filter), and they seem to me to be the logical first step in developing state-space filtering theory.

There are several special features of this text that I would like to point out to the reader. Chapter 3 contains a derivation of the maximum entropy principle assuming differentiability of the entropy function; ostensibly a bit restrictive, but it is fast. In Chapter 5, I give an overview and classification of general minimum variance estimation problems. In Chapters 5, 6, and 7, while doing the filter derivations, I keep the reader cognizant of the often blurred distinction between linear and affine estimates. Finally, Chapter 7 contains some new material about initializing Kalman filters using generalized Fisher estimates.

There are a number of individuals to whom I should express my sincere gratitude and appreciation. First, I would like to thank Professor Dave Foulis who taught me to do mathematics and gave me an appreciation for the interplay between so-called pure and applied mathematics. Next, my thanks to and my admiration for the late Professor Charles Randall. Charley, as he liked to be called, taught me more statistics over a glass of beer than I ever learned in a classroom. It was Charley who made me realize that statistics is a vibrant, living discipline that is as much rooted in philosophy as it is in mathematics. This realization has served me well in both academic studies and industrial consulting. Special thanks are given to my former colleague, Charles Hutchinson, now Dean of the Thayer School of Engineering at Dartmouth College. When he was a professor of electrical engineering at the University of Massachusetts, he and I ran many joint seminars and courses wherein we both enhanced our understanding of state-space analysis and its ramifications.

In the past decade, I have been fortunate to have had some industrial experience in Kalman filtering and other related topics. In this regard, I would like to thank Dr. Steven Alter, Dr. Joseph D'Appolito, Jack Fagan, Dr. James Foltz, Bob Geddes, Dr. Michael Geyer, Dick Healy, Dr. John Ladik, Bill O'Halloran, Paul Olinski, and Ron Warren, all of whom, at one time or another, have been industrial colleagues of mine at TASC and have provided me with "hands on" experience. Very special thanks go to Dr. Ray Nash, Dr. Thomas Mottl, and especially Dr. Robert Pyle for making the arrangements that enabled me to teach this material at TASC in 1984, an opportunity for which I am very, very grateful.

I would also like to thank all of my students during the past decade. There are too many to mention all of them, but each helped me in his or her own way. One of them, Gary Noseworthy, read a good bit of this manuscript and made very helpful suggestions—special thanks to him.

Finally, my very special thanks go to Peg Bombardier who typed the manuscript with accuracy, speed, and an abundance of good humor. Words cannot really express my gratitude to her.

Amherst, Massachusetts
October 1987

Donald E. Catlin

Contents

1

Basic Probability

1.1 Definitions

We begin by introducing the notion of a probability measure, and then we will discuss the intuitive interpretation of this definition.

1.1.1. **Definition.** Let $(\Omega, \mathcal{E}, P)$ be a *measure space,* that is, Ω is a set, P is a countably additive measure on Ω, and $\mathcal{E}$ is the collection of P-measurable sets. If it is the case that

$$P(\Omega) = 1, \tag{1.1-1}$$

then P is called a ***probability measure,*** the measurable sets $\mathcal{E}$ are called ***events,*** the points in Ω are called ***outcomes,*** and the triple $(\Omega, \mathcal{E}, P)$ is called a ***probability space.***

Recall from the theory of real variables that the following properties must hold for the measure P (see Appendix A)

(1) $P(\phi) = 0$

(2) For each $A \in \mathcal{E}$, $\quad P(A) \geq 0$

(3) If $\{A_i \mid i = 1, 2, \ldots\} \subset \mathcal{E}$, then if $i \neq j \Rightarrow A_i \cap A_j = \phi$, then

$$P\left(\bigcup_{i=1}^{\infty} A_i\right) = \sum_{i=1}^{\infty} P(A_i). \tag{1.1-2}$$

We can easily construct a probability space as follows.

1.1.2. **Example.** Let $\{p_i\}$ be a finite or countably infinite collection of nonnegative real numbers such that

$$\sum_i p_i = 1.$$

Define $\Omega = \{p_i\}$, and if $E \subset \Omega$, define

$$P(E) = \sum_{p_i \in E} p_i; \quad P(\phi) = 0.$$

Then P is a probability measure and in this case $\mathcal{E} = 2^{\Omega}$, the set of all subsets of Ω. We leave it to the interested reader to check that relations (1.1-2) hold.

The probability space in Example 1.1.2 is called a discrete probability space, since the measure P is concentrated on a discrete set of points. Although our primary concern will be continuous probability measures (definition to come later), discrete probability spaces will be useful to us for conceptualizing certain ideas in that simple setting. This practice must be done with care, however, since discrete spaces enjoy certain properties that general probability spaces do not; for instance, in Example 1.1.2, we have $\mathcal{E} = 2^{\Omega}$, whereas in general this is untrue (see Appendix A).

If $E \in \mathcal{E}$, then $P(E)$ is read "the probability that an outcome is in event E" or simply "the probability that event E occurs." Most people have some vague notion about what such a statement means, their interpretation generally involving synonyms such as chance, odds, uncertainty, prevalence, risk, and expectancy. It would be nice if we could, at this point, tell the reader that we are going to give $P(E)$ a precise meaning. Unfortunately, we are unable to do this, and despite vigorous assertions to the contrary, no one else seems to be able to do so either. Let us elaborate on this for a moment.

On the one hand, probability theory is simply a branch of mathematics. More specifically, it is a study of measure spaces with the constraint (1.1-1). On the other hand, if probability theory is to be useful, it must have some viable interpretation. The branch of human endeavor devoted to making this interpretation is known as statistics, and its practitioners, statisticians. We could generalize this a bit and define statistics as the science (or art—take your pick) of drawing plausible or reasonable inferences from information containing some degree of uncertainty. The generalization here is the introduction of the words plausible and reasonable. Again, there is a certain vagueness in our language: what are we willing to accept as plausible and reasonable? Clearly, we cannot begin to address this issue until we know what we mean by the phrase "some degree of uncertainty," and, of course, this is just another way of asking what is meant by $P(E)$. Statisticians build elaborate theories using very fancy, very deep mathematics (in fact you are going to see such a theory), but the ultimate interpretation of their theory and its reasonableness lie rooted in their beliefs as to the meaning of probability, an inherently philosophical issue.

There are at least eight distinct interpretations of probability of which we are aware. A complete discussion of these would take us too far afield (and would be complicated by the fact that there are some interpretations we simply do not understand). Thus, we will content ourselves with the following discussion (which, at least, we do understand).

There are two extreme (our word) interpretations of probability, the so-called objective school and the subjective school. The objective school holds that probabilities are long-run relative frequencies. This means that one should compute a probability by taking the number of favorable outcomes of an experiment, divide it by the total number of trials, and then take the limit as the number of trials becomes large. The first part of the preceding sentence is the interpretation of "relative frequency," the second part is the meaning of "long run." Most statisticians have no trouble accepting the notion of relative frequency as just stated. However, many have trouble with the concept of "long run" as we have stated it. The problem is, simply, that it is physically impossible to experimentally take the limit of a sequence of trials; there just is not enough time. The philosopher/statistician John Maynard Keynes couched the criticism in a rather humorous vein when he said, "In the long run we will all be dead." Nevertheless, the belief that the limit of a finite relative frequency exists, whether or not we can physically construct it, seems to be a rational belief. For, if one does a large number of trials and plots relative frequency versus the number of trials, the resulting curve does suggest that the limit exists. The concept is also self-consistent in that there are theorems (the laws of large numbers for example) asserting that this limiting behavior is true for existing probability measures. Still, we must emphasize that the interpretation of probability as long-run relative frequency is in the final analysis a belief.

The other extreme, the subjective school of probability, holds that probabilities are subjective assignments based on combining rational thought with available information. We should point out at once that there are subjective statisticians who would argue with our "definition." The subjective school is really a number of different philosophical points of view, and some of the differences between these viewpoints are quite subtle; some, quite frankly, we do not even understand. Some subjective probabilists interpret probabilities as betting rates. This interpretation of probability requires that one establish (payoff) odds on a set of outcomes in such a way that the person setting the odds be willing to take either side of the bet. Professional odds-makers (bookies) assign probabilities in precisely this fashion. Others simply interpret probabilities as degrees of credibility or degrees of belief associated with alternative outcomes. Whatever the interpretation, it seems to us that subjective probability is based on the belief that if two rational human beings are faced with the same information, they will arrive at the same assignment of probabilities. Here, of course, "rational human being" effectively means a rational or reasonable inference procedure, and "information" means data in its broadest sense. As vague and imprecise as this notion seems to be, there have been some remarkable efforts, and results thereof, toward making the notion precise. The real breakthrough was done by Claude Shannon [25], who gave a precise definition of the measure of information contained in an assignment of probabilities. The rational inference procedure one uses in conjunction with the idea of information is to assign probabilities in such a way that the resulting distribution contains no more information than is inherent in the data. This principle can be stated quite precisely and is known as the *maximum entropy principle.* For an especially lucid explanation of this idea vis-a-vis subjective probabilities, the reader can refer to the work of E. T. Jaynes, especially the papers in References [12] and [13]. In Chapter 3, we will give a rapid treatment of the maximum entropy principle and thus will delay any detailed discussion of this idea until then.

In addition to the strictly objective and strictly subjective schools of probability previously described, there is another school known as the Bayesian school. The Bayesian school is polytheistic in that it regards both the objective and subjective viewpoints as valid. A Bayesian statistician assigns probability distributions by first choosing a prior distribution based on some form of subjective reasoning and then adjusts his or her beliefs as subsequent data are accumulated. The word Bayesian is derived from the fact that the inference procedure, often used to adjust prior beliefs in the presence of data, utilizes a formula known as Bayes rule. Bayes rule, in turn, is so named because it has a formal resemblance to Bayes Theorem from elementary probability theory. A detailed understanding of Bayes rule is really a peripheral issue as far as these lectures are concerned and hence will not be discussed. What is *not* a peripheral issue as far as these lectures are concerned is the philosophical notion of Bayesian estimation. In the course of these lectures, we are going to develop a particular estimation technique known as the Kalman filter. The Kalman filter is an inherently Bayesian procedure. Rather than recursively assigning a probability distribution as described, the Kalman filter addresses the seemingly less ambitious problem of recursively estimating the values taken on by a certain random vector (definition to be given shortly), without estimating the underlying probability distribution. As we will see, this procedure solves a huge class of extremely important and difficult problems.

1.1.3. **Definition.**

(a) By a *random variable* on Ω, we mean a real valued measurable function from Ω to R (see Appendix C).

(b) By a *random vector* on Ω, we mean a measurable function from Ω to R^n.

1.1.4. **Definition.**

(a) Let $\mathcal{F}$ denote the collection of closed subsets of R^n. By the collection of *Borel sets* in R^n, denoted B_n, we mean the minimal (with respect to inclusion) σ field containing $\mathcal{F}$.

(b) Let $\mathbf{X} : \Omega \to \mathrm{R}^n$ be a random vector. We then define a function

$$P_{\mathbf{X}} : B_n \to \mathrm{R}$$

via

$$P_{\mathbf{X}}(B) \triangleq P(\mathbf{X}^{-1}(B)).$$

Here, $P_{\mathbf{X}}$ is called the probability measure on R^n induced by $\mathbf{X}$.

Note that $\omega \in \mathbf{X}^{-1}(B)$ if and only if $\mathbf{X}(\omega) \in B$. For this reason, $P_{\mathbf{X}}(B)$ is interpreted as the probability that the random vector X takes on values in the Borel set B. This interpretation certainly justifies the introduction of the construction in 1.1.4(b); but why the restriction to Borel sets? We will address this issue momentarily.

It has been said that a random variable is like the Holy Roman Empire—it wasn't holy, it wasn't Roman, and it wasn't an empire. A random variable is neither random nor variable, it is simply a function. The values it takes on, however, are both random and variable because they are generated by $(\Omega, \mathcal{E}, P)$. Before we proceed, it will be helpful to illustrate this with a simple example.

1.1.5. **Example.** Let $D = \{1, 2, 3, 4, 5, 6\}$. Here, D will represent the outcome resulting from a single roll of a die. If we next consider the outcomes Ω resulting from a single roll of a pair of dice, it is easily seen that

$$\Omega = D \times D.$$

Here, of course, there are 36 outcomes of the form (n_1, n_2), where n_1 is the result shown on die 1 and n_2 is the result shown on die 2. If both dice are considered fair, one assigns a probability measure P on Ω such that

$$P((n_1, n_2)) = \frac{1}{36}. \tag{1.1-3}$$

Note that the assignment in (1.1-3) is a naive instance of the maximum entropy principle. We assign each point in Ω the same value since there is no information available that suggests we do otherwise. In fact, the word fair indicates that probabilities are to be assigned in exactly this fashion.

In many games involving the roll of a pair of dice, one is not interested in the specific outcome on each die. Rather, one is interested in the total number of spots on both. We can represent this mathematically by introducing the random variable

$$X : \Omega \to \mathrm{R}$$

defined by

$$X(x, y) = x + y. \tag{1.1-4}$$

It is easily seen that the range of X is concentrated on the integers $\{2, 3, \ldots, 11, 12\}$ and hence by 1.1.4(b), P_X is concentrated on this set. Let us calculate $P_X(\{6\})$. By 1.1.4(b), we have

$$\begin{aligned} P_X(\{6\}) &= P(X^{-1}(\{6\})) \\ &= P((1,5),\ (2,4),\ (3,3),\ (4,2),\ (5,1)) \\ &= \frac{5}{36}. \end{aligned}$$

This example clearly illustrates the reason for introducing the construction in 1.1.4(b). Very soon we are going to prove that P_X has the properties of a probability measure. However, as mentioned earlier, this measure is restricted to the Borel sets in R^n, and we have yet to address the reason for doing this. We will explain this now.

In order to illustrate the necessity for introducing the Borel sets, it is necessary to have at our disposal a probability space $(\Omega, \mathcal{E}, P)$ and a function

$$X : \Omega \rightarrow \mathrm{R}$$

such that

(1) X is a measurable function, that is, a random variable,

(2) there is a measurable set $M \subset \mathrm{R}$ such that M is not a Borel set, and

(3) $X^{-1}(M)$ is not a measurable set.

The previous dice example is useless for making such a construction because in that case every subset is measurable. A bit of thought will convince the reader that any example satisfying (1), (2), and (3) will involve some moderately sophisticated constructions. Because of this, we will relegate the construction of such an example to Appendix C and proceed with our explanation under the assumptions of (1), (2), and (3). The reader might also consult reference [11].

We have already noted that P_X measures the probability with which the random variable X takes on certain values in R. Keeping in mind the way one generally records experimental data, that is, a nominal value plus or minus an error, it seems reasonable to us that one would like to have $P_X([\alpha, \beta])$ make sense for any closed interval $[\alpha, \beta]$. In other words, every closed interval must be P_X measurable. It follows, of course, that if P_X is indeed a measure, then all of the sets in the smallest σ field containing the closed intervals are also P_X measurable. Now, the closed intervals form a (topological) basis for the closed sets in R, so that each closed set must be P_X measurable. Finally, it follows that the Borel sets must be a collection of P_X-measurable sets.

Very well then, if we want closed intervals to be P_X measurable, then all Borel sets must be P_X measurable. Why not just declare that all Lebesgue measurable subsets of R are to be P_X measurable and be done with it? This is where the aforementioned example is useful. Let X be the random variable described in item (1) and let M be the measurable set described in item (2). If we try to form $P(X^{-1}(M))$, we see at once that by item (3) we have a problem, for $X^{-1}(M)$ is not measurable. Thus, M is not P_X measurable.

In summary, we have shown that B_1 is the smallest σ field of P_X-measurable sets. However, by the above example, there is no assurance that a non-Borel set is P_X measurable so that P_X is also the largest σ field that we can safely assume contains all P_X-measurable sets. Thus, B_1 is just the right size!

We have tried to show why P_X is so defined and why it is restricted to the σ field of Borel sets. The first point is essential to our work. The second is simply a technical point; we need Borel sets so that the mathematics is correct. But in point of fact, it is very doubtful that any practitioner of the theory we develop here would ever encounter a non-Borel set in practice.

Very well, let's continue with the development of our theory.

1.1.6. **Theorem.**

(a) *Let* $\mathbf{X} : \Omega \to \mathrm{R}^n$ *be any function and let* $X_i \triangleq \pi_i \circ \mathbf{X}$, *where* π_i *is the ith projection function on* R^n $[\pi_i(\alpha_1, \alpha_2, \ldots, \alpha_n) = \alpha_i]$. *Then* $\mathbf{X}$ *is a random vector if and only if* X_i *is a random variable for each* i.

(b) $P_{\mathbf{X}}$ *as defined in* 1.1.4 *is a well-defined probability measure on* (R^n, B_n).

Proof. (a) Let $\mathbf{X}$ be a random vector, U an open subset of R. Then,

$$X_i^{-1}(U) = (\pi_i \circ \mathbf{X})^{-1}(U) = \mathbf{X}^{-1}(\pi_i^{-1}(U)).$$

However, π_i is continuous, so $\pi_i^{-1}(U)$ is open in R^n. Hence, $\mathbf{X}^{-1}(\pi_i^{-1}(U)) \in \mathcal{E}$ as required. Thus, X_i is a random variable.

Conversely, suppose each X_i is a random variable. It then suffices to show that $\mathbf{X}^{-1}(U) \in \mathcal{E}$ for any open set $U \subset \mathrm{R}^n$. But since $\mathbf{X}^{-1}$ preserves unions, it is sufficient to assume that U is a basic open set. Now, a common (topological) basis for the open sets in R^n consists of rectangular open sets, that is, sets of the form

$$U = U_1 \times U_2 \times \cdots \times U_n, \tag{1.1-5}$$

where each U_i is open in R. Now,

$$U = U_1 \times \cdots \times U_n = \pi_1^{-1}(U_1) \cap \pi_2^{-1}(U_2) \cap \cdots \cap \pi_n^{-1}(U_n),$$

the first equality being (1.1-5) and the second being obtained from elementary set theory. Thus,

$$\begin{aligned}
\mathbf{X}^{-1}(U) &= X^{-1}\left(\bigcap_{i=1}^{n} \pi_i^{-1}(U_i)\right) \\
&= \bigcap_{i=1}^{n} X^{-1}(\pi_i^{-1}(U_i)) \\
&= \bigcap_{i=1}^{n} (\pi_i \circ X)^{-1}(U_i) \\
&= \bigcap_{i=1}^{n} X_i^{-1}(U_i).
\end{aligned}$$

However, since each X_i is measurable, $X_i^{-1}(U_i)$ is in $\mathcal{E}$ hence so is the above intersection, that is, $\mathbf{X}^{-1}(U) \in \mathcal{E}$.

(b) From the discussion preceding this theorem, we know that for any Borel set B, $P_{\mathbf{X}}(B)$ makes sense. Thus, assuming that P is a probability measure on Ω, we will show that (P_X, B_n, R^n) is a probability space, that is, $P_{\mathbf{X}}$ satisfies (1.1-2) on the Borel sets. We will also show that (1.1-1) holds.

(1) $P_{\mathbf{X}}(\phi) = P(\mathbf{X}^{-1}(\phi)) = P(\phi) = 0$.

(2) $P_{\mathbf{X}}(B) = P(\mathbf{X}^{-1}(B)) \geq 0$.

(3) Let $\{B_i, i = 1, 2, \ldots\}$ be a disjoint family of Borel sets. Then, $\{\mathbf{X}^{-1}(B_i),\ i = 1, 2, \ldots\}$ is a disjoint family in $\mathcal{E}$ since

$$\mathbf{X}^{-1}(B_i) \cap \mathbf{X}^{-1}(B_j) = \mathbf{X}^{-1}(B_i \cap B_j) = \mathbf{X}^{-1}(\phi) = \phi$$

whenever $i \neq j$. Thus,

$$\begin{aligned} P_{\mathbf{X}}\left(\bigcup_{i=1}^{\infty} B_i\right) &= P\left(\mathbf{X}^{-1}\left(\bigcup_{i=1}^{\infty} B_i\right)\right) \\ &= P\left(\bigcup_{i=1}^{\infty} \mathbf{X}^{-1}(B_i)\right) \\ &= \sum_{i=1}^{\infty} P(\mathbf{X}^{-1}(B_i)) \\ &= \sum_{i=1}^{\infty} P_{\mathbf{X}}(B_i). \end{aligned}$$

The third step here uses the fact that P is a measure.

Finally, we note that (1.1-1) holds since

$$P_{\mathbf{X}}(\mathrm{R}^n) = P(\mathbf{X}^{-1}(\mathrm{R}^n)) = P(\Omega) = 1. \qquad \square$$

Before leaving this section, it should be pointed out that one often sees the notation $P(X \in B)$ instead of $P_X(B)$. The reason for such notation is simply that

$$P_{\mathbf{X}}(B) = P(\mathbf{X}^{-1}(B)) = P(\{\omega \mid \mathbf{X}(\omega) \in B\}).$$

While we prefer the more formal notation $P_{\mathbf{X}}$, we must admit that the notation $P(X \in B)$ is intuitively suggestive and often enhances clarity.

In the next few sections we will sometimes work with R^2 rather than R^n in order to simplify notation. However, we will only do this in circumstances where we feel the generalization to R^n is clear. Finally, note that the construction in 1.14(b) can be depicted as in Figure 1.1.

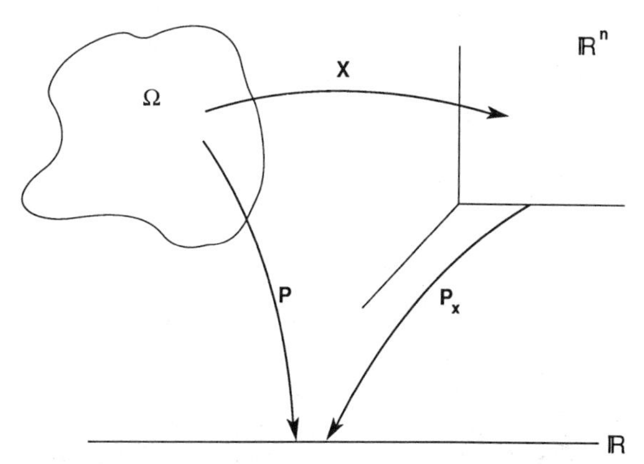

FIGURE 1.1. The induced probability measure.

1.2 Probability Distributions and Densities

1.2.1. **Definition.** Let $\mathbf{X}$ be a random vector. By the *cumulative probability distribution,* $F_{\mathbf{X}}$, we will mean the function

$$F_{\mathbf{X}} : \mathrm{R}^n \to \mathrm{R}$$

defined by

$$F_{\mathbf{X}}(x_1, x_2, \ldots, x_n) \triangleq P_X((-\infty, x_1] \times \cdots \times (-\infty, x_n]). \qquad (1.2\text{-}1)$$

Clearly, $F_{\mathbf{X}}$ measures the probability that the random vector $\mathbf{X} = (X_1, X_2, \ldots, X_n)$ takes on (vector) values $\mathbf{X}(\omega)$ such that $X_i(\omega) \leq x_i$ for each $i = 1, 2, \ldots, n$. For this reason, one often sees the right-hand side of (1.2-1) written as $P(X_1 \leq x_1, \ldots, X_n \leq x_n)$.

1.2.2. **Theorem** (Properties of $F_{\mathbf{X}}$ in R^2).

(a) $0 \leq F(x_1, x_2) \leq 1$,

(b) *F is right continuous and monotone increasing in x_1 and x_2,*

(c) $\lim_{x_1 \to -\infty} F(x_1, x_2) = 0$; $\lim_{x_1 \to -\infty} F(x_1, x_2) = 0$,

(d) $\lim_{x_1 \to \infty} F(x_1, x_2) = P_X((-\infty, \infty) \times (-\infty, x_2]) = P(X_2 \leq x_2)$, *and*

(e) $F(+\infty, +\infty) = 1$ *(means* $\lim_{\substack{x_1 \to \infty \\ x_2 \to \infty}} F(x_1, x_2)$*).*

(f) *If $a_1 \leq a_2$ and $b_1 \leq b_2$, then*

$$F(a_2, b_2) - F(a_2, b_1) - F(a_1, b_2) + F(a_1, b_1) \geq 0.$$

Proof. This is an exercise in measure theory. Here, (a)–(e) depend on the theorem that if E_1, $E_2, \cdots$ is an increasing (respectively decreasing) sequence of measurable sets, then

$$\lim_{n\to\infty} \mu(E_n) = \mu\left(\bigcup_{n=1}^{\infty} E_n\right) \quad \left(\text{respectively } \lim_{n\to\infty} \mu(E_n) = \mu\left(\bigcap_{n=1}^{\infty} E_n\right)\right).$$

Here, (f) is an application of finite additivity of measures (a picture is very helpful). □

1.2.3. **Theorem.** *If F satisfies* (a)–(f) *of Theorem* 1.2.2, *then F determines a unique probability measure on* R^2 *by taking the class of all half open rectangles as a sequential covering class whose covering value (measure) is*

$$\tau((a_1, b_1] \times (a_2, b_2]) \triangleq F(a_2, b_2) - F(a_2, b_1) - F(a_1, b_2) + F(a_1, b_1)$$

and constructing outer measures in the usual way. Moreover, rectangles turn out to be measurable. (See Appendix A.)

Proof. See reference [19] for this construction; also see reference [23], and Appendix A. □

1.2.4. **Definition.** By a *Baire* function we mean a Borel measurable function on R^n, that is, the inverse image of open sets are Borel sets.

1.2.5. **Definition.** Let g be a Baire function on R^n. We define

$$\underbrace{\int\int\cdots\int}_{E} g(x)\, dF_{\mathbf{X}}(\mathbf{x}) \triangleq \underbrace{\int\int\cdots\int}_{E} g\, dP_{\mathbf{X}},$$

that is, the left-hand side is common notation for the right-hand side.

The reason for the introduction of the notation in 1.2.5 is simply that, if g is continuous, hence certainly a Baire function, then the left-hand side of the equality exists as a (multiple) Riemann–Steiltjes integral and the definition is a theorem. We will not pursue the proof of this fact here.

1.2.6. **Theorem.** *If g is a Baire function on* R^n, $\mathbf{X}$ *a random vector on* Ω, *then $g \circ \mathbf{X}$ is a random variable on* R^n.

Proof. Exercise. □

1.2.7. **Theorem** (Change of Variables). *If g is a Baire function on* R^n, $\mathbf{X}$ *a random vector on* Ω, *then for any Borel set* $B \subset \mathrm{R}^n$,

$$\underbrace{\int\int\cdots\int}_{B} g\, dP_{\mathbf{X}} = \int_{\mathbf{X}^{-1}(B)} (g \circ \mathbf{X})\, dP. \tag{1.2-2}$$

Proof. First let $g = C_E$, $E \in B_n$, where C_E denotes the characteristic function of E.[1] Then

$$g \circ \mathbf{X} = C_{\mathbf{X}^{-1}(E)} \tag{1.2-3}$$

because

$$1 = (C_E \circ \mathbf{X})(\omega) \Longleftrightarrow \mathbf{X}(\omega) \in E \Longleftrightarrow \omega \in \mathbf{X}^{-1}(E)$$

or

$$1 = (C_E \circ \mathbf{X})(\omega) \Longleftrightarrow 1 = C_{\mathbf{X}^{-1}(E)}(\omega).$$

Now, the left-hand side of (1.2-2) is given by

$$\begin{aligned} \text{L.H.S.} &= \underbrace{\int\int\cdots\int}_{B} C_E \, dP_{\mathbf{X}} \\ &= \underbrace{\int\int\cdots\int}_{B\cap E} dP_{\mathbf{X}} \\ &= P_{\mathbf{X}}(B \cap E). \end{aligned}$$

Similarly, the right-hand side is [by (1.2.-3)]

$$\begin{aligned} \text{R.H.S.} &= \int_{\mathbf{X}^{-1}(B)} C_{\mathbf{X}^{-1}(E)} \, dP \\ &= \int_{\mathbf{X}^{-1}(B)\cap\mathbf{X}^{-1}(E)} dP \\ &= \int_{\mathbf{X}^{-1}(B\cap E)} dP \\ &= P(\mathbf{X}^{-1}(B \cap E)) \\ &= P_{\mathbf{X}}(B \cap E). \end{aligned}$$

Thus, the theorem holds for characteristic functions. If g is a simple function, then by the same calculation done to show (1.2-3), one can show that $g \circ \mathbf{X}$ is a simple function. Using the result for characteristic functions, the linearity of the integral implies that the theorem holds for simple functions.

Now, if g is a nonnegative Baire function, that is, Borel measurable, then from real variable theory, one can always construct a sequence of simple functions $g_n \uparrow g$. In this case, however, $g \circ \mathbf{X}$ is nonnegative and $g_n \circ \mathbf{X} \uparrow g \circ \mathbf{X}$. From the definition of integral, the theorem thus holds for nonnegative, integrable Baire functions, whence for all integrable Baire functions. (See Appendix D, Definition 8.) □

---

[1] $C_E(\mathbf{x}) = 0$ if $x \notin E$; $C_E(\mathbf{x}) = 1$ if $\mathbf{x} \in E$.

1.3 Expected Value, Covariance

1.3.1. Definition.

(a) Let X be a random variable on Ω, $X \in \mathcal{L}_1(\Omega, P)$.[2] By the expected value of X (or mean of X), written either $E(X)$ or μ_X, we mean the number

$$E(X) \triangleq \int_\Omega X(\omega)\, dP(\omega).$$

(b) If $\mathbf{X} = (X_1, X_2, \ldots, X_n)$ is a random vector, then

$$\mu_{\mathbf{X}} \triangleq E(\mathbf{X}) \triangleq (E(X_1), E(X_2), \ldots, E(X_n)).$$

1.3.2. Theorem. *If X is a random variable,*

$$E(X) = \int_{-\infty}^{\infty} x\, dP_X.$$

Proof. In theorem 1.2.7, let $n = 1$ and $g =$ identity function; then $\Omega = X^{-1}(\mathrm{R})$ and $X = id \circ X$, so

$$\begin{aligned} E(X) &= \int_\Omega X(\omega)\, dP(\omega) = \int_{X^{-1}(\mathrm{R})} (id \circ X)(\omega)\, dP(\omega) \\ &= \int_{\mathrm{R}} id\, dp_X \\ &= \int_{\mathrm{R}} x\, dP_X. \qquad \square \end{aligned}$$

1.3.3. Theorem. *Let $\mathbf{X}_1, \mathbf{X}_2, \ldots, \mathbf{X}_n$ be random vectors; let $\alpha_1, \alpha_2, \ldots \alpha_n$ be scalars. Then,*

$$E\left(\sum_{i=1}^n \alpha_i \mathbf{X}_i\right) = \sum_{i=1}^n \alpha_i E(\mathbf{X}_i). \tag{1.3-1}$$

Proof. Exercise. $\square$

Recall that since $P(\Omega) = 1$, it follows that the constant function $X(\omega) = 1$ is integrable. Using this fact together with the inequality

$$\frac{Y^2 + X^2}{2} \geq |YX| \tag{1.3-2}$$

[just use $(Y \pm X)^2 \geq 0$], it follows that $\mathcal{L}_2(\Omega, P)$[3] $\subset \mathcal{L}_1(\Omega, P)$. Hence the following are well-defined.

---

[2]This is simply notation for the collection of all integrable functions on (Ω, ϵ, P).

[3]See Appendix E for definition of $\mathcal{L}_2(\Omega, P)$.

1.3.4. **Definitions.**

(a) Let X and Y be random variables in $\mathcal{L}_2(\Omega, P)$. By the *covariance* of X and Y, we mean the number

$$\begin{aligned}\operatorname{cov}(X, Y) &\triangleq E((X - \mu_X) \cdot (Y - \mu_Y)) \\ &= \int_\Omega (X - \mu_X)(Y - \mu_Y)\, dP.\end{aligned}$$

(b) By the *variance* of X, we mean the number

$$V(X) \triangleq \operatorname{cov}(X, X).$$

(c) By the *standard deviation* of X, we mean the number

$$\sigma_X \triangleq \sqrt{V(X)}.$$

(d) By the *correlation* between X and Y, we mean the number

$$\operatorname{cor}(X, Y) \triangleq E(X \cdot Y).$$

(e) A *standardized random variable* Z is one with the property that

$$E(Z) = 0, \quad V(Z) = 1.$$

Before stating the next theorem, we will introduce the following terminology. If X and Y are random variables, then the pair (X, Y) is a random vector (see 1.1.6). The probability measure on R^2 induced by this vector, namely, $P_{(X,Y)}$, is called the *joint measure* for X and Y. The proof of Theorem 1.1.6 suggests the reason for this terminology.

1.3.5. **Theorem.** *Let X and Y be random variables in $\mathcal{L}_2(\Omega, P)$. Then*

$$\operatorname{cov}(X, Y) = \int\int_{\mathrm{R}^2} (x - \mu_X)(y - \mu_Y)\, dP_{(X,Y)}.$$

Proof. This follows from Theorem 1.2.7 letting $n = 2$ and $g(x, y) = x \cdot y$. Note that $(X, Y)^{-1}(\mathrm{R}^2) = \Omega$. The details are left to the reader. □

1.3.6. **Lemma.**

(a) *If X is any random variable in $\mathcal{L}_2(\Omega, P)$, then*

$$Z \triangleq \frac{X - \mu_X}{\sigma_X}$$

is a standardized random variable.

(b) *If Z_1 and Z_2 are standardized random variables, then*

$$|\mathrm{cov}(Z_1, Z_2)| \leq 1.$$

Proof.

(a)

$$\begin{aligned} E(Z) &= E\left(\frac{X-\mu_X}{\sigma_X}\right) = \frac{1}{\sigma_X}E(X-\mu_X) \\ &= \frac{1}{\sigma_X}[E(X) - E(\mu_X)] \\ &= \frac{1}{\sigma_X}[\mu_X - \mu_X] \\ &= 0. \end{aligned}$$

Also,

$$\begin{aligned} V(Z) &= E\left(\left(\frac{X-\mu_X}{\sigma_X}\right)^2\right) \\ &= E\left(\frac{1}{\sigma_X^2}(X-\mu_X)^2\right) \\ &= \frac{1}{\sigma_X^2}E((X-\mu_X)^2) \\ &= \frac{1}{V(X)} \cdot V(X) \\ &= 1. \end{aligned}$$

(b) $E(Z_1 \pm Z_2)^2 \geq 0$. Hence,

$$E(Z_1^2 \pm 2Z_1Z_2 + Z_2^2) \geq 0,$$

and so,

$$E(Z_1^2) + E(Z_2^2) \geq \pm 2E(Z_1Z_1).$$

Hence, by the definition of standardized random variables

$$2 \geq \pm 2 \ \mathrm{cov}(Z_1, Z_2)$$

or

$$1 \geq \pm\mathrm{cov}(Z_1, Z_2).$$

The result follows. □

1.3.7. **Corollary.** $|\mathrm{cov}(Z_1, Z_2)| = 1 \iff Z_1 = \pm Z_2$ (a.e.)

1.3.8. **Theorem.** *Let* $X_1, X_2 \in \mathcal{L}_2(\Omega, P)$, $\alpha \in \mathrm{R}$. *Then,*

(a) $V(\alpha X_i) = \alpha^2 V(X_i)$,

(b) $V(X_i) = E(X_i^2) - [E(X_i)]^2$,

(c) $V(X_1 + X_2) = V(X_1) + 2\,\mathrm{cov}(X_1, X_2) + V(X_2)$,

(d) $|\mathrm{cov}(X_1, X_2)| \leq \sigma_{X_1}\sigma_{X_2}$,

(e) $V(X_1 + X_2) \leq [\sigma_{X_1} + \sigma_{X_2}]^2$, *and*

(f) $\mathrm{cov}(X_1, X_2) = \mathrm{cor}(X_1, X_2) - E(X_1) \cdot E(X_2)$.

Proof. We will do part (d); the rest are left as exercises.
(d) Let

$$Z_i \triangleq \frac{X_i - \mu_{X_i}}{\sigma_{X_i}}.$$

By 1.3.6, $|\mathrm{cov}(Z_1, Z_2)| \leq 1$, that is,

$$\left| E\left[\frac{(X_1 - \mu_{X_1})(X_2 - \mu_{X_2})}{\sigma_{X_1}\sigma_{X_2}}\right]\right| \leq 1.$$

It follows that

$$|E[(X_1 - \mu_{X_1})(X_2 - \mu_{X_2})]| \leq \sigma_{X_1}\sigma_{X_2},$$

which is equivalent to the result. □

1.3.9. **Definitions.** Let $X_1, X_2 \in \mathcal{L}_2(\Omega, P)$.

(a) X_1 and X_2 are said to be *uncorrelated* if and only if $\mathrm{cov}(X_1, X_2) = 0$.

(b) $\rho(X_1, X_2) \triangleq \dfrac{\mathrm{cov}(X_1, X_2)}{\sigma_{X_1}\sigma_{X_2}}$ is called a *correlation coefficient.*

1.3.10. **Corollary** (To 1.3.7 and 1.3.8).

(a) $|\rho(X_1, X_2)| \leq 1$

(b) $|\rho(X_1, X_2)| = 1$ *if and only if*

$$X_1 - \mu_{X_1} = \pm\frac{\sigma_{X_1}}{\sigma_{X_2}}(X_2 - \mu_{X_2}),$$

where the choice of sign is the sign of $\rho(X_1, X_2)$.

Part (b) of 1.3.10 says that the correlation coefficient of two random variables is a measure of the degree to which they are linearly correlated. We will have occasion to return to this remark later when we study estimation.

1.3.11. **Theorem.** *The following are equivalent.*

(a) X_1 *and* X_2 *are uncorrelated.*

(b) $E(X_1 \cdot X_2) = E(X_1) \cdot E(X_2)$.

(c) $\mathrm{var}(X_1 + X_2) = \mathrm{var}(X_1) + \mathrm{var}(X_2)$.

Proof. The equivalence of (a) and (c) follows from part (c) of 1.3.7, and the equivalence of (b) and (c) follows from part (f) of the same theorem. □

1.3.12. **Definition.** Let

$$\mathbf{Y} = \begin{bmatrix} Y_1 \\ \vdots \\ Y_n \end{bmatrix}, \qquad \mathbf{Z} = \begin{bmatrix} Z_1 \\ \vdots \\ Z_m \end{bmatrix}$$

be random vectors. By the *covariance matrix* of Y and Z, we mean the matrix

$$\mathrm{cov}(\mathbf{Y}, \mathbf{Z}) \triangleq E((\mathbf{Y} - \boldsymbol{\mu}_{\mathbf{Y}})(\mathbf{Z} - \boldsymbol{\mu}_{\mathbf{Z}})^T) = [\mathrm{cov}(Y_i, Z_j)]$$

(recall 1.3.1(b)).

1.3.13. **Theorem.** *Let A and B be constant matrices. Then,*

(a) $E(A\mathbf{Y}) = AE(\mathbf{Y})$,

(b) $E(\mathbf{Y}B) = E(\mathbf{Y})B$,

(c) $\mathrm{cov}(A\mathbf{Y}, B\mathbf{Z}) = A\,\mathrm{cov}(\mathbf{Y}, \mathbf{Z})B^T$.

Proof.

(a) Let $A = [\alpha_{ij}]$. Then

$$E(A\mathbf{Y}) = E\begin{bmatrix} \sum_j \alpha_{ij} Y_j \\ \vdots \\ \sum_j \alpha_{nj} Y_j \end{bmatrix} = \begin{bmatrix} E\left(\sum_j \alpha_{ij} Y_j\right) \\ \vdots \\ E\left(\sum_j \alpha_{nj} Y_j\right) \end{bmatrix}$$

$$= \begin{bmatrix} \sum_j \alpha_{ij} E(Y_j) \\ \vdots \\ \sum_j \alpha_{nj} E(Y_j) \end{bmatrix} = AE(\mathbf{Y}).$$

(b) Similar to (a).

(c) If X is a matrix of random variables, then by partititioning X into columns, we can write

$$\begin{aligned} E(A\mathbf{X}) &= E(A[X_1|X_2|\cdots|X_n]) \\ &= E([AX_1|AX_2|\cdots|AX_n]) \\ &= [E(AX_1)|E(AX_2)|\cdots|E(AX_n)] \\ &= [AE(X_1)|AE(X_2)|\cdots|AE(X_n)] \quad \text{[part (a)]} \\ &= A[E(X_1)\cdots E(X_n)] \\ &= AE(\mathbf{X}). \end{aligned}$$

Similarly, one can show that

$$E(\mathbf{X}C) = E(\mathbf{X})C$$

for C, a matrix of constants.

This done, we have

$$\begin{aligned} \operatorname{cov}(A\mathbf{Y}, B\mathbf{Z}) &= E(A(\mathbf{Y} - \boldsymbol{\mu}_{\mathbf{Y}})(B(\mathbf{Z} - \boldsymbol{\mu}_{\mathbf{Z}}))^T) \\ &= E(A(\mathbf{Y} - \boldsymbol{\mu}_{\mathbf{Y}})(\mathbf{Z} - \boldsymbol{\mu}_{\mathbf{Z}})^T B^T) \\ &= AE((\mathbf{Y} - \boldsymbol{\mu}_{\mathbf{Y}})(\mathbf{Z} - \boldsymbol{\mu}_{\mathbf{Z}})^T)B^T \\ &= A\operatorname{cov}(\mathbf{Y}, \mathbf{Z})B^T. \qquad \square \end{aligned}$$

It is clear, from the definition of a covariance matrix, that for a random vector $\mathbf{Y}$, $\operatorname{cov}(\mathbf{Y}, \mathbf{Y})$ is a symmetric matrix. Now, from linear algebra recall that a symmetric matrix Q is called positive definite (respectively semidefinite) providing the associated quadratic form is positive (respectively nonnegative). This means, by definition, that for any (column) vector $\mathbf{a}$, $\mathbf{a} \neq \mathbf{0}$ then $\mathbf{a}^T Q\mathbf{a} > 0$ (respectively $\mathbf{a}^T Q\mathbf{a} \geq 0$). From part (c) of the above theorem, letting $A = \mathbf{a}^T$ and $B = \mathbf{a}$, we have

$$\mathbf{a}^T \operatorname{cov}(\mathbf{Y}, \mathbf{Y})\mathbf{a} = \mathbf{a}^T \operatorname{cov}(\mathbf{Y}, \mathbf{Y})\mathbf{a}^{TT} = \operatorname{cov}(\mathbf{a}^T\mathbf{Y}, \mathbf{a}^T\mathbf{Y}).$$

But $\mathbf{a}^T\mathbf{Y}$ is scalar valued, that is, it is a random variable, and so

$$\mathbf{a}^T \operatorname{cov}(\mathbf{Y}, \mathbf{Y})\mathbf{a} = \int_\Omega (\mathbf{a}^T(\mathbf{Y} - \boldsymbol{\mu}_{\mathbf{Y}}))^2\, dP \geq 0.$$

In other words, $\operatorname{cov}(\mathbf{Y}, \mathbf{Y})$ is a positive semidefinite matrix. Furthermore, the only way the above expression can be zero is if $\mathbf{a}^T(\mathbf{Y} - \boldsymbol{\mu}_{\mathbf{Y}}) = 0$ almost everywhere, that is, if the components of $\mathbf{Y}$ form a dependent set in $\mathcal{L}_2(\Omega, P)$. Using the contrapositive, these observations prove the following.

1.3.14. **Theorem.**

(a) *If* $Y_1, Y_2, \ldots Y_n$ *are linearly independent in* $\mathcal{L}_2(\Omega, P)$ *and*

$$\mathbf{Y} = \begin{bmatrix} Y_1 \\ \vdots \\ Y_n \end{bmatrix},$$

then

$$\operatorname{cor}(\mathbf{Y}, \mathbf{Y}) \triangleq E(\mathbf{Y}\mathbf{Y}^T)$$

is positive definite and conversely.

(b) *If* $Y_1 - \mu_{Y_1}, Y_2 - \mu_{Y_2}, \ldots, Y_n - \mu_{Y_n}$ *are linearly independent in* $\mathcal{L}_2(\Omega, P)$, *then* $\operatorname{cov}(\mathbf{Y}, \mathbf{Y})$ *is positive definite and conversely.*

1.4 Independence

1.4.1. **Definition.** Let $(\Omega, \mathcal{E}, P)$ be a probability space and let A and B be events, that is, elements of $\mathcal{E}$. Then, A and B are said to be *independent* provided that

$$P(A \cap B) = P(A) \cdot P(B). \tag{1.4-1}$$

This simple definition is of great philosophic importance in statistics, and hence of great importance, period. To explain this remark, we begin with an example.

1.4.2. **Example.** Let $\Omega = \{\omega_1, \omega_2, \ldots, \omega_n\}$ be a discrete outcome set with a uniform probability measure, that is, $p(\omega_i) = 1/n$ for each i. Suppose A and B are subsets of Ω such that

(1) A contains k points,

(2) B contains s points, and

(3) $A \cap B$ contains p points.

If $(\Omega, 2^\Omega, P)$ represents some experiment, and we are told that upon performing the experiment, we secured a result in B, what is the probability that the result was also in A? Since the distribution is uniform, that is, each outcome is equally likely, the probability in question, which we write as $P(A \mid B)$ is given by

$$P(A \mid B) = \frac{\text{number of points in } A \cap B}{\text{number of points in } B}$$

$$P(A \mid B) = \frac{p}{s}.$$

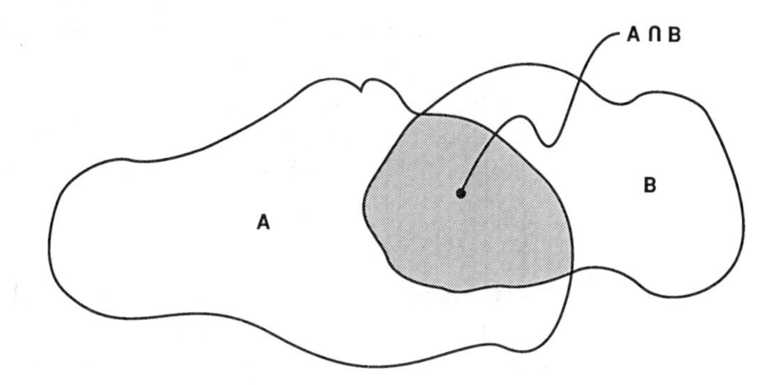

FIGURE 1.2. The domains used in constructing the conditional probability measure.

Now, if we divide the numerator and denominator of this fraction by n, we obtain

$$P(A \mid B) = \frac{p/n}{s/n} = \frac{P(A \cap B)}{P(B)}.$$

With a bit of imagination one can see that the above formula for $P(A \mid B)$ makes sense even if the distribution is not uniform (or discrete for that matter). Referring to Figure 1.2, we see that $P(A \mid B)$ is just the ratio of the probability measure of the event in question, $A \cap B$, to the probability measure of the secured event B. We thus make the following definition.

1.4.3. **Definition.** By the *conditional probability* of A given B, written $P(A \mid B)$, we mean the probability measure

$$P(A \mid B) = \frac{P(A \cap B)}{P(B)}, \qquad P(B) \neq 0.$$

We will (for the time being) leave $P(A \mid B)$ undefined if $P(B) = 0$. The fact that $P(A \mid B)$, as a function of A, is a measure is left as an exercise for the reader.

Very well, we have a notion of conditional probability $P(A \mid B)$, which measures the probability that event A has occurred given that it is known that B occurred, and $P(B) \neq 0$ (note that $P(B)$ can occur even if $P(B) = 0$). In a later section, we will generalize this idea, so for now we will not worry about the restriction that $P(B) \neq 0$. Suppose now that it turns out that knowledge of the fact that B has occurred in no way influences the probability that A has also occurred, that is to say,

$$P(A \mid B) = P(A). \tag{1.4-2}$$

In this case, it seems reasonable to say that A is independent from B. But Definition 1.4.3 applied to (1.4-2) yields equation (1.4-1), which was the condition we used to define independence.

The above discussion is provided to make Definition 1.4.1 seem plausible. Certainly in the case where A and B are past events (as in the above discussion), we have interpreted Definition 1.4.1 in a reasonable fashion using Definition 1.4.3. In general, if A and B are events that are possibly temporally related, or related in some other complicated fashion, the relationship

$$P(A \cap B) = P(A) \cdot P(B)$$

has the feature that if $P(B)$ is known, then the joint probability $P(A \cap B)$ can be calculated with simply the additional information $P(A)$ and nothing else about B and its relation to A. Intuitively, then, this equation seems to capture the idea of causally independent events, or at least the joint probability of such events.

With the introduction of the notion of "causally," we would be remiss if we did not discuss the way independence is used in practice. This is the "great philosophic importance" we aluded to in the remark immediately following Definition 1.4.1. Certainly Definition 1.4.1 is a precise mathematical definition (whether or not $P(B) = 0$ by the way) and so, in principle, one could always use (1.4-1) to check and see if two events A and B are independent or not. With the exception of simple textbook exercises, this is never done. Rather, there is a metaprinciple that is seldom (if ever) explicitly stated, but is generally (if not always) invoked.

Metaprinciple. If two events are causally independent, then they are mathematically independent.

Note that this principle is not a theorem—it is a belief based on experience. It cannot be proved, or disproved, and as we have said, it is seldom explicitly stated. Of course, in the world of applied mathematics, it is always necessary to draw mathematical inferences from observations of the "real world," and such inferences are always really based on human belief. When we define velocity as a time derivative of position, we are expressing our belief that the time derivative does indeed convey the essence of the physical notion of velocity. Of course, since the definition looks so reasonable, even obvious, the fact that human belief is involved in its execution is usually not mentioned. However, it is there!

In this book, we chose to explicitly state the above metaprinciple about causality and independence, not because it is any more or less correct than any other metaprinciple used to interpret the world around us, but rather to emphasize the way that mathematical independence is utilized. Except in very trivial examples, it is not a conclusion that one deduces from other mathematical hypotheses. Instead, it is usually a hypothesis itself, an assumption based on the belief that the physical events represented by our mathematical events are indeed causally independent.

It is probably natural for us to wonder if the converse of the above principle works, that is, if two events are mathematically independent are they causally independent? Clearly, it is impossible to decide this unless we have an explicit definition of causal independence, something we don't have and perhaps don't even want. As in all such metaprinciples, the best we can hope for is some illustration of plausibility. The next example, from Feller [6], seems to convince some that the converse does not hold. Fortunately, whether it does or not doesn't seem to really matter, for the question seldom (if ever) comes up in practical situations.

1.4.4. **Example.** Let Ω_n = the set of families having exactly n children and let A be the event that a family has children of both sexes, B the event that a family has at most one girl. If we assume that boys and girls are equally likely, then

$$P(A) = \frac{2^n - 2}{2^n} = \frac{2^{n-1} - 1}{2^{n-1}}$$

$$P(B) = \frac{n+1}{2^n}$$

and

$$P(A \cap B) = \frac{n}{2^n}.$$

When $n = 3$, we have

$$P(A \cap B) = \frac{3}{8} = \frac{3}{4} \cdot \frac{1}{2} = P(A) \cdot P(B).$$

However, when $n = 2$ or 4 the events A and B are not independent.

The notion of independence of two events as given in 1.4.1 easily generalizes as follows.

1.4.5. **Definition.** Let $\mathcal{E}_1$ and $\mathcal{E}_2$ be two σ fields (presumably sets of events associated with two different statistical situations). We say that $\mathcal{E}_1$ and $\mathcal{E}_2$ are *independent from each other* if and only if for each $E_1 \in \mathcal{E}_1$ and each $E_2 \in \mathcal{E}_2$, E_1 and E_2 are independent events, that is,

$$P(E_1 \cap E_2) = P(E_1) \cdot P(E_2).$$

Definition 1.4.5 allows us to easily say what we mean by two random variables being independent. First, however, we need a technical definition and a simple lemma.

1.4.6. **Definition.** Let X be a random variable (vector) on $(\Omega, \mathcal{E}, P)$. By $B(X)$, we mean

$$B(X) = \{X^{-1}(B) \mid B \in B_1 \quad (\text{resp. } B \in B_n)\}.$$

1.4.7. **Lemma.**

(a) *$B(X)$ is a σ field, $B(X) \subset \mathcal{E}$.*

(b) *$B(X)$ is the smallest σ field such that X is measurable.*

Proof.

(a) This follows at once from the fact that X^{-1} preserves arbitrary unions, intersections, and complements.

(b) The word smallest in the statement of the theorem means that if $\mathcal{F}$ is any σ field, $\mathcal{F} \subset \mathcal{E}$, such that $X^{-1}(B) \in \mathcal{F}$ for every $B \in B_1$, then $B(X) \subset \mathcal{F}$. Of course, this is obvious. □

1.4.8. **Examples.**

(a) Let $X = C_E$, $E \in \mathcal{E}$. Then

$$B(X) = \{\phi, E, E', \Omega\}.$$

(b) Let $X = \alpha_1 C_{E_1} + \alpha_2 C_{E_2}$; $E_1, E_2 \in \mathcal{E}$. Then

$$B(X) = \{\phi, E_1, E_2, E_1', E_2', E_1 \cup E_2, E_1 \cup E_2', E_2 \cup E_1',$$
$$E_1 \cap E_2, E_1 \cap E_2', E_1' \cap E_2, E_1' \cap E_2', \Omega\}.$$

1.4.9. **Definition.** Let X_1 and X_2 be random variables (vectors); then X_1 and X_2 are said to be *independent* if and only if $B(X_1)$ and $B(X_2)$ are independent σ fields.

1.4.10. **Theorem.** *Let X_1 and X_2 be random variables, $\mathbf{X} = (X_1, X_2)$; then the following are equivalent.*

(a) *X_1 and X_2 are independent.*

(b) *$P_{\mathbf{X}}(A \times B) = P_{X_1}(A) \cdot P_{X_2}(B)$ for all $A, B \in B_1$.*

(c) *$F_{\mathbf{X}}(x_1, x_2) = F_{X_1}(x_1) \cdot F_{X_2}(x_2)$.*

Proof. (a) $\Rightarrow$ (b):

$$\begin{aligned} P_{\mathbf{X}}(A \times B) &= P(X^{-1}(A \times B)) \\ &= P(X_1^{-1}(A) \cap X_2^{-1}(B)) \\ &= P(X_1^{-1}(A) \cdot P(X_2^{-1}(B)) \\ &= P_{X_1}(A) \cdot P_{X_2}(B). \end{aligned}$$

(b) $\Rightarrow$ (c): In (b), simply let $A = (-\infty, x_1]$ and $B = (-\infty, x_2]$.
(c) $\Rightarrow$ (a): From the property of additivity of the measure $P_{\mathbf{X}}$, we have

$$\begin{aligned} P_{\mathbf{X}}((-\infty, b_1] \times (-\infty, a_2]) &= P_{\mathbf{X}}((-\infty, a_1] \cup (a_1, b_1] \times (-\infty, a_2]) \\ &= P_{\mathbf{X}}((-\infty, a_1] \times (-\infty, a_2]) + P_{\mathbf{X}}((a_1, b_1] \\ &\quad \times (-\infty, a_2]). \end{aligned}$$

Hence, from 1.2.1,

$$P_{\mathbf{X}}((a_1, b_1)] \times (-\infty, a_2]) = F_{\mathbf{X}}(b_1, a_2) - F_{\mathbf{X}}(a_1, a_2). \tag{1.4-3}$$

Similarly,

$$P_{\mathbf{X}}((-\infty, a_1] \times (a_2, b_2]) = F_{\mathbf{X}}(a_1, b_2) - F_{\mathbf{X}}(a_1, a_2). \tag{1.4-4}$$

Also,

$$\begin{aligned} P_{\mathbf{X}}((-\infty, b_1] \times (-\infty, b_2]) &= P_{\mathbf{X}}((-\infty, a_1] \cup (a_1, b_1]) \times ((-\infty, a_2] \\ &\quad \cup (a_2, b_2])) \\ &= P_{\mathbf{X}}((-\infty, a_1] \times (-\infty, a_2] \cup (a_1, b_1] \\ &\quad \times (-\infty, a_2] \\ &\quad \cup (-\infty, a_1] \times (a_2, b_2) \cup (a_1, b_1] \times (a_2, b_2). \end{aligned}$$

Since the four sets in this last expression are disjoint, we have (again applying 1.2.1)

$$\begin{aligned} F_{\mathbf{X}}(b_1, b_2) &= F_{\mathbf{X}}(a_1, a_2) + P_{\mathbf{X}}((a_1, b_1] \times (-\infty, a_2]) \\ &\quad + P_{\mathbf{X}}((-\infty, a_1] \times (a_2, b_2]) + P_{\mathbf{X}}((a_1, b_1] \times (a_2, b_2]). \end{aligned}$$

Solving this last expression for $P_{\mathbf{X}}((a_1, b_1] \times (a_2, b_2])$ and using (1.4-1) and (1.4-2), we have

$$P_{\mathbf{X}}((a_1, b_1] \times (a_2, b_2]) = F_{\mathbf{X}}(b_1, b_2) - F_{\mathbf{X}}(a_1, b_2) - F_{\mathbf{X}}(b_1, a_2) + F_{\mathbf{X}}(a_1, a_2).$$ [4]

Applying the hypothesis (c) to this expression,

$$\begin{aligned} P_{\mathbf{X}}((a_1, b_1] \times (a_2, b_2]) &= F_{X_1}(b_1) \cdot F_{X_2}(b_2) - F_{X_1}(a_1) \cdot F_{X_2}(b_2) \\ &= F_{X_1}(b_1) \cdot F_{X_2}(b_2) + F_{X_1}(a_1) \cdot F_{X_2}(a_2) \\ &= (F_{X_1}(b_1) - F_{X_1}(a_1))(F_{X_2}(b_2) - F_{X_2}(a_2)) \\ &= P_X((a_1, b_1]) \cdot P_{X_2}((a_2, b_2]). \end{aligned}$$

Hence,

$$P(X_1^{-1}(a_1, b_1] \cap X_2^{-1}(a_2, b_2]) = P(X_1^{-1}(a_1, b_1]) \cdot P(X_2^{-1}(a_2, b_2]).$$

The result now follows from the structure of Borel sets in R and Definition 1.4.6. □

1.4.11. **Theorem.** *If X_1 and X_2 are independent random variables, then they are uncorrelated.*

---

[4] This calculation explains the reason for the particular construction in Theorem 1.2.3.

Proof. We first show that

$$E(X_1 X_2) = E(X_1) \cdot E(X_2).$$

However, from the construction of the (Lebesgue) integral, it suffices to show that this relation holds when X_1 and X_2 are simple. Thus, suppose that E_1 and E_2 are arbitrary measurable sets and let

$$X_1 = C_{E_1}; \qquad X_2 = C_{E_2}.$$

Then

$$X_1 \cdot X_2 = C_{E_1} \cdot C_{E_2} = C_{E_1 \cap E_2},$$

and so,

$$\begin{aligned} E(X_1 X_2) &= \int_\Omega C_{E_1 \cap E_2}\, dP \\ &= P(E_1 \cap E_2) \\ &= P(E_1) \cdot P(E_2) \\ &= \int_\Omega C_{E_1}\, dP \cdot \int_\Omega C_{E_2}\, dP \\ &= E(X_1) \cdot E(X_2). \end{aligned}$$

From this calculation, it follows easily that

$$E(X_1 X_2) = E(X_1) \cdot E(X_2)$$

holds when X_1 and X_2 are simple, hence when they are in $\mathcal{L}_2(\Omega, P)$. It now follows from part (f) of Theorem 1.3.8 that

$$\operatorname{cov}(X_1, X_2) = E(X_1 X_2) - E(X_1) \cdot E(X_2) = 0,$$

whence by part (a) of Definition 1.3.9 that X_1 and X_2 are uncorrelated. □

1.4.12. **Example.** The converse of 1.4.11 fails as we will now show. Suppose $X = (X_1, X_2)$ is uniformly distributed on the unit disk D, that is,

$$P_X(E) = \frac{\mu(E \cap D)}{\pi},$$

where μ is the Lebesgue measure on $\mathbf{R}^2$. Then, letting

$$g(x_1, x_2) = x_1 \cdot x_2,$$

we have

$$E(X_1 X_2) = \int_\Omega X_1(\omega) X_2(\omega)\, dP(\omega)$$

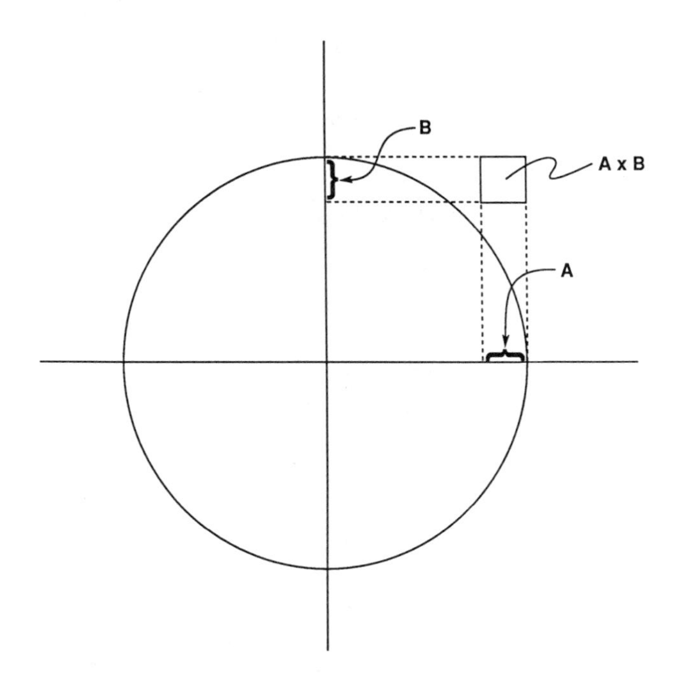

FIGURE 1.3. Uncorrelated random variables need not be independent.

$$
\begin{aligned}
&= \int_\Omega g \circ X(\omega)\, dP(\omega) \\
&= \int_{\mathbb{R}^2} g(x)\, dP_X \\
&= \int\int_{\mathbb{R}^2} x_1 x_2\, dP_X \\
&= \frac{1}{\pi}\int\int_D x_1 x_2\, dx_1\, dx_2 \\
&= 0.
\end{aligned}
$$

Since $E(X_1) = E(X_2) = 0$, it follows that X_1 and X_2 are uncorrelated. However, X_1 and X_2 are not independent. This can easily be seen by looking at Figure 1.3. If A and B are chosen as shown, both $P_{X_1}(A)$ and $P_{X_2}(B)$ are nonzero, whereas $P_X(A \times B) = 0$.

1.5 The Radon–Nikodym Theorem

In this section we will state and prove a very important theorem in mathematics, and in particular in probability theory. This theorem, the Radon–Nikodym theorem, will be central to our discussion of conditional expectation later in this chapter. The proof we will present here is from von

Neuman and is as slick an argument as we have seen. It also provides us with some practice at utilizing Hilbert spaces.

1.5.1. **Definitions.** Let S be an arbitrary set.

(a) A measure μ on S is said to be *bounded* providing $\mu(S) < \infty$.

(b) Let μ_1 and μ_2 be two measures defined on the same σ field $\mathcal{M}$ of sets in S. Define $\mu_1 + \mu_2$ via

$$(\mu_1 + \mu_2)(E) = \mu_1(E) + \mu_2(E)$$

for all $E \in \mathcal{M}$.

(c) Let μ_1 and μ_2 be as in (b). Then μ_1 is said to be *absolutely continuous* with respect to μ_2, written $\mu_1 << \mu_2$, provided that for every $E \in \mathcal{M}$

$$\mu_2(E) = 0 \Rightarrow \mu_1(E) = 0.$$

1.5.2. **Lemma.** *If μ_1 and μ_2 are as above, then $\mu_1 + \mu_2$ is a measure on S also.*

1.5.3. **Theorem.** *Let μ_1 and μ_2 be measures defined on the σ field $\mathcal{M}$. Then, if f is integrable with respect to both μ_1 and μ_2, f is integrable with respect to $\mu_1 + \mu_2$ and for any $E \in \mathcal{M}$*

$$\int_E f d(\mu_1 + \mu_2) = \int_E f\, d\mu_1 + \int_E f\, d\mu_2.$$

Proof. This is left as an optional exercise. (Hint: show the theorem for simple functions and then simply use the definition of the integral.) □

1.5.4. **Theorem.** *Let λ and μ be bounded measures on S that are defined on the same σ field $\mathcal{M}$, and suppose that $\lambda << \mu$. Then there exists a unique h (up to sets of measure zero), integrable with respect to μ, such that for any $E \in \mathcal{M}$*

$$\lambda(E) = \int_E h\, d\mu.$$

Proof. Define

$$\phi = \lambda + \mu.$$

Then ϕ is a bounded measure on S. Note that

$$\int_S |f| d\phi = \int_S |f| d\lambda + \int_S |f| d\mu > \int_S |f| d\lambda,$$

so if $f \in \mathcal{L}_2(S, \mathcal{M}, \phi)$,[5] the Cauchy–Schwarz inequality yields

$$\left| \int_S f\, d\lambda \right| \le \int_S |f| d\lambda \le \int_S |f| d\phi \le \|f\| \cdot \|1\| = \|f\| \sqrt{\phi(s)} < \infty.$$

[5]We include $\mathcal{M}$ here to emphasize that f is only $\mathcal{M}$-square integrable, i.e., f is an $\mathcal{M}$-measurable function.

This proves that the mapping

$$f \mapsto \int_S f\, d\lambda$$

is a bounded (hence continuous) linear functional on $\mathcal{L}_2(S, \mathcal{M}, \phi)$. By the Riesz representation theorem, there is a unique $g \in \mathcal{L}_2(S, \mathcal{M}, \phi)$ such that for all $f \in \mathcal{L}_2(S, \mathcal{M}, \phi)$

$$\int_S f\, d\lambda = \int_S f \cdot g\, d\phi. \tag{1.5-1}$$

Note that, as usual, g is unique as an element of $\mathcal{L}_2(S, \mathcal{M}, \phi)$, but as a function on S is only determined almost everywhere.

Now, put

$$f = C_E,$$

where $E \in \mathcal{M}$ is any set such that $\phi(E) > 0$. Then using Equation (1.5-1),

$$\lambda(E) = \int_E g\, d\phi.$$

Since $0 \leq \lambda(E) \leq \phi(E)$, we have

$$0 \leq \int_E g\, d\phi \leq \phi(E),$$

and since $\phi(E) > 0$, this can be written

$$0 \leq \frac{1}{\phi(E)} \int_E g\, d\phi \leq 1. \tag{1.5-2}$$

Since $E \in \mathcal{M}$ was an arbitrary set satisfying $\phi(E) > 0$, it follows from (1.5-2) that

$$0 \leq g(x) \leq 1 \quad \text{(a.e.) on } S.$$

(Suppose not—you can then easily find an E contradicting (1.5-2).) Thus, we can redefine g on a set of ϕ-measure zero so that

$$0 \leq g(x) \leq 1 \quad \text{for all} \quad x \in S, \tag{1.5-3}$$

and (1.5-1) still holds for this redefined g. If we rewrite (1.5-1) as

$$\int_S f\, d\lambda = \int_S f \cdot g\, d(\lambda + \mu) = \int_S f \cdot g\, d\lambda + \int_S f \cdot g\, d\mu, \tag{1.5-4}$$

we obtain

$$\int_S (1 - g) f\, d\lambda = \int_S f \cdot g\, d\mu. \tag{1.5-5}$$

Next, define $A = g^{-1}(1) = \{x \mid g(x) = 1\}$. Clearly, $A \in \mathcal{M}$ (since $g \in \mathcal{L}^2(S, \mathcal{M}, \phi)$ and is thus $\mathcal{M}$ measurable), so we can let $f = C_A$ in (1.1-5) and obtain

$$\int_A (1-g)\,d\lambda = \int_A g\,d\mu,$$

whence by the definition of A

$$0 = \mu(A).$$

By hypothesis, $\lambda << \mu$, so it follows that $\lambda(A) = 0$. Hence, $\phi(A) = 0$, and so we can redefine g on a set of ϕ-measure zero (namely A) so that

$$0 \leq g(x) < 1 \quad \text{for all} \quad x \in S \tag{1.5-6}$$

and (1.5-5) still holds. Since g is bounded, $g^n \in \mathcal{L}_2(S, \mathcal{M}, \phi)$ for all n, so in (1.5-5) we can replace f by $(1 + g + g^2 + g^3 + \cdots g^n)C_E$, where $E \in \mathcal{M}$ is arbitrary, and obtain

$$\int_E (1 - g^{n+1})\,d\lambda = \int_E (g + g^2 + g^3 + \cdots + g^{n+1})\,d\mu. \tag{1.5-7}$$

Now $g < 1$ by (1.5-6), so $1 - g^{n+1}$ converges monotonically upward to 1, while $g + g^2 + \cdots + g^{n+1}$ is a monotonically increasing sequence of $\mathcal{M}$-measurable functions that converge to

$$h = \frac{g}{1-g}.$$

Hence, by the Lebesgue monotone convergence theorem and (1.5-7), we obtain

$$\int_E d\lambda = \int_E h\,d\mu$$

or

$$\lambda(E) = \int_E h\,d\mu. \tag{1.5-8}$$

Finally, this holds for all $E \in \mathcal{M}$, so letting $E = S$, we have

$$\int_S h\,d\mu = \lambda(S) < \infty,$$

and thus, h is μ integrable. This established, (1.5-8) is our desired conclusion. □

1.5.5. **Definition.** A measure μ on S is said to be σ *finite* provided that

(1) there exists a countable collection of sets in $\mathcal{M}$, $E_1, E_2, \ldots$, pairwise disjoint, such that

$$S = \bigcup_{i=1}^{\infty} E_i, \quad \text{and}$$

(2) $\mu(E_i) < \infty$ for each i.

1.5.6. **Theorem** (Radon–Nikodym). *If λ, μ are measures defined on the same σ field $\mathcal{M}$, λ bounded, μ σ finite, $\lambda << \mu$, then there is an h, integrable with respect to μ, such that for all $E \in \mathcal{M}$*

$$\lambda(E) = \int_E h \, d\mu.$$

Proof. Use 1.5.4 to show

$$\lambda(E \cap E_i) = \int_{E \cap E_i} h_i \, d\mu,$$

where h_i is defined on E_i. Let $h = \sum_{i=1}^{\infty} h_i$ and use the countable additivity of λ and μ—details left to the reader. □

1.5.7. **Example.** Lebesgue measure on the plane is an example of a σ-finite but unbounded measure. For

$$\mathrm{R}^2 = \bigcup_{n=-\infty}^{\infty} \bigcup_{m=-\infty}^{\infty} (n, n+1] \times (m, m+1],$$

and each square $(n, n+1] \times (m, m+1]$ has measure (area) 1. However, $\mu(\mathrm{R})^2 = \infty$.

The h in Theorem 1.5.6 is often called the Radon–Nikodym derivative of λ with respect to μ and one formally writes $h = d\lambda/d\mu$.

1.6 Continuously Distributed Random Vectors

Up to now, all of our examples have been of a rather trivial nature, primarily discrete spaces. In this section, we will introduce the notion of a continuous distribution, and as a consequence, we will begin to introduce those examples that will be germane to our study.

1.6.1. **Definition.** Let $\mathbf{X}$ be a random vector. X is called *continuously distributed* provided $P_{\mathbf{X}} << \mu_n$, μ_n being Lebesgue measure on R^n. Note that, in this case, $P_{\mathbf{X}}$ cannot be concentrated on a point, hence it cannot be a discrete measure.

1.6.2. **Theorem.** *Let $\mathbf{X}$ be a continuously distributed random vector* (c.d.r.v.). *Then there exists a Baire function $f_{\mathbf{X}}$ such that*

$$P_{\mathbf{X}}(B) = \int_B f_{\mathbf{X}} \, d\mu_n$$

for all $B \in B_n$.

Proof. This is the Radon–Nikodym theorem. □

1.6.3. **Definition.** Let X be a c.d.r.v. Then the function $f_{\mathbf{X}}$ in 1.6.2 is called a *probability density function* (p.d.f.)

1.6.4. **Theorem.** *Let* $\mathbf{X}$ *be a random vector with* p.d.f. f_X. *If* $f_{\mathbf{X}}$ *is continuous* a.e., *then*

$$f_{\mathbf{X}}(x_1, \ldots, x_n) = \frac{\partial^n F_X}{\partial x_1 \, \partial x_2, \ldots, \partial x_n}.$$

Proof.

$$F_{\mathbf{X}}(x_1, x_2, \ldots, x_n) = \int_{-\infty}^{x_1} \int_{-\infty}^{x_2} \cdots \int_{-\infty}^{x_n} f_{\mathbf{X}}(z_1, \ldots, z_n) \, dz_1 \cdots dz_n.$$

Use the fundamental theorem of integral calculus. □

1.6.5. **Theorem.** *Let* X *be a random vector with* p.d.f. $f_{\mathbf{X}}$. *Then for all* $A \in B(\mathbf{X})$,

$$\int_A g \, dP_{\mathbf{X}} = \int_A g \cdot f_{\mathbf{X}} \, d\mu_n \tag{1.6-1}$$

for any Baire function g.

Proof. The proof is the usual scenario utilizing the definition of integral. Namely, if $g = C_E$, E Borel, then the above reduces to

$$P_{\mathbf{X}}(A \cap E) = \int_{A \cap E} f_{\mathbf{X}} \, d\mu_n,$$

which is true by 1.6.2. From this result, it easily follows that (1.6-1) holds for simple functions. Finally, if g_n is a sequence of simple functions such that $g_n \uparrow g$, we have

$$\int_A g_n \, dP_{\mathbf{X}} = \int_A g_n \cdot f_{\mathbf{X}} \, d\mu,$$

whence

$$\int_A g \, dP_{\mathbf{X}} = \int_A g \cdot f_{\mathbf{X}} \, d\mu,$$

the left-hand side following from the definition of integral and the right-hand side following from the Lebesgue monotone convergence theorem (Appendix D). □

With (1.6-1) established, one can formally write

$$dP_{\mathbf{X}} = dF_{\mathbf{X}} = f_{\mathbf{X}} \, d\mu_n$$

with impunity, that is, formal substitutions under integral signs will produce correct formulas. Note that (1.6-1) and (1.2-2) together imply the special case

$$\int_{a_1}^{b_1} \cdots \int_{a_n}^{b_n} g(z_1, \ldots, z_n) f_{\mathbf{X}}(z_1, \ldots, z_n) \, dz_1 \cdots dz_n$$

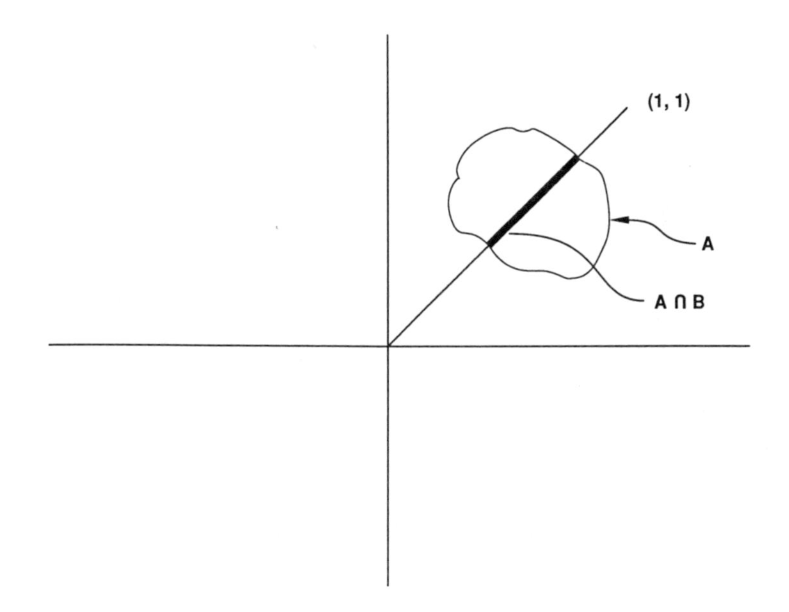

FIGURE 1.4. A counterexample showing that the converse of Theorem 1.6.6 fails.

$$= \int_{X^{-1}(\prod_{i=1}^{n}(a_i,b_i])} (g \circ \mathbf{X})\, dp. \qquad (1.6\text{-}2)$$

The reader might be interested in rereading the remark following Definition 1.2.5.

1.6.6. **Theorem.** *If* $\mathbf{X} = (X_1, X_2)$ *is a* c.d.r.v., *then so are* X_1 *and* X_2.

Proof. Suppose $A \subset \mathrm{R}$ and $\mu(A) = 0$. Then $\mu_2(A \times \mathrm{R}) = 0$ and so by hypothesis $P_{\mathbf{X}}(A \times \mathrm{R}) = 0$. Thus,

$$\begin{aligned} 0 &= P_{\mathbf{X}}(A \times \mathrm{R}) = P(\mathbf{X}^{-1}(A \times \mathrm{R})) = P(X_1^{-1}(A) \cap X_2^{-1}(\mathrm{R})) \\ &= P(X_1^{-1}(A) \cap \mathrm{R}^2) = P(X_1^{-1}(A)) = P_{X_1}(A). \end{aligned}$$

Hence, we have shown that $\mu(A) = 0$ implies $P_{X_1}(A) = 0$. By symmetry this also holds for X_2. □

1.6.7. **Example.** The converse of 1.6.6 does not hold. Let L be the line segment from (0,0) to (1,1) in R^2 and define $P_X(A) = 1/\sqrt{2}$ length of $(A \cap L)$ (one-dimensional Lebesgue measure of $A \cap L$) (see Figure 1.4). Then, if $A \subset L$, $\mu_2(A) = 0$ and A can be chosen so that $P_X(A) \neq 0$ (take $A = L$ for instance). However,

$$\begin{aligned} P_{X_1}(A) = P_X(A \times \mathrm{R}) = \frac{1}{\sqrt{2}}\mu((A \times \mathrm{R}) \cap L) &= \frac{1}{\sqrt{2}} \cdot \sqrt{2}\,\mu(A \cap [0,1]) \\ &= \mu(A \cap [0,1]). \end{aligned}$$

1.6.8. **Theorem.** *Suppose $(S, \mathcal{M}, \mu)$ is a measure space, $\Im \subset \mathcal{M}$ is a σ field, and f_1 and f_2 are $\Im$-measurable functions. If*

$$\int_A f_1 \, d\mu = \int_A f_2 \, d\mu$$

for all $A \in \Im$, then $f_1 = f_2$ (a.e.).

Proof. Let $f = f_1 - f_2$. Then f is $\Im$ measurable, and the theorem is equivalent to showing that

$$\int_A f \, d\mu = 0, \quad \text{for all} \quad A \in \Im \Rightarrow f = 0 \quad \text{(a.e.)}.$$

Suppose not. Then there is some $M \in \mathcal{M}$, $\mu(M) > 0$, such that $f(x) \neq 0$ on M. Define

$$\begin{aligned} A_1 &= \{x \mid f(x) > 0\} = f^{-1}((0, \infty)) \\ A_2 &= \{x \mid f(x) < 0\} = f^{-1}((-\infty, 0)). \end{aligned}$$

Then A_1 and A_2 are elements of $\Im$ since f is $\Im$ measurable and moreover

$$A_1 \cup A_2 \supset M.$$

Hence,

$$\mu(A_1) + \mu(A_2) \geq \mu(M) > 0,$$

so either $\mu(A_1) > 0$ or $\mu(A_2) > 0$. Thus, either $\int_{A_1} f \, d\mu \neq 0$ or $\int_{A_2} f \, d\mu \neq 0$, so we have established the contrapositive. □

1.6.9. **Example.** Theorem 1.6.8 would fail if f_1 and f_2 were simply hypothesized to be measurable rather than $\Im$ measurable. As an illustration, suppose $\Im = \{\phi, [-1, 0], (0, 1], [-1, 1]\}$ where $S = \{-1, 1\}$ and f is as shown in Figure 1.5. Then f is certainly Lebesgue measurable but not $\Im$ measurable. Moreover, $f \neq 0$, yet

$$\int_{[-1,0]} f(x) \, dx = \int_{(0,1]} f(x) \, dx = 0,$$

whence

$$\int_{-1}^{1} f(x) \, dx = 0,$$

so $\int_A f(x) \, dx = 0$ for every $A \in \Im$.

1.6.10. **Theorem.** *Suppose X_1 and X_2 are* c.d.r.v. *and $\mathbf{X} = (X_1, X_2)$. Then X_1 and X_2 are independent if and only if*

$$f_{\mathbf{X}} = f_{X_1} \cdot f_{X_2}.$$

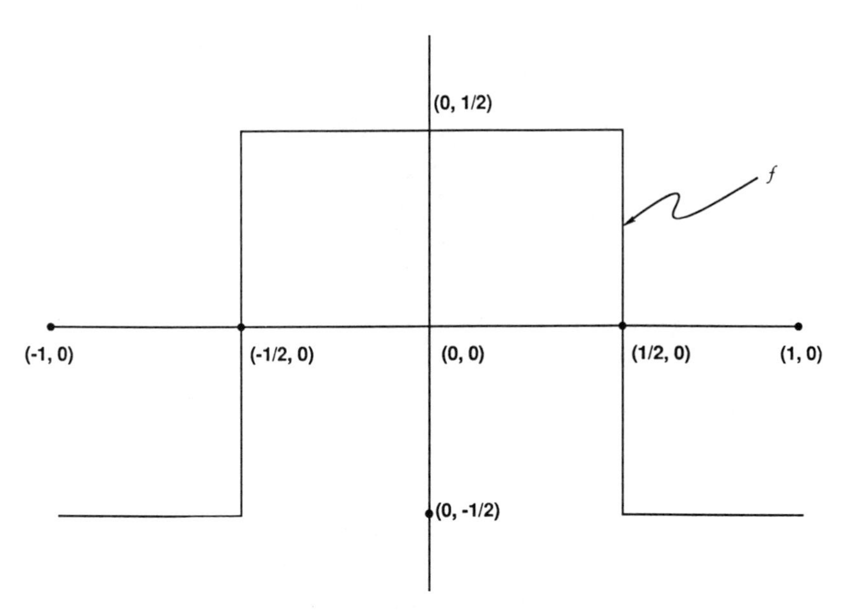

FIGURE 1.5. An example showing the necessity of assuming f_1 and f_2 to be $\mathcal{A}$ measurable in Theorem 1.6.8.

Proof. Let us define

$$f(x_1, x_2) = f_{X_1}(x_1) \cdot f_{X_2}(x_2).$$

Since both f_{X_1} and F_{X_2} are Baire functions, that is, are Borel measurable, so is f. (Use arguments similar to those in Appendix C.)

Now, by Theorem 1.4.10, X_1 and X_2 are independent if and only if for every $A, B \in B_1$

$$P_{\mathbf{X}}(A \times B) = P_{X_1}(A) \cdot P_{X_2}(B).$$

But this condition is equivalent to

$$\int_{A\times B} f_{\mathbf{X}}\, d\mu_2 = \int_{A\times B} f_{X_1}\, d\mu_1 \int_B f_{X_2}\, d\mu_2$$

or

$$\int_{A\times B} f_{\mathbf{X}}\, d\mu_2 = \int_{A\times B} f\, d\mu_2 \quad \text{for all} \quad A, B \in B_1.$$

But B_2, the Borel sets in R^2, are generated by sets of the form $A \times B$; $A, B \in B_1$, so

$$\int_E f_{\mathbf{X}}\, d\mu_2 = \int_E f\, d\mu_2$$

for all $E \in B_2$. By 1.6.8, $f_X = f$ (a.e.). □

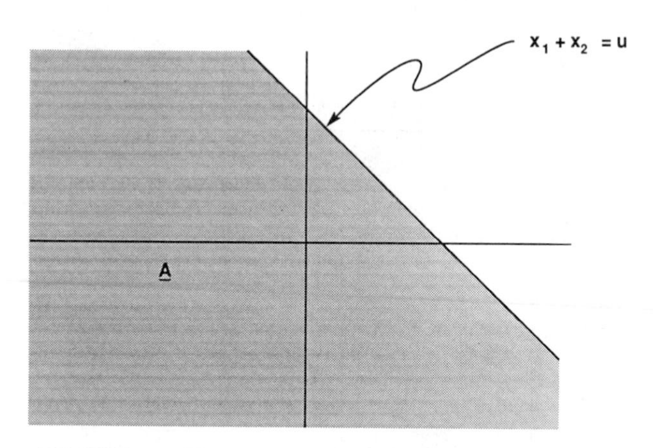

FIGURE 1.6. The set A in the proof of Theorem 1.6.11.

1.6.11. **Theorem.** *Suppose X_1 and X_2 are independent and have continuous* p.d.f.'s *f_{X_1} and f_{X_2}, respectively. Then $X_1 + X_2$ is a* c.d.r.v. *and has* p.d.f.

$$f_{X_1+X_2} = f_{X_1} * f_{X_2},$$

where $*$ represents convolution.

Proof. Let $\mathbf{X} = (X_1, X_2)$ and define

$$H(u) \triangleq P(X_1 + X_2 \le u).$$

Then

$$H(u) = P_{\mathbf{X}}(A),$$

where

$$A = \{(x_1, x_2) \mid x_1 + x_2 \le u\}.$$

This set is shown in Figure 1.6.

Next, define the set $A(a)$ shown in Figure 1.7. Then

$$H(u) = \lim_{a \to \infty} P_{\mathbf{X}}(A(a))$$

since $A(a)$ is an increasing sequence of nested sets. (See Exercise 1.10.3). Hence, using Theorem 1.6.10

$$\begin{aligned}
H(u) &= \lim_{a \to \infty} \int\int_{A(a)} f_{\mathbf{X}}(x_1, x_2)\, dx_1 dx_2 \\
&= \lim_{a \to \infty} \int\int_{A(a)} f_{X_1}(x_1) f_{X_2}(x_2)\, dx_1\, dx_2 \\
&= \lim_{a \to \infty} \int_{-a}^{u+a} \left(\int_{-a}^{u-x_1} f_{X_2}(x_2)\, dx_2 \right) f_{X_1}(x_1)\, dx_1
\end{aligned}$$

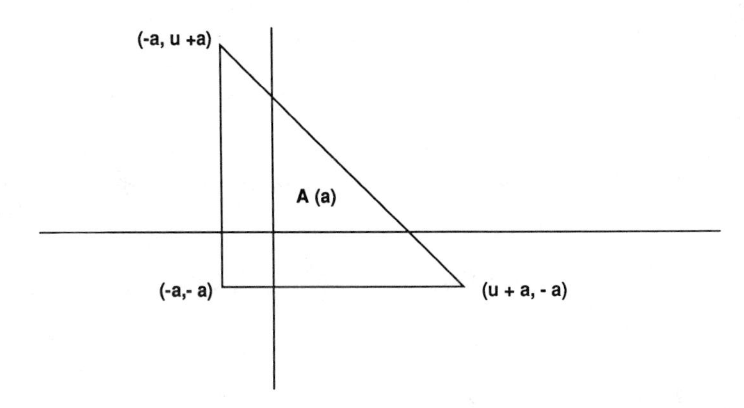

FIGURE 1.7. The domain used to construct the integral in the proof of Theorem 1.6.11.

$$
\begin{aligned}
&= \int_{-\infty}^{\infty} \int_{-\infty}^{u-x_1} f_{X_2}(x_2) f_{X_1}(x_1)\, dx_2\, dx_1 \\
&= \int_{-\infty}^{\infty} F_{X_2}(u-x) f_{X_1}(x)\, dx.
\end{aligned}
$$

Thus,

$$H'(u) = \int_{-\infty}^{\infty} F'_{X_2}(u-x) f_{X_1}(x)\, dx$$

or

$$
\begin{aligned}
f_{X_1+X_2}(u) &= \int_{-\infty}^{\infty} F_{X_2}(u-x) f_{X_1}(x)\, dx \\
&= f_1 * f_2. \qquad \square
\end{aligned}
$$

1.6.12. **Theorem.** *Let X be a random variable with* p.d.f. f. *Then*

(a) $\mu = E(X) = \int_{-\infty}^{\infty} x f(x)\, dx$, *and*

(b) $\sigma^2 = V(X) = \int_{-\infty}^{\infty} (x-\mu)^2 f(x)\, dx$.

Proof.

(a) This is just Theorem 1.6.5 applied to the result in Theorem 1.3.2. Specifically,

$$E(X) = \int_{-\infty}^{\infty} x\, dP_X = \int_{-\infty}^{\infty} x f(x)\, dx.$$

(b) From Definition 1.3.4, parts (a) and (b), we have

$$V(X) = \int_\Omega (X-\mu)^2\, dP.$$

From the change of variables formula (Theorem 1.2.7) and the observation that $\Omega = X^{-1}(\mathrm{R})$, we obtain

$$V(X) = \int_{\mathrm{R}} (x-\mu)^2\, dP_X.$$

Finally, using 1.6.5, we obtain

$$V(X) = \int_{-\infty}^{\infty} (x-\mu)^2 f(x)\, dx. \qquad \square$$

1.6.13. **Example.** Let

$$f(x) = \frac{1}{\sqrt{2\pi}\,\sigma} e^{-x^2/2\sigma^2}.$$

We will show that

$$\int_{-\infty}^{\infty} f(x)\, dx = 1,$$

so that $f(x)$ is a possible p.d.f.

Let

$$I = \int_{-\infty}^{\infty} e^{-x^2}\, dx.$$

Then,

$$\begin{aligned} I^2 &= \int_{-\infty}^{\infty}\int_{-\infty}^{\infty} e^{-x^2} e^{-y^2}\, dx\, dy \\ &= \int_{-\infty}^{\infty}\int_{-\infty}^{\infty} e^{-(x^2+y^2)}\, dx\, dy \\ &= \int_0^{\infty}\int_0^{2\pi} e^{-r^2}\, r\, dr\, d\theta \\ &= 2\pi \int_0^{\infty} e^{-r^2}\, r\, dr \\ &= 2\pi \left[-\frac{e^{-r^2}}{2} \Big|_0^{\infty} \right] \\ &= \pi. \end{aligned}$$

Thus, $I = \sqrt{\pi}$, that is,

$$\sqrt{\pi} = \int_{-\infty}^{\infty} e^{-x^2}\, dx.$$

Making a change of variables via $x = (1/\sqrt{2}\,\sigma)z$, we obtain

$$\sqrt{\pi} = \int_{-\infty}^{\infty} e^{-z^2/2\sigma^2} \frac{1}{\sqrt{2}\,\sigma} dz$$

or

$$1 = \frac{1}{\sqrt{2\pi}\,\sigma} \int_{-\infty}^{\infty} e^{-z^2/2\sigma^2} dz$$

as required.

Next note that if $\int_{-\infty}^{\infty} f(x)\,dx = 1$, then $\int_{-\infty}^{\infty} f(x-\mu)\,dx = 1$ so that the function defined by

$$f(x) = \frac{1}{\sqrt{2\pi}\,\sigma} e^{-(x-\mu)^2/2\sigma^2}$$

is also a p.d.f.

1.6.14. **Lemma.** *Let f be a function satisfying $f(x) \geq 0$ and*

(a) $\displaystyle\int_{-\infty}^{\infty} f(x)\,dx = 1$

(b) $\displaystyle\int_{-\infty}^{\infty} x f(x)\,dx = 0$

(c) $\displaystyle\int_{-\infty}^{\infty} x^2 f(x)\,dx = 1.$

Then, $g(x) \triangleq (1/\sigma)f(x-\mu/\sigma)$ is a p.d.f. *with mean μ and variance σ^2.*

Proof. Letting $x = \sigma z + \mu$, we have

$$\int_{-\infty}^{\infty} g(x)\,dx = \int_{-\infty}^{\infty} \frac{1}{\sigma} f\left(\frac{x-\mu}{\sigma}\right)\,dx = \int_{-\infty}^{\infty} f(z)\,dz = 1,$$

so g is a p.d.f. Next, note

$$\begin{aligned} \int_{-\infty}^{\infty} x g(x)\,dx &= \int_{-\infty}^{\infty} (\mu + \sigma z) f(z)\,dz \\ &= \mu \int_{-\infty}^{\infty} f(z)\,dz + \sigma \cdot 0 \\ &= \mu \end{aligned}$$

so that g has mean μ.

Finally,

$$\int_{-\infty}^{\infty} (x-\mu)^2 g(x)\,dx = \int_{-\infty}^{\infty} \sigma^2 z^2 f(z)\,dz = \sigma^2. \qquad \square$$

1.6.15. **Example** (1.6.13 continued). From 1.6.13 it follows that

$$f(x) = \frac{1}{\sqrt{2\pi}} e^{-x^2/2}$$

satisfies (a) of 1.6.14. It is obvious that (b) of 1.6.14 holds and (c) can be easily checked using (a) and (b) and integration by parts. Hence,

$$g(x) = \frac{1}{\sqrt{2\pi}\,\sigma} e^{-(x-\mu)^2/2\sigma^2}$$

is a p.d.f. with mean μ and variance σ^2.

1.6.16. **Definition.** A random variable X is said to be *normally distributed* with mean μ and variance σ^2 providing X is continuously distributed with p.d.f.

$$f_X(x) = \frac{1}{\sqrt{2\pi}\,\sigma} e^{-(x-\mu)^2/2\sigma^2}.$$

We write this as $X \sim N(\mu, \sigma^2)$.

1.6.17. **Lemma.** *Let*

$$g_i(x) = f_i(x - \mu_i); \qquad i = 1, 2.$$

Then

$$(g_1 * g_2)(y) = (f_1 * f_2)(y - \mu_1 - \mu_2).$$

Proof. Exercise. □

1.6.18. **Theorem.** *If X_1 and X_2 are independent, $X_i \sim N(\mu_i, \sigma_i^2)$, for $i = 1, 2$, then*

$$\alpha_1 X_1 + \alpha_2 X_2 \sim N(\alpha_1\mu_1 + \alpha_2\mu_2, (\alpha_1\sigma_1)^2 + (\alpha_2\sigma_2)^2).$$

Proof. Exercise. □

Next we are going to study the generalization of the normal distribution to random vectors, the so-called multivariate normal distribution. Our emphasis on this distribution is intentional as it arises quite naturally in estimation theory. Before we do this, however, we must do a bit of matrix algebra.

1.7 The Matrix Inversion Lemma

1.7.1. **Theorem.** *Let A be a square matrix partitioned as*

$$A = \left[\begin{array}{c|c} A_{11} & A_{12} \\ \hline A_{21} & A_{22} \end{array}\right]; \qquad A_{11}, A_{22} \text{ square.}$$

If A is nonsingular, let

$$A^{-1} = \begin{bmatrix} F_{11} & F_{12} \\ F_{21} & F_{22} \end{bmatrix},$$

where F_{ij} and A_{ij} are the same size. Then,

(a) *if A_{22}^{-1} exists,*

$$\begin{aligned} F_{11} &= (A_{11} - A_{12}A_{22}^{-1}A_{21})^{-1} \\ F_{12} &= -F_{11}A_{12}A_{22}^{-1} \\ F_{21} &= -A_{22}^{-1}A_{21}F_{11} \\ F_{22} &= A_{22}^{-1} + A_{22}^{-1}A_{21}F_{11}A_{12}A_{22}^{-1}; \end{aligned}$$

(b) *and if A_{11}^{-1} exists,*

$$\begin{aligned} F_{11} &= A_{11}^{-1} + A_{11}^{-1}A_{12}F_{22}A_{21}A_{11}^{-1} \\ F_{12} &= -A_{11}^{-1}A_{12}F_{22} \\ F_{21} &= -F_{22}A_{21}A_{11}^{-1} \\ F_{22} &= (A_{22} - A_{21}A_{11}^{-1}A_{12})^{-1}. \end{aligned}$$

Proof. Direct calculation using block multiplication. □

1.7.2. **Corollary.** *If the indicated inverses exist, then*

$$(A_{11} - A_{12}A_{22}^{-1}A_{21})^{-1} = A_{11}^{-1} + A_{11}^{-1}A_{12}(A_{22} - A_{21}A_{11}^{-1}A_{12})^{-1}A_{21}A_{11}^{-1}.$$

1.7.3. **Corollary.**

$$[I + A]^{-1} = I + [I + A^{-1}]^{-1}.$$

Proof. Let $A_{11} = I$, $A_{12} = I$, $A_{21} = I$, and $A_{22} = A$. □

1.7.4. **Lemma.** *Let A be partitioned as in* 1.7.1. *Then*

$$\det(A) = \det(A_{11} - A_{12}A_{22}^{-1}A_{21}) \cdot \det(A_{22}).$$

Proof. Here, we exploit the invariance of determinants under the addition of linear combinations of rows and columns to other rows and columns.

$$\begin{aligned} \det \begin{bmatrix} A_{11} & A_{12} \\ A_{21} & A_{22} \end{bmatrix} &= \det \begin{bmatrix} A_{11} - A_{12}A_{22}^{-1}A_{21} & A_{12} \\ 0 & A_{22} \end{bmatrix} \\ &= \det \begin{bmatrix} A_{11} - A_{12}A_{22}^{-1}A_{21} & 0 \\ 0 & A_{22} \end{bmatrix}. \end{aligned}$$

The result now follows. □

1.7.5. **Lemma.** *Let A be partitioned as in* 1.7.1. *Then*

$$\operatorname{tr}(A) = \operatorname{tr}(A_{11}) + \operatorname{tr}(A_{22}).$$

1.8 The Multivariate Normal Distribution

Suppose that $X_1, X_2, \ldots, X_n$ are pairwise independent, normally distributed random variables. Then, for any matrix A of constants, A nonsingular (for technical reasons that will be apparent later), we can define a new random vector $\mathbf{Y}$ by the simple matrix equation

$$\mathbf{Y} = \begin{bmatrix} Y_1 \\ Y_2 \\ \vdots \\ Y_n \end{bmatrix} = A \begin{bmatrix} X_1 \\ X_2 \\ \vdots \\ X_n \end{bmatrix}.$$

Using simple mathematical induction, one can generalize Theorem 1.6.18 to a finite number of random variables, and the result so obtained implies that each Y_i will be normally distributed. In general, however, the Y_i's will not be independent since they can be correlated. It is thus natural to ask the following. If a random vector $\mathbf{Y}$ has normally distributed marginals Y_i that are correlated, does there exist a set of independent, normally distributed random variables $X_1, \ldots, X_n$ and a matrix A such that (1.8-1) holds? The answer is provided by the following theorem. Before reading this rather lengthy theorem and proof, the reader is advised to review the latter part of Section 1.3 as well as Appendix F on the spectral theorem for symmetric matrices.

1.8.1. **Theorem.** *Let $\mathbf{Y}^T = (Y_1, Y_2, \ldots, Y_n)$ be a random vector such that for each i, $Y_i \sim N(\mu_i, \sigma_i^2)$. Moreover, let*

$$\Gamma = \operatorname{cov}(Y, Y)$$

and

$$\boldsymbol{\mu} = \begin{bmatrix} \mu_1 \\ \mu_2 \\ \vdots \\ \mu_n \end{bmatrix},$$

both of which are assumed known. Then, if Γ is nonsingular (see 1.3.15),

(a) *the function $f : \mathrm{R}^n \to \mathrm{R}$ defined by*

$$f(\mathbf{Z}) = \frac{1}{(2\pi)^{n/2}\sqrt{\det(\Gamma)}} e^{-\langle \Gamma^{-1}(\mathbf{Z}-\boldsymbol{\mu}), \mathbf{Z}-\boldsymbol{\mu}\rangle/2} \tag{1.8-1}$$

($\langle \cdot, \cdot \rangle$ *denotes the inner product in* R^n) *is a possible*[6] p.d.f. *for* Y, *and*

[6]Meaning that f produces means, variances, and marginals identical to those for Y.

(b) *there exists an orthogonal matrix V such that the random vector X defined by*

$$X = VY$$

has marginals that are pairwise independent and normally distributed.

Proof. Parts (a) and (b) will be proved together.

Using the spectral theorem for symmetric matrices (Appendix E), there exists an orthogonal matrix V that diagonalizes Γ, that is,

$$D \triangleq \begin{bmatrix} \lambda_1 & 0 & \dots & 0 \\ 0 & \lambda_2 & & \\ \vdots & \vdots & \ddots & \vdots \\ 0 & 0 & 0 & 0 \\ 0 & 0 & 0 & \lambda_n \end{bmatrix} = V\Gamma V^T. \tag{1.8-2}$$

Then, letting

$$X = VY, \tag{1.8-3}$$

we have

$$\begin{aligned} \mathrm{cov}(X,) &= \mathrm{cov}(VY, VY) \\ &= V\,\mathrm{cov}(Y, Y)V^T \\ &= V\Gamma V^T \\ &= D, \end{aligned}$$

so

$$\lambda_i = \mathrm{cov}(X_i, X_i) \triangleq d_i^2, \tag{1.8-4}$$

where the assumption that λ_i can be written in the form d_i^2 follows from 1.3.15. Also note that

$$D^{-1} = (V\Gamma V^T)^{-1} = V\Gamma^{-1}V^T \tag{1.8-5}$$

and

$$\det(\Gamma) = \det(D) = d_1^2 d_2^2 \cdots d_n^2. \tag{1.8-6}$$

Define

$$\boldsymbol{\nu} \triangleq \begin{bmatrix} E(X_1) \\ E(X_2) \\ \vdots \\ E(X_n) \end{bmatrix} = \begin{bmatrix} \nu_1 \\ \nu_2 \\ \vdots \\ \nu_n \end{bmatrix}. \tag{1.8-7}$$

Then

$$\boldsymbol{\nu} = V\boldsymbol{\mu} \tag{1.8-8}$$

by Theorem 1.3.13. Letting

$$S_n \triangleq (-\infty, x_1] \times (-\infty, x_2] \cdots (-\infty, x_n], \tag{1.8-9}$$

we can express the cumulative distribution function for X as follows.

$$\begin{aligned} F_{\mathbf{X}}(x_1, x_2, \ldots, x_n) &= P(\mathbf{X} \leq (x_1, x_2, \ldots, x_n)) \\ &= P(\mathbf{X} \in S_n) \\ &= P(V\mathbf{Y} \in S_n) \end{aligned}$$

$$V_{\mathbf{X}}(x_1, x_2, \ldots, x_n) = P(\mathbf{Y} \in V^{-1}(S_n)). \tag{1.8-10}$$

Assume for the moment that f is, in fact, a p.d.f. that is compatible with $\mathbf{Y}$ (this will be shown later in the proof). Then, using (1.8-10),

$$\begin{aligned} F_{\mathbf{X}}(x_1, \ldots, X_n) &= \int_{V^{-1}(S_n)} \cdots \int \frac{1}{(2\pi)^{n/2}\sqrt{\det(\Gamma)}} e^{-\langle \Gamma^{-1}(\mathbf{Y}-\boldsymbol{\mu}), \mathbf{Y}-\boldsymbol{\mu}\rangle/2} \\ &\quad \times dy_1 \cdots dy_n \\ &= \frac{1}{(2\pi)^{n/2}\sqrt{\det(\Gamma)}} \\ &\quad \times \int_{V^{-1}(S_n)} \cdots \int e^{-\langle \Gamma^{-1}V^T V(\mathbf{Y}-\boldsymbol{\mu}), V^T V(\mathbf{Y}-\boldsymbol{\mu})\rangle/2} \\ &\quad \times dy_1 \cdots dy_n \\ &= \frac{1}{(2\pi)^{n/2}\sqrt{\det(\Gamma)}} \\ &\quad \times \int_{(S_n)} \cdots \int e^{-\langle \Gamma^{-1}V^T(\mathbf{X}-\boldsymbol{\nu}), V^T(\mathbf{X}-\boldsymbol{\nu})\rangle/2} \\ &\quad \times J(V)\, dx_1 \cdots dx_n, \end{aligned} \tag{1.8-11}$$

where $J(V)$ is the determinant of the Jacobian of the transformation V and $V(\mathbf{Y} - \boldsymbol{\mu}) = X - \boldsymbol{\nu}$ follows from (1.8-3) and (1.8-8). Hence, since V is orthogonal, $J(V) = 1$, and we have

$$\begin{aligned} F_X(x_1, \ldots, x_n) &= \frac{1}{(2\pi)^{n/2}\sqrt{\det(\Gamma)}} \int_{S_n} \cdots \int e^{-\langle V\Gamma^{-1}V^T(\mathbf{X}-\boldsymbol{\nu}), \mathbf{X}-\boldsymbol{\nu}\rangle/2} \\ &\quad \times dx_1 \cdots dx_n \\ &= \frac{1}{(2\pi)^{n/2}\sqrt{\det(\Gamma)}} \int_{S_n} \cdots \int e^{-\langle D^{-1}(\mathbf{X}-\boldsymbol{\nu}), \mathbf{X}-\boldsymbol{\nu}\rangle/2} \\ &\quad \times dx_1 \cdots dx_n \\ &= \frac{1}{(2\pi)^{n/2}\sqrt{\det(\Gamma)}} \\ &\quad \times \int_{-\infty}^{x_1} \cdots \int_{-\infty}^{x_n} e^{-[((x_1-\nu_i)/d_1)^2 + \cdots + ((x_n-\nu_n)/d_n)^2]/2} \\ &\quad \times dx_1 dx_2 \cdots dx_n. \end{aligned}$$

But the integral on the right can be expressed as an iteration of n integrals as

$$F_{\mathbf{X}}(x_1, \ldots, x_n) = \frac{1}{\sqrt{2\pi}\, d_1} \int_{-\infty}^{x_1} e^{-((x_1-\nu_i)/d_1)^2/2} dx_1$$

$$\cdots \frac{1}{\sqrt{2\pi}\, d_n} \int_{-\infty}^{x_n} e^{-((x_1-\nu_n)/d_n)^2/2} dx_n. \tag{1.8-12}$$

By replacing each of the upper limits of integration except x_i with $+\infty$, we obtain from Example 1.6.15 and Theorem 1.2.2 that for each i

$$F_{X_i}(x_i) = \frac{1}{\sqrt{2\pi}\, d_i} \int_{\infty}^{x_i} e^{-((x_i-\nu_n)/d_i)^2/2}\, dx_i,$$

whence each X_i is normally distributed and

$$F_{\mathbf{X}}(x_1, \ldots, x_n) = F_{X_i}(x_1) \cdots F_{X_n}(x_n). \tag{1.8-13}$$

Of course, (1.8-13) implies that $X_1, \ldots, X_n$ are independent. Thus, the theorem has been proved provided that f is indeed a p.d.f. that is compatible with $\mathbf{Y}$. We now show this.

First note that the right-hand side of Equation (1.8-11) is simply

$$\int_{V^{-1}(S_n)} \cdots \int f(\mathbf{y})\, d\mathbf{y},$$

so if we replace S_n by R^n and equate the right-hand sides of (1.8-11) and (1.8-12), we obtain

$$\begin{aligned} \int_{\mathrm{R}^n} \cdots \int f(\mathbf{y})\, d\mathbf{y} \quad &= \quad \frac{1}{\sqrt{2\pi}\, d_1} \int_{-\infty}^{\infty} e^{-((x_1-E(X_1))/d_1)^2/2} dx_1 \\ &\times \cdots \frac{1}{\sqrt{2\pi}\, d_n} \int_{-\infty}^{\infty} e^{-((x_n-E(X_n))/d_n)^2/2} dx_n \end{aligned}$$

and by Example 1.6.15 that

$$\int_{\mathrm{R}^n} \cdots \int f(\mathbf{y})\, d\mathbf{y} = 1. \tag{1.8-14}$$

Hence, f is indeed a p.d.f. If we now show that f produces the correct mean, covariance matrix, and marginal distributions for Y, we will be finished.

For the covariance matrix, note that

$$\begin{aligned} C \quad &\triangleq \quad \frac{1}{(2\pi)^{n/2}\sqrt{\det(\Gamma)}} \int_{\mathrm{R}^n} \cdots \int_{\mathbf{Y}-\boldsymbol{\mu}} (\mathbf{Y}-\boldsymbol{\mu})^T e^{-\langle \Gamma^{-1}(\mathbf{Y}-\boldsymbol{\mu}), \mathbf{Y}-\boldsymbol{\mu}\rangle/2} \\ &\quad \times dy_1 \cdots dy_n \end{aligned}$$

$$\begin{aligned}
&= V^T \left[\frac{1}{(2\pi)^{n/2}\sqrt{\det(\Gamma)}} \int_{\mathrm{R}^n} \right. \\
&\quad \left. \cdots \int V(\mathbf{Y}-\boldsymbol{\mu})(\mathbf{Y}-\boldsymbol{\mu})^T V^T e^{-\langle\cdot,\cdot\rangle/2} dy_1 \cdots dy_n \right] V \\
&= V^T \left[\frac{1}{(2\pi)^{n/2} d_1 \cdots d_n} \int_{\mathrm{R}^n} \cdots \int (\mathbf{X}-\boldsymbol{\nu})(\mathbf{X}-\boldsymbol{\nu})^T e^{-\langle D^{-1}(X-\boldsymbol{\nu}), X-\boldsymbol{\nu}\rangle/2} \right. \\
&\quad \left. \times\ dx_1 \cdots dx_n \right] V.
\end{aligned}$$

Now, the (i,j)th entry of the matrix in the above brackets is given by

$$\begin{aligned}
\alpha_{ij} &= \frac{1}{2\pi\, d_i d_j} \int_{-\infty}^{\infty} \int_{-\infty}^{\infty} (x_i - \nu_i)(x_j - \nu_j) e^{-[((x_i-\nu_i)/d_i)^2 + ((x_j-\nu_j)/d_j)^2]/2} \\
&\quad \times dx_i\, dx_j \quad \text{for} \quad i \neq j
\end{aligned}$$

and

$$\alpha_{ii} = \frac{1}{\sqrt{2\pi}\, d_i^2} \int_{-\infty}^{\infty} (x_i - \nu_i)^2 e^{-((x_i-\nu_i)/d_i)^2/2} dx_i.$$

Clearly, $\alpha_{ij} = 0$ when $i \neq j$ and from 1.6.15

$$\alpha_{ii} = d_i^2.$$

Thus, the above calculation becomes

$$C = V^T D V$$

or

$$C = \Gamma.$$

Therefore, f produces the correct covariance matrix.

In a similar fashion, we can show that

$$\frac{1}{(2\pi)^{n/2}\sqrt{\det(\Gamma)}} \int_{\mathrm{R}^n} \cdots \int Y e^{-1/2\langle\Gamma^{-1}(Y-\boldsymbol{\mu}),(Y-\boldsymbol{\mu})\rangle} dy_1 \cdots dy_n = \boldsymbol{\mu},$$

and we leave this to the reader as an exercise.

Finally, to check the marginal distributions, we proceed as follows, checking the result for Y_1 for notational convenience. Define

$$H(y_1) = \int_{-\infty}^{y_1} \int_{-\infty}^{\infty} \cdots \int_{-\infty}^{\infty} \frac{1}{(2\pi)^{n/2}\sqrt{\det(\Gamma)}} e^{-\langle\Gamma^{-1}(Y-\boldsymbol{\mu}), Y-\boldsymbol{\mu}\rangle/2} dy_1 \cdots dy_n. \tag{1.8-15}$$

If f works as we hope, (1.8-15) should produce the cumulative distribution for Y_1 (see 1.2.2). Next, partition Γ as

$$\Gamma = \begin{bmatrix} \Gamma_{11} & \Gamma_{12} \\ \Gamma_{21} & \Gamma_{22} \end{bmatrix},$$

where Γ_{11} is the scalar

$$\Gamma_{11} = \text{cov}(Y_1, Y_1).$$

Then define

$$\begin{aligned} \mathbf{Z}_1 &\triangleq Y_1 - \mu_1 \\ \mathbf{Z}_2 &\triangleq \begin{bmatrix} Y_2 - \mu_2 \\ \vdots \\ Y_n - \mu_n \end{bmatrix} \end{aligned} \tag{1.8-16}$$

so that

$$\mathbf{Y} - \boldsymbol{\mu} = \begin{bmatrix} Z_1 \\ Z_2 \end{bmatrix}.$$

Replacing A_{ij} in Theorem 1.7.1 by Γ_{ij} and then defining F_{ij} via 1.7.1(b), we have

$$\begin{aligned} \langle \Gamma^{-1}(\mathbf{Y} - \boldsymbol{\mu}), \mathbf{Y} - \boldsymbol{\mu} \rangle &= \langle \Gamma^{-1} \begin{bmatrix} Z_1 \\ Z_2 \end{bmatrix}, \begin{bmatrix} Z_1 \\ Z_2 \end{bmatrix} \rangle \\ &= [Z_1, Z_2^T] \begin{bmatrix} F_{11} & F_{12} \\ F_{21} & F_{22} \end{bmatrix} \begin{bmatrix} Z_1 \\ Z_2 \end{bmatrix} \\ &= Z_1 F_{11} Z_1 + Z_1 F_{12} Z_2 + Z_2^T F_{21} Z_1 + Z_2^T F_{22} Z_2 \\ &= Z_1 \Gamma_{11}^{-1} Z_1 + Z_1 \Gamma_{11}^{-1} \Gamma_{12} F_{22} \Gamma_{21} \Gamma_{11}^{-1} Z_1 \\ &\quad - Z_1 \Gamma_{11}^{-1} \Gamma_{12} F_{22} Z_2 \\ &\quad - Z_2^T F_{22} \Gamma_{21} \Gamma_{11}^{-1} Z_1 + Z_2^T F_{22} Z_2 \\ &= Z_1 \Gamma_{11}^{-1} Z_1 \\ &\quad + [Z_2^T - (\Gamma_{21} \Gamma_{11}^{-1} Z_1)^T] F_{22} [Z_2 - \Gamma_{21} \Gamma_{11}^{-1} Z_1]. \end{aligned} \tag{1.8-17}$$

From 1.7.4, we have that

$$\det(\Gamma) = \det(\Gamma_{11}) \det(\Gamma_{22} - \Gamma_{21} \Gamma_{11}^{-1} \Gamma_{12}) = \Gamma_{11} \cdot \det(F_{22}^{-1})$$

and from (1.8-7), that

$$\langle \Gamma^{-1}(\mathbf{Y} - \boldsymbol{\mu}), (\mathbf{Y} - \boldsymbol{\mu}) \rangle = \Gamma_{11}^{-1} Z_1^2 + \langle F_{22}(\mathbf{Z}_2 - \Gamma_{21} \Gamma_{11}^{-1} Z_1), \mathbf{Z}_2 - \Gamma_{21} \Gamma_{11}^{-1} Z_1 \rangle,$$

so that

$$\begin{aligned} f(\mathbf{y}) = &\frac{1}{(2\pi)^{n/2} \sqrt{\Gamma_{11}} \sqrt{\det(F_{22}^{-1})}} e^{-(Z_1^2/\Gamma_{11})/2} \\ &\cdot e^{-\langle F_{22}(Z_2 - \Gamma_{21} \Gamma_{11}^{-1} Z_1, Z_2 - P_{21} P_{11}^{-1} Z_1 \rangle /2}. \end{aligned} \tag{1.8-18}$$

Defining the new variables

$$\mathbf{S} = \begin{bmatrix} S_2 \\ \vdots \\ S_n \end{bmatrix} = \mathbf{Z}_2 - \Gamma_{21} \Gamma_{11}^{-1} Z_1, \tag{1.8-19}$$

we then have from (1.8-18) that

$$f(\mathbf{y}) = \frac{1}{\sqrt{2\pi}\sqrt{\Gamma_{11}}} e^{-(Z_1^2/\Gamma_{11})/2} \cdot \frac{1}{(2\pi)^{(n-1/2)}\sqrt{\det(F_{22}^{-2})}} e^{-\langle F_{22}\mathbf{S},\mathbf{S}\rangle/2}.$$

If we then change variables in (1.8-15), we note that the last $n-1$ iterated integrals do not involve y_1, hence Z_1, as a variable and so $dS_i = dy_i$. Hence, (1.8-15) becomes

$$\begin{aligned} H(y_1) &= \int_{-\infty}^{y_1} \frac{1}{\sqrt{2\pi}\sqrt{\Gamma_{11}}} e^{-((y_1-\mu_1)/\sqrt{\Gamma_{11}})^2/2} dy_1 \\ &\times \int_{-\infty}^{\infty} \cdots \int_{-\infty}^{\infty} \frac{1}{(2\pi)^{(n-1/2)}\sqrt{\det(F_{22}^{-1})}} e^{-\langle F_{22}\mathbf{S},\mathbf{S}\rangle/2} dS_2 \cdots dS_n \end{aligned}$$

and from our previous results the last $n-1$ integrals produce unity. Thus,

$$H(y_1) = \int_{-\infty}^{y_1} \frac{1}{\sqrt{2\pi}\sqrt{\Gamma_{11}}} e^{-((y_1-\mu_1)/\sqrt{\Gamma_{11}})^2/2} dy_1$$

and so

$$H'(y) = \frac{1}{\sqrt{2\pi\Gamma_{11}}} e^{-((y-\mu_1/\sqrt{\Gamma_{11}})^2/2}. \tag{1.8-20}$$

Since $\Gamma_{11} = \text{cov}(Y_1, Y_1)$ and $\mu_1 = E(Y_1)$, it follows that $H'(y)$ is indeed the p.d.f. for Y_1 and so the p.d.f. f does indeed produce the correct marginal distributions for the random vector Y.

This completes the proof. □

1.8.2. **Definition.** The distribution of $\mathbf{Y}$ in 1.8.1 is called the *multivariate normal* distribution, and its p.d.f. is the function f we defined. When a random variable $\mathbf{Y}$ is so distributed, we write

$$Y \sim N(\boldsymbol{\mu}, \Gamma).$$

Note that any such random vector has the decomposition given in part (b) of 1.8.1.

As mentioned at the end of Section 1.6, the multivariate normal distribution arises quite naturally in estimation theory and will occupy a central role in the theory we develop. The explicit connection will be made in Chapter 3.

1.9 Conditional Expectation

In this section, we discuss the problem of calculating the expected value of a random variable X given that another random variable Y has already taken on the value y_0. If X and Y are causally, hence mathematically, independent, we would expect the answer to simply be $E(X)$. If, however, X and Y are dependent, we would expect the calculation to involve y_0 and the statistical structure of X and Y. It turns out that we have already studied a particular case of conditional expectation. For, if $X = C_A$, then $E(X) = P(A)$, and so $P(A \,|\, Y^{-1}(B))$ is the same as the expected value of X given that $Y \in B$. In the case where $P(Y^{-1}(B)) > 0$, Definition 1.4.3 provided us with a means of calculating this number, namely,

$$P(A \,|\, Y^{-1}(B)) = \frac{P(A \cap Y^{-1}(B))}{P(Y^{-1}(B))}. \tag{1.9-1}$$

In the case where $B = \{y_0\}$, we would have

$$P(A \,|\, Y = y_0) = \frac{P(A \cap Y^{-1}(y_0))}{P(Y^{-1}(y_0))}. \tag{1.9-2}$$

However, if Y were a continuously distributed random variable, $Y^{-1}(y_0)$ would have measure zero (check this) and so (1.9-2) would become

$$P(A \,|\, Y = y_0) = \frac{0}{0},$$

which is nonsense. However, the question of calculating $P(A \,|\, Y = y_0)$ is not meaningless. For instance, suppose X and Y are uniformly distributed on the unit disk D, that is,

$$P_{YY}(A) = \frac{1}{\pi}\mu_2(A \cap D). \tag{1.9-3}$$

If we fix $Y = y_0$, we then have that X is uniformly distributed on the line ℓ shown in Figure 1.8. It seems intuitively clear from this figure that

$$P(X \in S \,|\, Y = y_0) = \frac{\text{one dimensional measure of } S \cap \ell}{\text{length of } \ell},$$

so that the notion does make sense even though Y is continuously distributed. How can we describe the solution in the general case? Here is a first attempt.

Suppose (X, Y) is distributed on R^2 with p.d.f. $f_{XY}(x, y)$ (so that X is also continuously distributed) and let $A \subset \mathrm{R}$. We can easily define

$$P(X \in A \,|\, Y \in (y_0 - \epsilon, y_0 + \epsilon)) = \frac{P(X \in A \text{ and } Y \in (y_0 - \epsilon, y_0 + \epsilon))}{P(Y \in (y_0 - \epsilon, y_0 + \epsilon))}$$

provided that the denominator of the right-hand expression is nonzero whenever $\epsilon > 0$. If such is the case, we then have

$$\begin{aligned} P(X \in A \,|\, Y \in (y_0 - \epsilon, y_0 + \epsilon)) &= \frac{P_{XY}(A \times (y_0 - \epsilon, y_0 + \epsilon))}{P_Y(y_0 - \epsilon, y_0 + \epsilon)} \\ &= \frac{\int_A \int_{y_0-\epsilon}^{y_0+\epsilon} f_{XY}(x, y)\, dy\, dx}{\int_{y_0-\epsilon}^{y_0+\epsilon} f_Y(y)\, dy} \end{aligned}$$

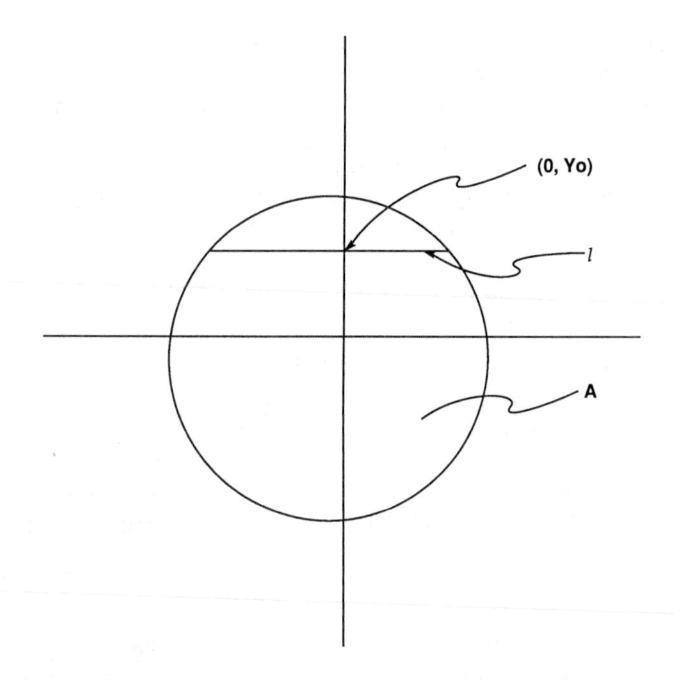

FIGURE 1.8. An example to motivate the existence of conditional densities in the case of continuous density functions.

where, of course,

$$f_Y(y) = \int_{-\infty}^{\infty} f_{XY}(x, y)\, dx.$$

Thus, from the mean value theorem for integrals

$$P(X \in A \mid Y \in (y_0 - \epsilon, y_0 + \epsilon)) = \frac{\int_A 2\epsilon f_{XY}(x, \xi_x)\, dx}{2\epsilon f_Y(\eta)},$$

where $\xi_x, \eta \in (y_0 - \epsilon, y_0 + \epsilon)$. From this expression, it is reasonable to define

$$P(X \in A \mid Y = y_0) \triangleq \int_A \frac{f_{XY}(x, y_0)}{f_Y(y_0)} dy. \tag{1.9-4}$$

Hence, if X and Y are (jointly) continuously distributed, and $f_Y(y_0) \neq 0$, it is possible to define the conditional density

$$f_{X \mid Y}(x \mid y_0) \triangleq \frac{f_{XY}(x, y_0)}{f_Y(y_0)}. \tag{1.9-5}$$

The condition that $f_Y(y_0) \neq 0$ is annoying in that $f_{XY}(x, y_0)$ might vanish whenever $f_Y(y_0)$ does, thereby suggesting that some description of $f_{X \mid Y}(x \mid y_0)$ might make physical sense at y_0 even though (1.9-5) does not.

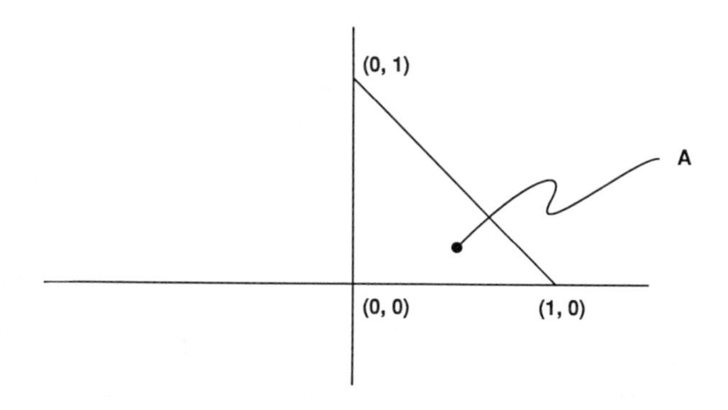

FIGURE 1.9. The domain of P_{XY} in Example 1.9.1.

It turns out that problems such as this will be avoided by appealing to the Radon–Nikodym theorem, although the resulting theory is rather abstract. Nevertheless, relation (1.9-4) is very useful in many, many situations, and we will have occasion to use it again.

Very well then, let us suppose for the moment that given two random variables X and Y, somehow we have been able to obtain a conditional density function $f_{X\,|\,Y}(x\,|\,y)$. The interpretation of this is that $f_{X\,|\,Y}(x\,|\,y_0)$ gives the probability distribution of the random variable X given that $Y = y_0$. One might appropriately refer to this as a conditional marginal; however, this is not standard terminology. Anyway, with this interpretation of $f_{X\,|\,Y}(x\,|\,y)$, it now seems reasonable to define

$$E(X\,|\,Y = y) \triangleq \int_{-\infty}^{\infty} x f_{X\,|\,Y}(x\,|\,y)\,dx \tag{1.9-6}$$

and refer to this as the (conditional) expected value of X given that $Y = y$. It is customary to write this as $E(X\,|\,Y)(y)$ to emphasize that this is indeed a function of y.

1.9.1. **Example.** Let (X, Y) be uniformly distributed on the triangular set A shown in Figure 1.9. Then,

$$P_{XY}(x, y) = 2\mu_2(E \cap A)$$

so

$$f_{XY}(x, y) = \begin{cases} 2 & \text{for} \quad (x, y) \in A \\ 0 & \text{otherwise.} \end{cases}$$

Hence,

$$f_Y(y) = \int_{-\infty}^{\infty} f_{XY}(x, y)\,dx = \int_0^{1-y} 2\,dx = 2(1 - y)$$

and so by (1.9-4),

$$f_{X\,|\,Y}(x\,|\,y) = \begin{cases} \frac{1}{1-y} & \text{for} \quad 0 \le x \le 1 - y \\ 0 & \text{otherwise,} \end{cases}$$

where $0 \le y < 1$. Applying (1.9-5)

$$\begin{aligned} E(X \mid Y)(y) &= \int_{-\infty}^{\infty} x f_{X \mid Y}(x \mid y)\, dx \\ &= \int_0^{1-y} x \cdot \frac{1}{1-y} dx \\ &= \frac{1-y}{2}. \end{aligned}$$

Note that from Figure 1.9, this result is quite reasonable.

Our motivation is finished, and we now turn to the formal machinery used to study conditional expectation. As mentioned earlier, the theory is somewhat abstract and will require some patience from the reader. We will define conditional expectation, study it, and finally make contact with the above motivation.

1.9.2. **Definition.** Let X be a random variable on $(\Omega, \mathcal{E}, P)$, $\mathcal{A}$ a σ field, $\mathcal{A} \subset \mathcal{E}$. By the *conditional expectation* of X given $\mathcal{A}$ we mean the random variable $E(X \mid \mathcal{A})(\omega)$ satisfying the following two conditions:

(a) $E(X \mid \mathcal{A})$ is $\mathcal{A}$-measurable.

(b) $E(C_A \cdot E(X \mid \mathcal{A})) = E(C_A X)$ for all $A \in \mathcal{A}$.
Condition (b) can also be written

(b′) $\displaystyle\int_A E(X \mid \mathcal{A})\, dP = \int_A X\, dP$ for all $A \in \mathcal{A}$.

Note that by Theorem 1.6.8, these two conditions define $E(X \mid \mathcal{A})$ uniquely up to a set of measure zero. One might be tempted to simply set $E(X \mid \mathcal{A})$ equal to X, but this ignores condition (a), that is, X may not be $\mathcal{A}$ measurable. Thus, we must first establish the existence of $E(X \mid \mathcal{A})$, and this is settled by the next theorem.

1.9.3. **Theorem.** *If X is any random variable, $\mathcal{A}$ a σ field, $\mathcal{A} \subset \mathcal{E}$, then $E(X \mid \mathcal{A})$ exists.*

Proof. Let $X = X^+ - X^-$ be the decomposition of X into its positive and negative parts.[7] Define

$$\begin{aligned} \nu^+(A) &\triangleq \int_A X^+\, dP \\ \nu^-(A) &\triangleq \int_A X^-\, dP \end{aligned}$$

[7] $X^+(\omega) = X(\omega)$ if $X(\omega) \ge 0$; $X^+(\omega) = 0$ if $X(\omega) < 0$. $X^-(\omega) = -X(\omega)$ if $X(\omega) < 0$; $X^-(\omega) = 0$ if $X(\omega) \ge 0$.

for every $A \in \mathcal{A}$. Then, from the properties of integrals, ν^+ and ν^- are measures on $(\Omega, \mathcal{A})$. Note also that $P|_{\mathcal{A}}$, that is, the restriction of P to sets in $\mathcal{A}$ is also a measure on $\mathcal{A}$. If $P(A) = 0$, for $A \in \mathcal{A}$, then clearly $\nu^+(A) = 0$ and $\nu^-(A) = 0$, so $\nu^+ << P|_{\mathcal{A}}$ and $\nu^- << P|_{\mathcal{A}}$. It follows from the Radon–Nikodym theorem that there exists unique $\mathcal{A}$-measurable functions h_1 and h_2 such that

$$\nu^+(A) = \int_A h_1 \, dP$$

$$\nu^-(A) = \int_A h_2 \, dP$$

for every $A \in \mathcal{A}$. Let

$$E(X \mid \mathcal{A}) \triangleq h_1 - h_2.$$

Then, clearly, $E(X \mid \mathcal{A})$ is $\mathcal{A}$ measurable since h_1 and h_2 are. Also,

$$\begin{aligned}
\int_A E(X \mid \mathcal{A}) \, dP &= \int_A (h_1 - h_2) \, dP \\
&= \int_A h_1 \, dP - \int_A h_2 \, dP \\
&= \int_A X^+ \, dP - \int_A X^- \, dP \\
&= \int_A (X^+ - X^-) \, dP \\
&= \int_A X \, dP. \qquad \square
\end{aligned}$$

1.9.4. **Example.** Let $\mathcal{A} = \{\phi, A, A', \Omega\}$ where $P(A) > 0$ and $P(A') > 0$. If a function h is $\mathcal{A}$ measurable, then it must be the case that for every open set $V \subset \mathrm{R}$, $h^{-1}(V) \in \mathcal{A}$. In the following, we will make use of the particular open sets

$$V_x \triangleq (-\infty, x).$$

We first claim that if $\omega_1, \omega_2 \in A$, then $h(\omega_1) = h(\omega_2)$. Suppose not. Then $\omega_1 \neq \omega_2$ and defining $x_1 = h(\omega_1)$, $x_2 = h(\omega_2)$, we can suppose without loss of generality that $x_1 < x_2$. Let

$$V = \left(-\infty, \frac{x_1 + x_2}{2}\right).$$

Then $x_1 \in V$ and $x_2 \neq V$. Hence, $\omega_1 \in h^{-1}(V)$ and $\omega_2 \notin h^{-1}(V)$. Now, $\omega_1 \in h^{-1}(V)$ implies that $h^{-1}(V) \cap A \neq \phi$ (since $\omega_1 \in A$), and since h is measurable this in turn implies that $h^{-1}(V) = A$ or $h^{-1}(V) = \Omega$. However, $\omega_2 \notin h^{-1}(V)$ implies that $h^{-1}(V) \neq \Omega$, that is, we must have $h^{-1}(V) = A$. However, $\omega_2 \in A$, so if this were true we would have $\omega_2 \in h^{-1}(V)$, a contradiction. Thus, h is constant on A. In a similar fashion, we can show that h is constant on A'.

From the above argument we see that $E(X \mid \mathcal{A}$ must have the form

$$E(X \mid \mathcal{A}) = \alpha_1 C_A + \alpha_2 C_{A'}.$$

From property (b') of 1.9.2, it follows that

$$\int_A X\,dP = \int_A (\alpha_1 C_A + \alpha_2 C_{A'})\,dP = \alpha_1 P(A)$$

or

$$\alpha_1 = \frac{1}{P(A)} \int_{A'} X\,dP.$$

A similar calculation holds for α_2, resulting in the formula

$$E(X \mid \mathcal{A}) = \left(\frac{1}{P(A)} \int_A X\,dP\right) C_A + \left(\frac{1}{P(A')} \int_{A'} X\,dP\right) C_{A'}.$$

This result generalizes to a so-called atomic σ field that is generated by a countable family of disjoint sets A_i, each of which has positive measure

$$E(X \mid \mathcal{A}) = \sum_{i=1}^{\infty} \left(\frac{1}{P(A_i)} \int_{A_i} X\,dP\right) C_{A_i}.$$

In this case, then, if $\omega \in A_i$, we have

$$E(X \mid \mathcal{A})(\omega) = \frac{1}{P(A_i)} \int_{A_i} X\,dP.$$

Writing the measure

$$P_{A_i}(E) \triangleq P(E \mid A_i) = \frac{P(E \cap A_i)}{P(A_i)}$$

and restricting X to A_i, we see that

$$E(X \mid \mathcal{A})(\omega) = \int_{A_i} X\,dP_{A_i},$$

which is the expected value of the random variable X restricted to A_i using the conditional probability measure P_{A_i}.

1.9.5. **Theorem.** *X and Y are random variables*

(a) *If $X \leq Y$, then $E(X \mid \mathcal{A}) \leq E(Y \mid \mathcal{A})$* [a.e.].

(b) $E(\alpha X + \beta Y \mid \mathcal{A}) = \alpha E(X \mid \mathcal{A}) + \beta E(Y \mid \mathcal{A})$ [a.e.].

(c) *If X is $\mathcal{A}$ measurable, then $E(X \mid \mathcal{A}) = X$.*

(d) *If Y is $\mathcal{A}$ measurable, X a random variable, then* $E(Y \cdot X \mid \mathcal{A}) = Y \cdot E(X \mid \mathcal{A})$ [a.e.].

(e) *If $\mathcal{A}_1$ and $\mathcal{A}_2$ are σ fields, $\mathcal{A}_1 \subset \mathcal{A}_2 \subset \mathcal{E}$, then* $E(E(X \mid \mathcal{A}_2) \mid \mathcal{A}_1) = E(X \mid \mathcal{A}_1)$ [a.e.].

Proof. (a) If $X \leq Y$, then for any $A \in \mathcal{A}$

$$\int_A X\, dP \leq \int_A Y\, dP.$$

It follows that

$$\int_A E(X \mid \mathcal{A})\, dP \leq \int_A E(Y) \mid \mathcal{A})\, dP$$

or

$$0 \leq \int_A [E(Y \mid \mathcal{A}) - E(X \mid \mathcal{A})]\, dP.$$

Let $g \triangleq E(Y \mid \mathcal{A}) - E(X \mid \mathcal{A})$. If it is not true that $g \geq 0$ [a.e.], then the set

$$B = g^{-1}(-\infty, 0) = \{\omega \mid g(\omega) < 0\}$$

is $\mathcal{A}$ measurable, has positive measure, and so

$$0 > \int_B g\, dP,$$

a contradiction. Hence, $g \geq 0$ [a.e.] as required.

(b) and (c) follow directly from the definition and Theorem 1.6.8.

(d) If Y is $\mathcal{A}$ measurable, then $Y \cdot E(X \mid \mathcal{A})$ is also $\mathcal{A}$ measurable and so part (a) of Definition 1.9.2 holds.

Now suppose $Y = C_B$, $B \in \mathcal{A}$. Then, for every $A \in \mathcal{A}$, we have

$$\begin{aligned}
E(C_A \cdot E(Y \cdot X \mid \mathcal{A})) &= \int_A E(C_B \cdot X \mid \mathcal{A})\, dP \\
&= \int_A C_B \cdot X\, dP \\
&= \int_{A \cap B} X\, dP \\
&= \int_{A \cap B} E(X \mid \mathcal{A})\, dP \\
&= \int_A C_B \cdot E(X \mid \mathcal{A})\, dP \\
&= E(C_A \cdot Y E(X \mid \mathcal{A})).
\end{aligned}$$

Thus,

$$Y \cdot E(X \mid \mathcal{A}) = E(Y \cdot X \mid \mathcal{A})$$

for $Y = C_B$, $B \in \mathcal{A}$. It follows that this equality holds whenever Y is a simple function.

Next, suppose $Y \geq 0$ and $X \geq 0$. We can always construct a sequence of simple functions Y_n with $Y_n \uparrow Y$. Thus, $Y_n \cdot X \uparrow Y \cdot X$. Likewise, $Y_n \cdot E(X \mid \mathcal{A}) \uparrow Y \cdot E(X \mid \mathcal{A})$. (Note that by part (a), $E(X \mid \mathcal{A}) \geq 0$.) By the Lebesgue monotone convergence theorem,

$$\begin{aligned} E(Y \cdot X \mid \mathcal{A})\, dP &= \int_A Y \cdot X\, dP \\ &= \lim_{n\to\infty} \int_A Y_n \cdot X\, dP \\ &= \lim_{n\to\infty} \int_A E(Y_n \cdot X \mid \mathcal{A})\, dP \\ &= \lim_{n\to\infty} \int_A Y_n \cdot E(X \mid \mathcal{A})\, dP \\ &= \int_A Y \cdot E(X \mid \mathcal{A})\, dP. \end{aligned}$$

The general case now follows by writing $X = X^+ - X^-$, $Y = Y^+ - Y^-$, and applying the result just obtained to the four terms in the product. Note that parts (a) and (b) are used here also.

(e) If $\mathcal{A}_1 \subset \mathcal{A}_2$, then for every $A \in \mathcal{A}_1$, C_A is both $\mathcal{A}_1$ and $\mathcal{A}_2$ measurable. Thus, for $A \in \mathcal{A}_1$,

$$E(C_A \cdot E(E(X \mid \mathcal{A}_2) \mid \mathcal{A}_1)) = E(E(C_A \cdot E(X \mid \mathcal{A}_2) \mid \mathcal{A}_1)),$$

which follows from part (d) above. Now, from this it follows that

$$\begin{aligned} E(C_A \cdot E(E(X \mid \mathcal{A}_2) \mid \mathcal{A}_1)) &= \int_\Omega E(C_A \cdot E(X \mid \mathcal{A}_2) \mid \mathcal{A}_1)\, dP \\ &= \int_\Omega C_A \cdot E(X \mid \mathcal{A}_2)\, dP \\ &\quad \text{(by (b) of Definition 1.9.2)} \\ &= \int_A E(X \mid \mathcal{A}_2)\, dP \\ &= \int_A X\, dP \\ &\quad \text{(since } A \in \mathcal{A}_2\text{)} \\ &= \int_A E(X \mid \mathcal{A}_1)\, dP \\ &\quad \text{(since } A \in \mathcal{A}_1\text{)} \\ &= E(C_A \cdot E(X \mid \mathcal{A}_1)). \end{aligned}$$

Thus, $E(E(X \mid \mathcal{A}_2) \mid \mathcal{A}_1)$ and $E(X \mid \mathcal{A}_1)$ are both $\mathcal{A}_1$ measurable and both satisfy part (b) of Definition 1.9.2. It follows that

$$E(E(X \mid \mathcal{A}_2) \mid \mathcal{A}_1) = E(X \mid \mathcal{A}_1) \quad \text{[a.e.]}. \qquad \square$$

Before reading the next definition, the reader is advised to review the definition of $B(X)$ given in 1.4.6.

1.9.6. **Definition.** Let X be a random variable, $\mathbf{Y}$ a random vector. By the *conditional expectation* of X based on $\mathbf{Y}$, we mean the random variable $E(X \mid \mathbf{Y})$ defined by

$$E(X \mid \mathbf{Y}) \triangleq E(X \mid B(\mathbf{Y})).$$

1.9.7. **Theorem.** *Let X be a random variable, $\mathbf{Y}$ a random vector. Then there exists a real valued Baire function g such that*

$$E(X \mid \mathbf{Y})(\omega) = g(\mathbf{Y}(\omega))$$

for all $\omega \in \Omega$.

Proof. Let

$$X = X^+ - X^- \tag{1.9-7}$$

be the decomposition of X into its positive and negative parts. Then define two measures ν_1 and ν_2 on B_n via

$$\nu_1(B) = \int_{\mathbf{Y}^{-1}(B)} X^+ \, dP$$

$$\nu_2(B) = \int_{\mathbf{Y}^{-1}(B)} X^- \, dP. \tag{1.9-8}$$

The fact that ν_1 and ν_2 are measures on B_n follows from the countable additivity of the integral and the preservation of unions by Y^{-1}. Next note that $P_{\mathbf{Y}}(B) = P(\mathbf{Y}^{-1}(B))$, so that if $P_{\mathbf{Y}}(B) = 0$, then $\nu_1(B) = \nu_2(B) = 0$, that is, $\nu_1 << P_Y$ and $\nu_2 << P_Y$. By the Radon–Nikodym theorem, there exists (Borel) measurable functions g^+ and g^- such that

$$\nu_1(B) = \int_B g^+ \, dP_{\mathbf{Y}}$$

$$\nu_2(B) = \int_B g^- \, dP_{\mathbf{Y}} \tag{1.9-9}$$

for all $B \in B_n$. Defining

$$g = g^+ - g^-, \tag{1.9-10}$$

we have from (1.9-6) through (1.9-9) that

$$\begin{aligned}\int_{\mathbf{Y}^{-1}(B)} X\,dP &= \int_{\mathbf{Y}^{-1}(B)} (X^+ - X^-)\,dP \\ &= \nu_1(B) - \nu_2(B) \\ &= \int_B g^+\,dP_{\mathbf{Y}} - \int_B g^-\,dP_{\mathbf{Y}} \\ &= \int_B g\,dP_{\mathbf{Y}} \\ &= \int_{\mathbf{Y}^{-1}(B)} g \circ \mathbf{Y}\,dP \quad \text{(Theorem 1.2.7).}\end{aligned}$$

From the fact that every $A \in B(Y)$ is of the form $Y^{-1}(B)$ for some $B \in B_n$, we have the result that

$$\int_A X\,dP = \int_A g \circ Y\,dP$$

for all $A \in B(Y)$. Hence, by Definition 1.9.2,

$$\int_A E(X \mid Y)\,dP = \int_A g \circ Y\,dP$$

for all $A \in B(Y)$. But both $E(X \mid Y)$ and $g \circ Y$ are $B(Y)$ measurable, so by 1.6.8

$$E(X \mid Y) = g \circ Y \quad \text{[a.e.].} \quad \square$$

We can now show the connection between Definition 1.9.2 and our motivation. Clearly, if g is any Baire function, then $g \circ \mathbf{Y}$ is $B(\mathbf{Y})$ measurable and so part (a) of Definition 1.9.2 holds for any such function. Now suppose that $\mathbf{Y}$ is a random vector, X a random variable, such that X and $\mathbf{Y}$ are jointly distributed with p.d.f. $f_{X\mathbf{Y}}$, and that

$$f_{X \mid \mathbf{Y}}(x \mid \mathbf{y}) \triangleq \frac{f_{X\mathbf{Y}}(x, \mathbf{y})}{f_{\mathbf{Y}}(\mathbf{y})}$$

is well defined for all $\mathbf{y} \in \mathrm{R}^n$. Let

$$g(\mathbf{y}) \triangleq \int_{-\infty}^{\infty} x f_{X \mid \mathbf{Y}}(x \mid \mathbf{y})\,dx. \tag{1.9-11}$$

If we suppose that $f_{X\mathbf{Y}}$ is continuous, then $f_{X \mid \mathbf{Y}}$ certainly is, whence g is a continuous function. Thus, g is a Baire function since open sets are Borel. We claim that for this g,

$$E(X \mid \mathbf{Y})(\omega) = g(\mathbf{Y}(\omega)), \tag{1.9-12}$$

and we will show this by demonstrating that 1.9.2(b) holds for $g \circ \mathbf{Y}$.

Let $A \in \mathcal{B}(\mathbf{Y})$ be arbitrary. Then $A = \mathbf{Y}^{-1}(B)$ for some Borel set $B \in \mathcal{B}_n$. Hence,

$$\begin{aligned}
E(C_A(g \circ \mathbf{Y})) &= \int_A g \circ \mathbf{Y}\, dP \\
&= \int_{\mathbf{Y}^{-1}(B)} g \circ \mathbf{Y}\, dP \\
&= \int_B g\, dP_{\mathbf{Y}} \\
&= \int_B g(\mathbf{y}) f_{\mathbf{Y}}(\mathbf{y})\, d\mathbf{y} \\
&= \int_B \int_{-\infty}^{\infty} x f_{X\,|\,\mathbf{Y}}(x\,|\,\mathbf{y}) f_{\mathbf{Y}}(\mathbf{y})\, dx\, d\mathbf{y} \\
&= \int_{\mathbf{R}^{n+1}} C_B(\mathbf{y}) x f_{X\mathbf{Y}}(x, \mathbf{y})\, dx\, d\mathbf{y} \\
&= \int_{\mathbf{R}^{n+1}} C_B(\mathbf{y}) x\, dP_{X\mathbf{Y}} \\
&= \int_{(X,\mathbf{Y})^{-1}(\mathbf{R}^{n+1})} C_B(\mathbf{Y}(\omega)) X(\omega)\, dP \\
&= \int_{\Omega} C_B(\mathbf{Y}(\omega)) X(\omega)\, dP \\
&= \int_{\Omega} C_{\mathbf{Y}^{-1}(B)}(\omega) X(\omega)\, dP \\
&= \int_{\mathbf{Y}^{-1}(B)} X(\omega)\, dP \\
&= \int_A X(\omega)\, dP.
\end{aligned}$$

Hence, we have shown that for all $A \in \mathcal{B}(Y)$,

$$E(C_A(g \circ \mathbf{Y})) = E(C_A(X). \tag{1.9-13}$$

Since $g \circ Y$ is $\mathcal{B}(Y)$ measurable and satisfies (b) of 1.9.2, it follows that

$$g \circ Y = E(X\,|\,Y).$$

as claimed.

From our motivation, we say that $g(\mathbf{y})$ can be interpreted as the expected value of X given that $\mathbf{Y}$ takes on the value $\mathbf{y}$. Thus, (1.9-11) says that $E(X\,|\,\mathbf{Y})(\omega)$ is the expected value of X given that $\mathbf{Y}$ takes on the value $\mathbf{Y}(\omega)$. In coordinate form, $g(y_1, \ldots, y_n)$ is the expected value of X given that $\mathbf{Y}_1(\omega) = y_1(\omega), \ldots Y_n(\omega) = \mathbf{y}_n$.

We can now clean up a "loose end," namely, the definition of *conditional probability.* The idea stems from the observation that

$$P(A) = \int_A dP = E(C_A). \tag{1.9-14}$$

Let X and Y be as above and let g be the Baire function satisfying

$$E(C_{X^{-1}(B)} \,|\, Y) = g \circ Y. \tag{1.9-15}$$

Such a g exists by the previous theorem. Then, from the above expression and the discussion in the previous paragraph, we know that $g(\mathbf{y})$ is the expected value of $C_{X^{-1}(B)}$ given that $\mathbf{Y}(\omega) = \mathbf{y}$. However, by (1.9-13), this is the same as saying that $g(\mathbf{y})$ is the probability that $X(\omega) \in B$ given that $\mathbf{Y}(\omega) = \mathbf{y}$. This is generally written

$$P(X \in B \,|\, \mathbf{Y} = \mathbf{y}) = g(\mathbf{y}),$$

where g satisfies (1.9-15).

We next rewrite Theorem 1.9.5 in the notation $E(X \,|\, \mathbf{Y})$.

1.9.8. **Theorem.** *X, Z are random variables, Y a random vector.*

(a) *If $X \leq Z$, then $E(X \,|\, \mathbf{Y}) \leq E(Z \,|\, \mathbf{Y})$,*

(b) *$E(\alpha X + \beta Z \,|\, \mathbf{Y}) = \alpha E(X \,|\, \mathbf{Y}) + \beta E(Z \,|\, \mathbf{Y})$, and*

(c) *$E(X \,|\, X) = X$;*

(d) *if X is $B(Y)$ measurable, then $E(XZ \,|\, \mathbf{Y}) = X \cdot E(Z \,|\, \mathbf{Y})$; and*

(e) *if $B(Y) \subset B(X)$, then $E(E(Z \,|\, X) \,|\, \mathbf{Y}) = E(Z \,|\, \mathbf{Y})$.*

We end this section with a lemma that will be of use in the next chapter.

1.9.9. **Lemma.** *If $X \in \mathcal{L}_2(\Omega, P)$, $\mathbf{Y}$ a random vector, and g a Baire function such that $g \circ \mathbf{Y} \in \mathcal{L}_2(\Omega, P)$, then*

$$[X - E(X \,|\, \mathbf{Y})] \perp [E(X \,|\, \mathbf{Y}) - g(\mathbf{Y})].$$

Proof. First, note that both $g \circ \mathbf{Y}$ and $E(X \,|\, \mathbf{Y})$ are $B(\mathbf{Y})$ measurable. Thus,

$$\begin{aligned}
\langle X &- E(X \,|\, \mathbf{Y}), E(X \,|\, \mathbf{Y}) - g(\mathbf{Y})\rangle \\
&= \int_{-\infty}^{\infty} [X - E(X \,|\, \mathbf{Y})][E(X \,|\, \mathbf{Y}) - g(\mathbf{Y})]\, dp \\
&= E(-Xg(\mathbf{Y}) + X \cdot E(X \,|\, \mathbf{Y}) - E(X \,|\, \mathbf{Y})^2 + g(\mathbf{Y}) \cdot E(X \,|\, \mathbf{Y})) \\
&= -E(X \cdot g(\mathbf{Y})) + E(X \cdot E(X \,|\, \mathbf{Y})) - E(E(X \,|\, \mathbf{Y}) \cdot E(X \,|\, \mathbf{Y})) \\
&\quad + E(g(\mathbf{Y}) \cdot E(X \,|\, \mathbf{Y}))
\end{aligned}$$

$$\begin{aligned}
&= -E(X \cdot g(\mathbf{Y})) + E(X \cdot E(X \mid \mathbf{Y})) - E(E(E(X \mid \mathbf{Y}) \cdot X \mid \mathbf{Y})) \\
&\quad + E(E(g(\mathbf{Y}) \cdot X \mid \mathbf{Y})) \\
&\quad \text{(part (d) of 1.9.8)} \\
&= -E(X \cdot g(\mathbf{Y})) + E(X \cdot E(X \mid \mathbf{Y})) - E(E(X \mid \mathbf{Y}) \cdot X) + E(g(\mathbf{Y}) \cdot X) \\
&= 0,
\end{aligned}$$

the next to last equality following from part (b) of Definition 1.9.2 with $A = \Omega$ and X replaced by $E(X \mid \mathbf{Y}) \cdot X$ and $g(Y) \cdot X$, respectively. □

1.10 Exercises

1. Let $\mathbf{X}_1$ and $\mathbf{X}_2$ be independent random vectors and let f and g be Baire functions. Show that $f \circ \mathbf{X}_1$ and $g \circ \mathbf{X}_2$ are independent random variables.

2. Prove Lemma 1.6.17 and Theorem 1.6.18.

3. Prove either (a) or (b). Let $(\Omega, \mathcal{F}, \mu)$ be a measure space.

   (a) If $A_1 \subset \ldots$, with $A_i \in \mathcal{F}$, then $\mu(\cup_{i=1}^{\infty} A_i) = \lim_{n\to\infty} \mu(A_n)$.

   (b) If $B_1 \supset B_2 \supset \ldots$, with $B_i \in F$, then $\mu(\cap_{i=1}^{\infty} B_i) = \lim_{n\to\infty} \mu(B_n)$.

   Hint: Let $F_1 = A_1$, $F_2 = A_2 \setminus A_1$, $F_3 = A_3 \setminus A_2$, and so on, and show

$$\bigcup_{i=1}^{\infty} F_i = \bigcup_{i=1}^{\infty} A_i.$$

4. Let

$$\Gamma(\alpha) = \int_0^{\infty} t^{\alpha-1} e^{-t}\, dt.$$

   (a) Using integration by parts, show that

$$\Gamma(\alpha + 1) = \alpha\Gamma(\alpha).$$

   (b) Using part (a), show that for $n \in \mathrm{N}$,

$$\Gamma(n + 1) = n!.$$

   (c) In $\Gamma(\alpha)$, let $t = \lambda x$. Show that

$$f(x) = \begin{cases} 0 & \text{for} \quad x \le 0 \\ \frac{\lambda^{\alpha}}{\Gamma(\alpha)} x^{\alpha-1} e^{-\lambda x} & \text{for} \quad x > 0 \end{cases}$$

   is a viable p.d.f.

(d) If $Y(\omega)$ has a p.d.f. of the form in (c), Then Y is said to be $\Gamma(\alpha, \lambda)$ distributed and we write $Y \sim \Gamma(\alpha, \lambda)$. Show that if $X \sim N(0, \sigma^2)$, then $(X/\sigma)^2 \sim \Gamma(\frac{1}{2}, \frac{1}{2})$.

5. Suppose that X_1 and X_2 have a joint p.d.f. f. If

$$Y_1 = g_1(X_1, X_2)$$

$$Y_2 = g_2(X_1, X_2)$$

can be inverted, that is, there exist h_1, h_2 such that

$$X_1 = h_1(Y_1, Y_2)$$

$$X_2 = h_2(Y_1, Y_2),$$

then show that Y_1 and Y_2 have a joint p.d.f. given by

$$w(y_1, y_2) = f(h_1(y_1, y_2), h_2(y_1, y_2)) \det\left(\frac{\partial(h_1, h_2)}{\partial(y_1, y_2)}\right).$$

Hint: This is a simple exercise in the change of variable formula for double integrals (which you may assume).

6. (a) Show that if $X_1 \sim \Gamma(\alpha_1, 1)$ and $X_2 \sim \Gamma(\alpha_2, 1)$, X_1 independent from X_2, then $X_1 + X_2 \sim \Gamma(\alpha_1 + \alpha_2, 1)$. Hint: Let $y_1 = x_1 + x_2$ and $y_2 = x_1/(x_1 + x_2)$ and calculate $w(y_1, y_2)$ using 1.10.5. Note in conclusion that Y_1 and Y_2 are independent.

(b) Show that if $X \sim \Gamma(\alpha, \lambda)$, then $\lambda X \sim \Gamma(\alpha, 1)$.

(c) Using (a) and (b), show that if $X_1, X_2, \ldots, X_n$ are independent, then $X_1 + X_2 + \cdots + X_n \sim \Gamma(\alpha_1 + \cdots + \alpha_n, \lambda)$. (Show for two generalization is clear.)

(d) Suppose $X_1, X_2, \ldots, X_n$ are independent and $X_i \sim N(0, \sigma_i^2)$ for each i. Then show that

$$\left(\frac{X_1}{\sigma_1}\right)^2 + \left(\frac{X_2}{\sigma_2}\right)^2 + \cdots + \left(\frac{X_n}{\sigma_n}\right)^2 \sim \Gamma\left(\frac{n}{2}, \frac{1}{2}\right).$$

This is called the chi-square distribution with n degrees of freedom.

7. Suppose X and Y are random vectors that are jointly distributed with p.d.f. f. Describe (using words if you wish) how you would construct $E(X \mid Y)$.

8. If X_1 and X_2 are jointly normal, show that $X_1 + X_2$ is normally distributed. Express the parameters for $X_1 + X_2$ in terms of the parameters for the joint distribution. Compare with 1.10.2.

2

Minimum Variance Estimation—How the Theory Fits

2.1 Theory Versus Practice—Some General Observations

Other than intellectual curiosity, we can only think of two reasons why a scientist or engineer might study mathematics. The first is to obtain procedures or algorithms to solve a problem or class of problems. The second is to clearly and precisely conceptualize an idea; to capture its essence. At first glance, the former may appear to be the more important of the two. After all, aren't solutions to problems what we are really after? Yes, indeed! But the two reasons above are not really dichotomous. Certainly, the history of probability and statistics shows that viable solutions to many problems were not forthcoming until people like Borel and Kolmororov laid the proper foundations so that others could technically formulate their problems and rigorously check them. For example, it is hard to imagine formulating and proving the Kalman theorem with the probability theory of 1850. It would be misleading, however, to suggest that the only reason for precisely formulating ideas is to obtain solutions to problems. There are other, very practical, reasons for doing so. Let us briefly explore this.

For one thing, precisely formulating ideas sometimes shows that a particular problem does not have a solution (in which case it is rather silly to look for it). Even more to the point, precise formulation sometimes shows that we are asking the wrong question, or even that we are asking something that only sounds like a question,[1] or that we have formulated a question so that in the context asked, it has no solution. In other words, clarity is a practical consequence of precise formulation.

Another possible consequence of precisely formulating a problem is that one may find that it is possible to idealistically describe the precise solution even though it is difficult, or impossible, to actually implement it. What good is that? Quite simply, it is a standard by which we can compare or judge "less than perfect" solutions to our problem, that is, those that we

[1] Which is farther, New York or by bus?

really can implement. We are going to see a vivid example of this idea in the context of *minimum variance estimation*, the subject of this chapter. So, let us begin at the beginning.

2.2 The Genesis of Minimum Variance Estimation

Let us begin by supposing that we have under consideration some "quantities" that take on real values, or at least that is our conceptualization, and that these values change in a somewhat random fashion. For convenience, we will give these quantities names: $Y_1, Y_2, \ldots, Y_n, X$. Le us further suppose that we have some notion, however vague, that all of these quantities are varying for the same set of reasons. Perhaps it is because the thermodynamic state of the ocean is changing, new prisons have been built, an election is about to be held, there is a sardine shortage—whatever. We call this collection of perceived circumstances Ω. The set Ω may be precisely described or be somewhat vague. There are only two things that we must believe about Ω.

(1) All of the circumstances affecting the quantities $Y_1, \ldots, Y_n, X$ are in Ω.

From this belief, we can now view $Y_1, \ldots, Y_n$ and X as functions on Ω, that is, if ω represents the conditions producing a particular circumstance in Ω, then $Y_1(\omega), \ldots, Y_n(\omega), X(\omega)$ are the corresponding quantities that result from this circumstance. This said, it will be convenient for us to introduce the vector function

$$\mathbf{Z} : \Omega \to \mathrm{R}^{n+1}$$

defined by

$$\mathbf{Z}(\omega) \triangleq \begin{bmatrix} Y_1(\omega) \\ Y_2(\omega) \\ \vdots \\ Y_n(\omega) \\ X(\omega) \end{bmatrix}.$$

(2) There is a probability measure P on Ω that measures the likelihood of the various circumstances in Ω ("likelihood" does not have a technical meaning here) and the events $\mathcal{E}$ associated with P comprise a large enough class that if R is any rectangular set in R^{n+1}, then $\mathbf{Z}^{-1}(R) \in \mathcal{E}$.

It is not necessary at this point to describe the measure P nor to build it, only to believe it is there (not your usual existence proof).[2] However, there

[2]This is a philosophical departure from some schools of probability and statistics. Strict objectivists would be disturbed with this because there are, at this

is a situation when we can indeed construct P; let us describe this.

Suppose that somehow or other (and "how" isn't important for the moment) we are able to find (describe) a probability measure $\tilde{P}$ on R^{n+1} such that rectangular sets, hence Borel sets, are $\tilde{P}$ measurable. Suppose, moreover, that we believe $\tilde{P}(B)$ is a measure of the probability that $Z \in B$, that is,

$$\operatorname{Prob}(Z \in B) = \tilde{P}(B)$$

for each $B \in B_n$. In this case, we can define

$$\mathcal{C} = \{Z^{-1}(B) \mid B \in B_{n+1}\}$$

and observe that $\mathcal{C}$ is a covering class for Ω. Moreover, we define

$$\tau(Z^{-1}(B)) \triangleq \tilde{P}(B)$$

for each element of $\mathcal{C}$, and then extend τ to sequential covers (see A3–A5 in Appendix A). Using this covering class, we can generate an outer measure P^* on Ω, hence a measure P and a collection of measurable sets $\mathcal{E}$ on Ω. By our very construction, $\mathcal{C}$ is regular and so the random vector $\mathbf{Z}$ is measurable, hence its components $Y_1, \ldots, Y_n, X$, are measurable also. It also follows that

$$P_{\mathbf{Z}} = \tilde{P}$$

since

$$P_{\mathbf{Z}}(B) = P(\mathbf{Z}^{-1}(B) = \tilde{P}(B).$$

Note the scenario we have described. Rather than define $P_{\mathbf{Z}}$ in terms of P (as the formal mathematics seems to infer), we have really defined P in terms of $P_{\mathbf{Z}}$. In practice, this is the typical situation. Anyway, we now have (2).

Given (1) and (2), it now makes sense to inquire about integrals over $(\Omega, \mathcal{E}, P)$. After all, to formulate integrals, all you need is a respectable measure space and some measurable functions. In particular, what about integrals of the form $\int_\Omega X\, dP$ or $\int_\Omega X^2\, dP$? Do they exist or not? At first glance it would seem that this question has already been decided. For, once we have $(\Omega, \mathcal{E}, P)$ and X, the question of whether or not $\int_\Omega X\, dP$ exists is

---

point, no data. Those of the strictly operational school would also object on the grounds that we are forming a statistical hypothesis, however vague, that is not necessarily testable. Finally, there is an even more fundamental problem. The assumption in (2) implies that all of the random variables under consideration are jointly measurable. If one were studying quantum mechanics, this would simply not be true (unless one accepts the belief in so-called hidden variables). Moreover, Foulis and Randall [7], [20] have shown that even in more general empirical systems, the existence of events that are not simultaneously measurable is a distinct possibility and must be handled with nonclassical sample spaces, that is, spaces that are much different than $(\Omega, \mathcal{E}, P)$.

a foregone conclusion (see Appendix D, Def. D8). Certainly, If P had been generated from $\tilde{P}$ as indicated above, then this is a case in which we can check the existence of $\int_\Omega X\,dP$ by checking the existence of $\int_R id\,dP_X$. In this case either $\int_\Omega X\,dP$ exists or it does not, and if it seems reasonable on physical grounds that it should exist, then one concludes that $\tilde{P}$ is a poor model if $\int_R id\,dP_X$ should fail to exist. In general, however, Ω and X are conceptualizations, X came to be a function on Ω by decree; it is an idea. The nature of this function X is based on physical grounds or on the basis of a finite set of data, or both. In the first case, the existence of $\int_\Omega X\,dP$ can be made on physical grounds; namely, does it seem sensible to speak of its average? In the second case, one would look at the data and decide if it appears that an average makes sense. Again, this is a judgement, for one cannot possibly collect enough data to determine X on a set of positive measures unless P is discrete (and it usually doesn't seem reasonable to suppose it is).

Thus, we suppose that $Y_1, Y_2, \ldots, Y_n, X$ are in $\mathcal{L}_2(\Omega, P)$, simply because Ω and P have been chosen so that this assumption is correct. Moreover, we suppose any other random variables of interest are also in $\mathcal{L}_2(\Omega, P)$. How can we say this? Simple! If it isn't true, we have misformulated things in the first place.

Note that at this point we have at our disposal the notions of expected value, variance, correlation, conditional expectation, and so on, as well as the theorems for manipulating these quantities. We also have the concept of a p.d.f. and the relation between the measure so generated and Lebesgue measure (absolute continuity). In short, we are now in the position to formulate our statistical estimation problem in precise language, even though we may not (at this point) be able to actually make any calculations. Before we proceed with the details, however, we should like to address a possible criticism, and by answering it, illustrate the sort of conceptual advantage we now have.

Suppose we have a couple of random variables X_1 and X_2 that are described by p.d.f.'s g_1 and g_2, respectively. Why not just work with g_1 and g_2 on R and forget all about these fancy measure theoretic constructions? Well, suppose g_1 is gaussian and g_2 is a gamma distribution. What is the expected value of $X_1 + X_2$? "Easy," says John Q. Critic, "you just add their expected values." "But," we ask, "how do you know to do that?" "It is a theorem," says he. "Can you prove it?" "Sure," says Mr. Critic, "just find g_3, the p.d.f. for $X_1 + X_2$, and show that

$$\int_{-\infty}^{\infty} x g_3(x)\,dx = \int_{-\infty}^{\infty} x g_1(x)\,dx + \int_{-\infty}^{\infty} x g_2(x)\,dx."$$

"Well," we say, "go ahead and do it." "Okay," says he, "are X_1 and X_2 independent?" "We don't know!" "Well, you'll have to tell me something more or I can't make the above calculation," complains Mr. Critic, and so on, and so on.

In fact you don't need to know g_3 to calculate $E(X_1 + X_2)$. You don't need to know if X_1 and X_2 are independent or not. All you really need are our two beliefs, (1) and (2), and most emphatically belief (1) that the random behavior of X_1 and X_2 is a consequence of the same set of circumstances. Our construction of $(\Omega, \mathcal{E}, P)$ reflects this belief, and as a simple consequence, we get the result that

$$E(X_1 + X_2) = E(X_1) + E(X_2)$$

just by noticing that the integral is additive (Theorem 1.3.3). In fact, without the above construction, it is not clear that we can even say what X_1+X_2 means let alone ask questions about it.

There are other observations we could make. Since our random variables are in $\mathcal{L}_2(\Omega, P)$ we have the Cauchy–Schwarz theorem available and so in the usual manner we have distance and orthogonality at our disposal. The additivity of the variances of uncorrelated random variables, for example, is simply the Pythagorean theorem, and so on. But for us, the most important observation is that $\mathcal{L}_2(\Omega, P)$ is a Hilbert space, that is, is complete, and thus we have the projection theorem at our disposal. At long last we have reached the subject in the title of this section. The projection theorem in $\mathcal{L}_2(\Omega, P)$ is the genesis of minimum variance estimation. In the next section, we finally formulate the problem of minimum variance estimation, and in the last section we solve it—at least in principle.

2.3 The Minimum Variance Estimation Problem

Let us now return to our random variables $Y_1, Y_2, \ldots, Y_n$ and X defined on $\mathcal{L}_2(\Omega, P)$. We wish to envision a situation wherein we have secured measurements of $Y_1, \ldots, Y_n$ but not of X. For example, suppose that Y_1 and Y_2 represent the position of a boat in the ocean (longitude and latitude) and Y_3 and Y_4 represent the north and east velocity relative to the ocean current. These are quantities that one could measure. If X represents north velocity relative to a set of earth-fixed coordinates, it is clear that unless one has precise information about ocean currents, there is not enough information in the vector (Y_1, Y_2, Y_3, Y_4) to calculate X. However, since Y_1, Y_2, Y_3, Y_4 and X are defined on the same space Ω, one might reasonably infer that if we know some statistical information about their relationship (ideally we know P), that we can use the measurements of Y_1, Y_2, Y_3, Y_4 to make an estimate of X. It is this idea we wish to pursue.

To begin with let us define

$$\mathbf{Y}(\omega) = \begin{bmatrix} Y_1(\omega) \\ \vdots \\ Y_n(\omega) \end{bmatrix}$$

and suppose that the above problem is completely solvable, that is, knowing $\mathbf{Y}(\omega)$ we can calculate $X(\omega)$ every time. This would mean that there is some function

$$g : \mathrm{R}^n \to \mathrm{R}$$

such that

$$X = g \circ \mathbf{Y}. \tag{2.3-1}$$

Note that since we are insisting that X be a random variable, that is, be measurable, it is necessary to insist that g be a Baire function, that is, be Borel measurable (see Theorem 1.2.6).

Now the equality (2.3-1) is more than one can reasonably expect. However, we certainly don't want to rule out its possibility, that is, we don't want to formulate our estimation problem so that the above equality is impossible. We thus come to our first formal definition in this chapter.

2.3.1. **Definition.** (a) Let X and $\mathbf{Y}$ be as above. By an *estimator* $\hat{X}$ for X given $\mathbf{Y}$, we mean a random variable of the form

$$\hat{X} = g \circ \mathbf{Y}, \tag{2.3-2}$$

where g is a Baire function. The function g is called an *estimating function* or *estimator function.*

(b) If $y_1 = Y_1(\omega), \ldots, y_n = Y_n(\omega)$ are measurements of $\mathbf{Y}$, then

$$\hat{x} = g(y_1, y_2, \ldots, y_n)$$

is called the *estimate* of X given $y_1, \ldots, y_n$. Equivalently,

$$\hat{x} = \hat{X}(\omega) = g(\mathbf{Y}(\omega))$$

when $\mathbf{Y}(\omega) = (y_1, \ldots, y_n)$.

This said, which Baire functions g do we pick? Well, that depends on our criteria, a subjective decision. One criterion that seems reasonable, and turns out to be computationally tractable, is the following.

2.3.2. **Definition.** An estimator $\hat{X}$ of X given $\mathbf{Y}$ is called a *minimum variance estimator* providing there is a Baire function g satisfying (2.3-2) and such that

$$\|\hat{X} - X\| \leq \|h \circ \mathbf{Y} - X\| \tag{2.3-3}$$

holds for all Baire functions h for which the right-hand side of (2.3-3) exists. In terms of integrals, (2.2-3) is equivalent to

$$\int_\Omega (\hat{X} - X)^2 \, dP \leq \int_\Omega (h \circ \mathbf{Y} - X)^2 \, dP. \tag{2.3-4}$$

Note that we now have a well-posed problem, the language of mathematics has made the criterion precise. We must now check existence and uniqueness and see if we can find a solution. The first two questions we will now settle, and the last will be done in the final section.

We begin by introducing some useful notation.

2.3.3. **Definition.**

$$\mathcal{M}(\mathbf{Y}) = \{g \circ \mathbf{Y} \mid g \quad \text{a Baire function}, \quad g \circ \mathbf{Y} \in \mathcal{L}_2(\Omega, P)\}.$$

2.3.4. **Theorem.**

(a) *$\mathcal{M}(\mathbf{Y})$ is a subspace of $\mathcal{L}_2(\Omega, P)$.*

(b) *$\mathcal{M}(\mathbf{Y})$ is closed in $\mathcal{L}_2(\Omega, P)$.*

Proof. (a) Certainly if g_1 and g_2 are Baire functions, then so is $\alpha_1 g_1 + \alpha_2 g_2$ a Baire function. If $g_1 \circ \mathbf{Y}$, and $g_2 \circ \mathbf{Y}$ are in $\mathcal{L}_2(\Omega, P)$, then $\alpha_1(g_1 \circ \mathbf{Y}) + \alpha_2(g_2 \circ \mathbf{Y}) \in \mathcal{L}_2(\Omega, P)$ (see Appendix E, Lemma E3-2). However,

$$\alpha_1(g_1 \circ \mathbf{Y}) + \alpha_2(g_2 \circ \mathbf{Y}) = (\alpha_1 g_1 + \alpha_2 g_2) \circ \mathbf{Y},$$

so this function is in $\mathcal{M}(Y)$.

(b) Suppose that g_n is a sequence of Baire functions, $g_n \circ \mathbf{Y} \in \mathcal{M}(\mathbf{Y})$ for each n, and $g_n \circ \mathbf{Y} \to h$ in $\mathcal{L}_2(\Omega, P)$. Since $g_n \circ \mathbf{Y}$ is convergent, it is Cauchy, and by Appendix E, there is a subsequence $g_{n_k} \circ \mathbf{Y}$ converging to h [a.e.]. Thus, from Theorem C8, Appendix C, h is $B(\mathbf{Y})$-measurable. Using the Radon–Nikodym theorem and the construction used in the proof of Theorem 1.9.7 (use h to generate a measure on R^n that is absolutely continuous with respect to $P_{\mathbf{Y}}$), it follows from this and 1.6.8 that $h = g_0 \circ \mathbf{Y}$ for some Baire function g_0. Thus, $h \in \mathcal{M}(\mathbf{Y})$, so $\mathcal{M}(\mathbf{Y})$ is closed. □

2.3.5. **Corollary.** *$\hat{X}$ is the projection of X onto the subspace $\mathcal{M}(\mathbf{Y})$. Thus, $\hat{X}$ exists, it is unique, and it is characterized by the condition $(\hat{X} - X) \perp \mathcal{M}(\mathbf{Y})$.*

Proof. Equation (2.3-3) can be rewritten as

$$\|\hat{X} - X\| \leq \|Z - X\|, \qquad \hat{X} \in \mathcal{M}(\mathbf{Y})$$

for all $Z \in \mathcal{M}(\mathbf{Y})$. But the solution of this problem is given by the projection theorem in $\mathcal{L}_2(\Omega, P)$; namely, it is the projection of X onto the closed subspace $\mathcal{M}(\mathbf{Y})$. □

This definitively settles existence and uniqueness. We now address the calculation of $\hat{X}$.

2.4 Calculating the Minimum Variance Estimator

We have already laid the groundwork for addressing the issue of calculating $\hat{X}$. Specifically, we can now easily prove the following.

2.4.1. **Theorem.** *The minimum variance estimator of X based on $\mathbf{Y}$ is given by*

$$\hat{X} = E(X \mid \mathbf{Y}).$$

Proof. Suppose g is any Baire function such that $g \circ Y \in \mathcal{M}(Y)$. Then, from the Pythagorean theorem in $\mathcal{L}_2(\Omega, P)$ and Lemma 1.9.9, we have the following calculation:

$$\begin{aligned} \|X - g \circ \mathbf{Y}\|^2 &= \|X - E(X \mid \mathbf{Y}) + E(X \mid \mathbf{Y}) - g \circ \mathbf{Y}\|^2 \\ &= \|X - E(X \mid \mathbf{Y})\|^2 + \|E(X \mid \mathbf{Y}) - g \circ \mathbf{Y}\|^2. \end{aligned}$$

Clearly, this expression is minimized when we take $g \circ \mathbf{Y} = E(X \mid \mathbf{Y})$, and this is possible by Theorem 1.9.7. □

From this result and Definition 2.3.1, it is clear that if we have secured measurements $y_1, \ldots, y_n$ of $\mathbf{Y}$, then to calculate the minimum variance estimate $\hat{x}$ of X, we need to know the function g. From the discussion following the proof of Theorem 1.9.7, we see that if we know the joint density $f_{X\mathbf{Y}}$ and if $f_{\mathbf{Y}}$ doesn't vanish at $y_1, \ldots, y_n$, then

$$\hat{x} = g(y_1, \ldots, y_n) = \int_{-\infty}^{\infty} x \frac{f_{X\mathbf{Y}}(x, y_1, \ldots, y_n)}{f_{\mathbf{Y}}(y_1, \ldots, y_n)} dx. \tag{2.4-1}$$

Thus, at this point we either need some method of determining $f_{X\mathbf{Y}}$ or we need to settle for a less ambitious optimization criteria in which $f_{X\mathbf{Y}}$ need not be known. We will address these issues in Chapter 3.

Before we leave this material, we wish to settle a technical issue that was raised at the beginning of Section 1.9.

2.4.2. **Corollary.** *If X and $\mathbf{Y}$ are independent, then $E(X \mid \mathbf{Y}) = E(X)$ (a constant function).*

Proof. If X and $\mathbf{Y}$ are independent, then so are $X - E(X)$ and $g \circ \mathbf{Y}$ (Exercise 1 in Section 1.10). Hence,

$$\begin{aligned} \langle X - E(X), g(\mathbf{Y}) \rangle &= E((X - E(X))g(\mathbf{Y})) \\ &= E(X - E(X)) \cdot E(g(\mathbf{Y})) \quad (1.4.11) \\ &= 0 \cdot E(g(\mathbf{Y})) \\ &= 0. \end{aligned}$$

Thus, $(X - E(X)) \perp \mathcal{M}(\mathbf{Y})$ and so $E(X) = E(X \mid \mathbf{Y})$. □

2.5 Exercises

1. Provide the details for the construction of g_0 in the proof of Theorem 2.3.4 (b).

2. An estimator $\hat{X}$ is called unbiased providing $E(X) = E(\hat{X})$. Describe a situation in which $\hat{X}$ will certainly be unbiased.

3. Let $\Omega = \{\omega_1, \omega_2, \omega_3, \omega_4, \omega_5\}$ and let the corresponding probabilities be 1/2, 1/4, 1/8, 1/16, and 1/16, respectively. Let X and Y be given by

$$X(\omega_1) = 1 \quad Y(\omega_1) = 2$$
$$X(\omega_2) = 2 \quad Y(\omega_2) = 0$$
$$X(\omega_3) = 3 \quad Y(\omega_3) = 0$$
$$X(\omega_4) = 2 \quad Y(\omega_4) = 1$$
$$X(\omega_5) = 3 \quad Y(\omega_5) = 2.$$

Find the estimator function g for the minimum variance estimate of X given Y.

3

The Maximum Entropy Principle

3.1 Introduction

In Section 1.1, we alluded to the principle that titles this chapter. The idea of this principle, as we said there, is to assign probabilities in such a way that the resulting distribution contains no more information than is inherent in the data. The first attempt to do this was by Laplace and was called the "Principle of Insufficient Reason." This principle said that two events should be assigned the same probability if there is no reason to do otherwise. This appears reasonable and is fine as far as it goes. However, Jaynes [12] said, "except in cases where there is an evident element of symmetry that clearly renders the events 'equally possible,' this assumption may appear just as arbitrary as any other that might be made." What is needed, of course, is a way to uniquely quantify the "amount of uncertainty" represented by a probability distribution. As mentioned in Section 1.1, this was done in 1948 by C. E. Shannon in a paper entitled *A Mathematical Theory of Communication* [25]. Shannon was specifically interested in the problem of sending data through a noisy, discrete transmission channel. However, his results, most notably a rigorous treatment of entropy, have had far-reaching consequences in the theory of probability and statistics. We are not going to even attempt to summarize the broad effects his work has had nor reference the many papers that have since been written on the subject; our purpose is to quickly develop the notion of entropy and get on with the estimation problem of the last chapter. For more information on the subject of entropy, the reader is referred to the text by Ellis [4].

3.2 The Notion of Entropy

From a philosophical point of view, the notion of entropy is a bit vague, or so it seems. Entropy is an attempt to quantify the notion of disorder or chaos within a system (the internal point of view) or the degree of uncertainty or ignorance of a system (the external point of view). Trying to precisely define terms such as these, or to decide if the internal and external viewpoints are really different or not, leads one into a hopeless philosophical quagmire. Shannon's idea was to avoid pitfalls such as these and define entropy in terms of mathematical properties upon which all viewpoints could agree. He was remarkably successful in this endeavor. What follows is a bit different than Shannon's original work but is certainly in the same spirit.

We begin with the discrete case, the generalization to the continuous case being rather obvious. Suppose $\Omega_n = \{\omega_1, \omega_2, \ldots, \omega_n\}$ is a sample space of outcomes and $\{p_1, \ldots, p_n\}$ is a set of probabilities on Ω_n, that is,

$$\sum_{i=1}^{n} p_i = 1. \tag{3.2-1}$$

For each such assignment, we wish to assign a number $H_n(p_1, p_2, \ldots, p_n)$ that represents some index of uncertainty or index of ignorance that is inherent in the particular assignment $p_1, \ldots, p_n$. Note that for purposes of this discussion we are taking the "external" view. The notation we have adopted implicitly suggests that the set Ω_n itself is unimportant except for its size and the particular choice of probabilities $p_1, \ldots, p_n$. It is clear, therefore, that we have in mind a collection of functions, $H_1, H_2, H_3, \ldots$, that are ostensibly related to each other. The plan is to write down a set of plausible properties that these functions should have and then use these properties to deduce the forms of H_n for any n.

To begin with, let us make a couple of observations that seem, to us at least, obvious. First of all, the condition

$$H_n(0, \ldots, 0, 1, 0, \ldots, 0) = 0 \tag{3.2-2}$$

should hold since there is no uncertainty here at all; everything is concentrated on the point whose probability is 1 (we are tacitly using 0 to represent no uncertainty). Next note that for $m > n$

$$H_m(p_1, \ldots, p_n, 0, \ldots 0) = H_n(p_1, \ldots, p_n) \tag{3.2-3}$$

since there is the same uncertainty present in the distribution $p_1, \ldots, p_n$, $0, \ldots, 0$ as there is in $p_1, \ldots, p_n$; the last $m - n$ probabilities, being zero, represent certainty, that is, there is certainly *no chance* of obtaining the last $m - n$ outcomes. Third, we would expect H_n to be symmetric, that is,

$$H_n(p_1, \ldots, p_i, \ldots, p_j, \ldots, p_n) = H_n(p_1, \ldots, p_j, \ldots, p_i, \ldots, p_n) \tag{3.2-4}$$

for all $i, j = 1, 2, \ldots, n$. Equation (3.2-4) is reasonable since the only difference between the two sides is a matter of changing labels and this should not affect the uncertainty present in this collection of p_i's. Fourth, we would expect that

$$H_n\left(\frac{1}{n}, \frac{1}{n}, \ldots, \frac{1}{n}\right) \quad \text{is a global maximum}$$

$$\text{on the simplex defined by (3.2-1)} \tag{3.2-5}$$

since this assignment of probabilities represents complete uncertainty. Finally, we think it reasonable to suppose that

$$H_n(p_1, \ldots, p_n) \quad \text{is continuous} \tag{3.2-6}$$

since slight perturbations in the p_i's should not have a large impact on the uncertainty of the probability distribution. We will have more to say about this momentarily.

We have yet to really consider the precise relationship that one might expect to see between H_n and H_{n+1}, although whatever it might be it certainly should respect (3.2-3). To gain a purchase on this relationship, suppose that $\Omega_n = \{\omega_1, \ldots, \omega_n\}$, with associated probabilities $p_1, \ldots, p_n$, has an uncertainty (or ignorance) index of $H_n(p_1, \ldots, p_n)$. Then suppose that we add a new point ω_{n+1} to the set Ω_n, obtaining $\Omega_{n+1} = \{\omega_1, \ldots, \omega_n, \omega_{n+1}\}$, and that we assign probabilities to Ω_{n+1} in such a way that the relative probabilities on Ω_n remain the same as before. Thus, the new distribution must necessarily be $\alpha p_1, \alpha p_2, \ldots, \alpha p_n, 1-\alpha$ where $0 \leq \alpha \leq 1$ is arbitrary. The uncertainty represented by this new probability assignment is, of course, $H_{n+1}(\alpha p_1, \ldots, \alpha p_n, 1-\alpha)$. This uncertainty stems from two sources. First, there is the uncertainty regarding whether or not a given outcome is in Ω_n or not. This, of course, is the same as the uncertainty associated with the probability distribution $\alpha, 1-\alpha$ on a two-element set, that is, $H_2(\alpha, 1-\alpha)$. Second, there is the original uncertainty $H_n(p_1, \ldots, p_n)$ associated with the probability distribution $p_1, \ldots, p_n$. Thus, $H_{n+1}(\alpha p_1, \ldots, \alpha p_n, 1-\alpha)$ should be a function of $H_n(p_1, \ldots, p_n)$ and $H_2(\alpha, 1-\alpha)$. But what sort of function? Let us try the least complicated function we can think of that is not utterly trivial, namely,

$$H_{n+1}(\alpha p_1, \ldots, \alpha p_n, 1-\alpha) = c_1 H_n(p_1, \ldots, p_n) + c_2 H_2(\alpha, 1-\alpha), \tag{3.2-7}$$

where c_1 and c_2 are constants. Now, if $(p_1, p_2, \ldots, p_n) = (1, 0, \ldots, 0)$, (3.2-7) becomes

$$H_{n+1}(\alpha, 0, \ldots, 0, 1-\alpha) = c_1 H_n(1, 0, \ldots, 0) + c_2 H_2(\alpha, 1-\alpha),$$

and from (3.2-4) and (3.2-3), the left-hand side is just $H_2(\alpha, 1-\alpha)$. But from (3.2-2), the above expression becomes

$$H_2(\alpha, 1-\alpha) = c_2 H_2(\alpha, 1-\alpha), \qquad 0 \leq \alpha \leq 1$$

so $c_2 = 1$. Thus, (3.2-7) becomes

$$H_{n+1}(\alpha p_1, \ldots, \alpha p_n, 1-\alpha) = c_1 H_n(p_1, \ldots, p_n) + H_2(\alpha, 1-\alpha). \tag{3.2-8}$$

Now, if $\alpha = 1$ in (3.2-8), we have

$$H_{n+1}(p_1, \ldots, p_n, 0) = c_1 H_n(p_1, \ldots, p_n) + H_2(1, 0),$$

which by (3.2-2) and (3.2-3) is the same as

$$H_n(p_1, \ldots, p_n) = c_1 H_n(p_1, \ldots, p_n)$$

or $c_1 = 1$. However, if $\alpha = 0$ in (3.2-8), we have

$$H_{n+1}(0, \ldots, 0, 1) = c_1 H_n(p_1, \ldots, p_n) + H_2(0, 1),$$

which reduces to

$$0 = c_1 H_n(p_1, \ldots, p_n)$$

or $c_1 = 0$. Thus, the constant c_1 is 0 when $\alpha = 0$ and 1 when $\alpha = 1$. If we set $c_1 = \alpha$, c_1 will be consistent with these two calculations and (3.2-8) will have the form

$$H_{n+1}(\alpha p_1, \alpha p_2, \ldots, \alpha p_n, 1 - \alpha) = \alpha H_n(p_1, \ldots, p_n) + H_2(\alpha, 1 - \alpha). \quad (3.2\text{-}9)$$

Certainly this "derivation" of (3.2-9) includes some unsubstantiated assumptions (as indeed it must), but the final form of (3.2-9) is consistent with our earlier observations and appears, to us, reasonable, in that the uncertainty index $H_n(p_1, \ldots, p_n)$ is scaled by α whenever the probabilities $p_1, p_2, \ldots, p_n$ are scaled by α. In any case, we are now ready to make our official declaration of assumptions about H_n. There are three of them.

(1) $H_n(p_1, \ldots, p_n)$ is symmetric and nonnegative. (3.2-10)

(2) If $\sum_{i=1}^{n} p_i = 1$, then

$$H_{n+1}(\alpha p_1, \ldots, \alpha p_n, 1 - \alpha) = \alpha H_n(p_1, \ldots, p_n) + H_2(\alpha, 1 - \alpha). \quad (3.2\text{-}11)$$

(3) $H_2(\alpha, 1 - \alpha)$ is twice differentiable on $(0, 1)$ and continuous on $[0, 1]$. (3.2-12)

The third condition is stronger than (3.2-6). We have introduced this stronger condition simply to save time. It turns out that we can obtain our desired result replacing (3.2-12) by the weaker condition (3.2-6). The interested reader can find this proof in Reference [5]. We wish to point out, however, that the class of functions specified in (3.2-12) is dense in the continuous functions[1] on $[0, 1]$, so assumption (3.2-12) is not really a disturbing shortcut.

In the theorems and lemmas that follow in this section, we will assume the hypotheses (3.2-10)–(3.2-12) and nothing else.

3.2.1. **Theorem.** $H_2(1, 0) = 0$.

[1] The Stone–Weierstrass theorem will suffice.

Proof. In (3.2-11) let $p_1 = 1$, $p_2 = 0$. Then

$$H_3(\alpha, 0, 1-\alpha) = \alpha H_2(1,0) + H_2(\alpha, 1-\alpha).$$

However, using (3.2-11) again and replacing p_1 by α and α by 1, we obtain

$$H_3(\alpha, 1-\alpha, 0) = H_2(\alpha, 1-\alpha) + H_2(1,0).$$

Equating the left-hand sides of these two expressions (using (3.2-10)), we obtain

$$H_2(1,0) = \alpha H_2(1,0)$$

for all α. The theorem follows. □

3.2.2. **Lemma.**

$$\begin{aligned} \alpha H_2(\beta, 1-\beta) + H_2(\alpha, 1-\alpha) &= (1-\alpha\beta) H_2\left(\frac{1-\alpha}{1-\alpha\beta}, \frac{\alpha-\alpha\beta}{1-\alpha\beta}\right) \\ &\quad + H_2(1-\alpha\beta, \alpha\beta) \end{aligned}$$

where $0 \le \alpha < 1$ *and* $0 \le \beta < 1$.

Proof. Note that $\alpha\beta$, $\alpha(1-\beta)$, $1-\alpha$ is a probability distribution. Hence, by (3.2-11),

$$H_3(\alpha\beta, \alpha(1-\beta), 1-\alpha) = \alpha H_2(\beta, 1-\beta) + H_2(\alpha, 1-\alpha). \tag{3.2-13}$$

Using (3.2-10), we can also write

$$H_3(\alpha\beta, \alpha(1-\beta), 1-\alpha) = H_3\left((1-\alpha\beta)\frac{1-\alpha}{1-\alpha\beta}, (1-\alpha\beta)\frac{\alpha-\alpha\beta}{1-\alpha\beta}, \alpha\beta\right),$$

and applying (3.2-11) to the right-hand side of this last expression, we obtain

$$H_3(\alpha\beta, \alpha(1-\beta), 1-\alpha) = (1-\alpha\beta) H_2\left(\frac{1-\alpha}{1-\alpha\beta}, \frac{\alpha-\alpha\beta}{1-\alpha\beta}\right) + H_2(1-\alpha\beta, \alpha\beta).$$

The lemma follows from this and (3.2-13). □

3.2.3. **Definition.** Define $f : [0,1] \to \mathrm{R}$ by

$$f(x) = H_2(x, 1-x).$$

By (3.2-12), f is twice differentiable on $(0,1)$ and continuous on $[0,1]$.

3.2.4. **Corollary** (to 3.2.2 and Definition 3.2.3).

$$\alpha f(\beta) + f(\alpha) = (1-\alpha\beta) f\left(\frac{1-\alpha}{1-\alpha\beta}\right) + f(\alpha\beta).$$

3.2.5. **Lemma.** $f(\beta) + f'(\alpha) - \beta f'(\beta) + f'\left(\frac{1-\alpha}{1-\alpha\beta}\right) = 0.$

Proof. Differentiate the expression in 3.2.4 first with respect to α and then with respect to β obtaining the two expressions

$$f(\beta) + f'(\alpha) = -\beta f\left(\frac{1-\alpha}{1-\alpha\beta}\right) + \frac{\beta - 1}{1-\alpha\beta} f'\left(\frac{1-\alpha}{1-\alpha\beta}\right) + \beta f'(\alpha\beta)$$

and

$$\alpha f'(\beta) = -\alpha f\left(\frac{1-\alpha}{1-\alpha\beta}\right) + \frac{\alpha(1-\alpha)}{1-\alpha\beta} f'\left(\frac{1-\alpha}{1-\alpha\beta}\right) + \alpha f'(\alpha\beta).$$

Multiplying the first equation by α, the second by β, and subtracting the second from the first, we obtain

$$\alpha[f(\beta) + f'(\alpha)] - \alpha\beta f'(\beta) = f'\left(\frac{1-\alpha}{1-\alpha\beta}\right) \frac{\alpha^2\beta - \alpha}{1-\alpha\beta}.$$

Dividing this by α we obtain the result. □

3.2.6. **Lemma.** $(x-1)xf''(x)$ *is constant on* $(0,1)$.

Proof. Differentiate the result in Lemma 3.2.5 first with respect to α and then with respect to β, and obtain the equations

$$f''(\alpha) + f''\left(\frac{1-\alpha}{1-\alpha\beta}\right) \frac{\beta - 1}{(1-\alpha\beta)^2} = 0$$

$$-\beta f''(\beta) + f''\left(\frac{1-\alpha}{1-\alpha\beta}\right) \frac{\alpha(1-\alpha)}{(1-\alpha\beta)^2} = 0.$$

Solving the first for $f''[(1-\alpha)/(1-\alpha\beta)]$ and substituting into the second, we obtain

$$-\beta f''(\beta) - \frac{(1-\alpha\beta)^2}{\beta - 1} f''(\alpha) \cdot \frac{\alpha(1-\alpha)}{(1-\alpha\beta)^2} = 0$$

or

$$\alpha(\alpha - 1) f''(\alpha) = \beta(\beta - 1) f''(\beta).$$

But α and β are arbitrary numbers in $(0,1)$, and this implies the result. □

3.2.7. **Theorem.** *The function* $f(x)$ *has the form*

$$f(x) = \begin{cases} -c[x \ln x + (1-x)\ln(1-x)] & x \in (0,1) \\ 0 \quad \text{for} \quad x = 0, 1 \end{cases}$$

for $x \in [0,1]$, $c > 0$.

Proof. On $(0,1)$ we have by Lemma 3.2.6 that

$$(x-1)xf''(x) = c,$$

where c is a constant. Thus,

$$f''(x) = c\left[\frac{1}{x-1} - \frac{1}{x}\right],$$

whence

$$f'(x) = c[\ell n|x-1| - \ell n|x|] + c_1.$$

Since $|x-1| = 1-x$, $|x| = x$,

$$f'(x) = c[\ell n(1-x) - \ell n\, x] + c_1.$$

Integrating one more time

$$f(x) = c[-(1-x)\,\ell n(1-x) + (1-x) - x\,\ell n\, x + x] + c_1 x + c_2. \quad (3.2\text{-}14)$$

Since $f(x) = f(1-x)$ for $0 < x < 1$, it follows from this expression that

$$c_1 x = c_1(1-x)$$

or

$$c_1(2x-1) = 0, \qquad 0 < x < 1.$$

It follows that $c_1 = 0$. We can then write (3.2-14) as

$$f(x) = -c[x\,\ell n\, x + (1-x)\,\ell n(1-x)] + c + c_2.$$

Since $\lim_{t\to 0^+} t\,\ell n\, t = 0$, we can make f continuous at 0 and 1 by letting $c_2 = -c$ and defining f as in the statement of the theorem. It remains to check the sign of c. Since by (3.2-9), $f(x) \geq 0$, it follows that $c > 0$. □

3.2.8. **Theorem.** *If* $\sum_{i=1}^{n} p_i = 1$, *then*

$$H_n(p_1, p_2, \ldots, p_n) = -c\sum_{i=1}^{n} p_i\,\ell n\, p_i,$$

with the convention that if $p_i = 0$, *then* $p_i\,\ell n\, p_i$ *is replaced by zero.*

Proof. We will use induction on n. Clearly, 3.2.7 establishes the result for $n = 2$. Thus, suppose the result is true for n. Then using (3.2-10), we have (setting $c = 1$)

$$H_{n+1}(p_1, \ldots, p_n, p_{n+1}) = qH_n\left(\frac{p_1}{q}, \frac{p_2}{q}, \ldots, \frac{p_n}{q}\right) + H_2(q, 1-q),$$

where $q \triangleq p_1 + p_2 + \ldots + p_n = 1 - p_{n+1}$ (replace α by q and p_i by p_i/q). By hypothesis,

$$H_{n+1}(p_1, \ldots, p_n, p_{n+1}) = q\left[-\sum_{i=1}^{n} \frac{p_i}{q}\ell n\frac{p_i}{q}\right] - q\,\ell n\, q - (1-q)\,\ell n(1-q),$$

and so

$$\begin{aligned} H_{n+1}(p_1,\ldots,p_n,p_{n+1}) &= -\sum_{i=1}^{n} p_i[\ell\text{n}\, p_i - \ell\text{n}\, q] - q\,\ell\text{n}\, q - p_{n+1}\,\ell\text{n}\, p_{n+1} \\ &= -\sum_{i=1}^{n} p_i\,\ell\text{n}\, p_i + q\,\ell\text{n}\, q - q\,\ell\text{n}\, q - p_{n+1}\,\ell\text{n}\, p_{n+1} \\ &= -\sum_{i=1}^{n+1} p_i\,\ell\text{n}\, p_i \end{aligned}$$

as claimed. □

We have shown that there is a function, namely, that given in Theorem 3.2.8, such that (3.2-10), (3.2-11), and (3.2-12) all hold. Moreover, as we now show, the additional requirement stipulated in (3.2-5) also is true.

3.2.9. **Theorem.** *The function given in Theorem* 3.2.8 *defined on the simplex*[2] $\sum_{i=1}^{n} x_i = 1$, $x_i \geq 0$ *obtains a global maximum at* $(1/n, 1/n, \ldots, 1/n)$.

Proof. Since the function is nonnegative and takes on values 0 at the vertices, the maximum clearly does not occur at a vertex. Since $H_{k+1}(1/(k+1),\ldots,1/(k+1)) \geq H_k(1/k,\ldots,1/k)$ for all $k \geq 2$ (Exercise 3.6.1(a)), it follows that if H_n obtains a local maximum at an interior point of the simplex, then it cannot obtain a maximum on a face (whatever the dimension) (Exercise 3.6.1(b)). Since H_n is differentiable on the interior of the simplex, any global maximum must be a stationary point. Hence, let

$$\begin{aligned} g(x_1,\ldots,x_{n-1}) &= H_n\left(x_1, x_2, \ldots, x_{n-2}, 1 - \sum_{i=1}^{n-1} x_i\right) \\ &= \sum_{i=1}^{n-2} f(x_i) + f\left(1 - \sum_{i=1}^{n-1} x_i\right), \end{aligned}$$

where $f(x) = -x\,\ell\text{n}\, x$, and note that

$$\frac{\partial g}{\partial x_i} = f'(x_i) - f'\left(1 - \sum_{i=1}^{n-1} x_i\right).$$

Then, since $f'(x) = -1 - \ell\text{n}\, x$, it follows that

$$\frac{\partial g}{\partial x_i} = -\ell\text{n}\, x_i + \ell\text{n}\left(1 - \sum_{i=1}^{n-1} x_i\right).$$

Setting

$$\frac{\partial g}{\partial x_i} = 0 \quad \text{for} \quad i = 1, 2, \ldots, n-1,$$

---

[2] See Definition 3.2.14 for the formal definition of this term.

we obtain

$$\ell n\, x_i = \ell n \left(1 - \sum_{j=1}^{n-1} x_j\right) \qquad \text{for} \quad i = 1, 2, \ldots, n-1.$$

In other words, if $(x_1, \ldots, x_{n-1})$ is a set of points satisfying this expression, we would have

$$\ell n\, x_1 = \ell n\, x_2 = \ldots = \ell n\, x_{n-1} = \ell n\, x_n,$$

where

$$x_1 + x_2 + \cdots + x_n = 1.$$

But $\ell n\, a = \ell n\, b$ if and only if $a = b$, so $x_i = 1/n$ for each i. Thus,

$$\frac{\partial g}{\partial x_i}\left(\frac{1}{n}, \frac{1}{n}, \ldots, \frac{1}{n}\right) = 0 \quad \text{for} \quad i = 1, 2, \ldots, n-1. \tag{3.2-15}$$

In addition, we have

$$\frac{\partial^2 g}{\partial x_i^2} = f''(x_i) + f''(x_n)$$

and

$$\frac{\partial^2 g}{\partial x_i\, \partial x_j} = f''(x_n),$$

so at $(1/n, 1/n, \ldots, 1/n)$, the Hessian of g is

$$\mathcal{H}(g) = f''\left(\frac{1}{n}\right) \begin{bmatrix} 2 & 1 & 1 & \ldots & 1 \\ 1 & 2 & 1 & \ldots & 1 \\ 1 & 1 & 2 & & \\ \vdots & \vdots & \vdots & \ddots & \\ 1 & 1 & 1 & & 2 \end{bmatrix}. \tag{3.2-16}$$

From calculations in Theorem 3.2.7, we have

$$f''\left(\frac{1}{n}\right) = c\left[\frac{n^2}{1-n}\right],$$

and since $c > 0$, $f''(1/n) < 0$. The matrix in (3.2-15) is positive definite. (Exercise 3.6.2), hence the Hessian is negative definite. This observation together with (3.2-15) implies that $(1/n, \ldots, 1/n)$ is a local maximum, and since it is the only one, it is the global maximum. □

3.2.10. **Corollary.** *The global maximum of* $H_n(x_1, \ldots, x_n)$ *on the simplex* $\sum_{i=1}^{n} x_i = 1$ *is* $c\,\ell n\, n$.

Proof. Direct calculation. □

With the proof of Theorem 3.2.9, we have exhibited a function H_n that satisfies all of the conditions in our motivation. In addition, we have shown that if (3.2-10), (3.2-11), and (3.2-12) hold, then this function is unique. Whether or not a function quantifying disorder, uncertainty, or ignorance should satisfy these three conditions is, of course, a subjective matter (remember, also, that (3.2-12) can be weakened to mere continuity). We have tried to make the assumptions seem reasonable, which is all one can really do. But if one accepts these assumptions as indeed the properties such a quantification should have, then H_n is the function we seek. Let us suppose this to be the case and study the consequences.

Here is our official definition.

3.2.11. **Definition.**

(a) By the *standard* $n-1$ *simplex* in R^n, we mean the set

$$S = \left\{ (x_1, \ldots, x_n) \mid \sum_{i=1}^{n} x_i = 1 \right\}.$$

(b) By the *entropy function* on S we mean the function $H_n : S \to \mathrm{R}$ defined by

$$H_n(x_1, \ldots, x_n) = -c \sum_{i=1}^{n} f(x_i),$$

where

$$f(x_i) \triangleq \begin{cases} x_i \ln x_i, & x_i \neq 0 \\ 0, & x_i = 0. \end{cases}$$

As shown above, $H_n(x_1, \ldots, x_n)$ is twice differentiable on the interior of S, is continuous on S, is zero at the vertices of S, and has a global maximum at $(1/n, 1/n, \ldots, 1/n)$ that is equal to $c \ln n$.

(c) Let $\Omega = \{\omega_1, \ldots, \omega_n\}$ be a finite set, and let $\{p_1, \ldots, p_n\}$ be an assignment of probabilities to Ω. By the *entropy of* $\{p_1, \ldots, p_n\}$, we mean the number $H_n(p_1, \ldots, p_n)$.

The notion of entropy can be extended to countably infinite sets with discrete measures in the obvious way, although convergence of the resulting series is not automatic. We will not discuss this since our real interest is in the generalization to the continuous case.

To generalize the above ideas to the case of continuous distributions we simply note that for the finite case

$$H_n(p_1, \ldots, p_n) = -cE(\ln p), \tag{3.2-17}$$

where $p : \Omega \to \mathrm{R}$ is the function $p(\omega_i) = p_i$ (again one must agree that $p_i \ln p_i = 0$ when $p_i = 0$). Thus, if ρ represents a p.d.f. on, for instance, R^n, we could define (analogous to (3.2-17))

$$H(\rho) = -cE(\ln \rho).$$

Specifically, we have the following.

3.2.12. **Definition.** Let ρ be a p.d.f. on R^n.

(a) By the *support* of ρ, we mean the subset $\mathrm{supp}(\rho)$ defined by

$$\mathrm{supp}(\rho) = \{\mathbf{x} \in \mathrm{R}^n \mid \rho(\mathbf{x}) \neq 0\}.$$

(b) By the *entropy* of ρ, $H(\rho)$, we mean

$$H(\rho) = -c \int_{\mathrm{supp}(\rho)} \rho \,\ell\mathrm{n}\, \rho \, d\mu_n$$

whenever this integral exists.

Unlike the discrete case, the function H in 3.2.13 need not have a maximum. This can easily be seen by taking ρ to be the normal distribution on R, that is,

$$\rho(x) = \frac{1}{\sqrt{2\pi}\sigma} e^{-x^2/2\sigma^2}.$$

Here (taking $c = 1$),

$$H(\rho) = -\int_{-\infty}^{\infty} \rho \left[-\ell\mathrm{n}\, \sqrt{2\pi}\,\sigma - \frac{x^2}{2\sigma^2} \right] dx$$

or

$$H(\rho) = \ell\mathrm{n} \sqrt{2\pi}\,\sigma + \frac{1}{2}. \tag{3.2-18}$$

Although H is defined for all σ, one can make $H(\rho)$ arbitrarily large by taking σ large. Note also that $H(\rho) \to -\infty$ as $\sigma \to 0$, so $H(\rho)$ can take on any real value, that is, it is not bounded. Nevertheless, (3.2-18) is consistent with our intuitive understanding of variance. If σ is very large, our distribution ρ is very "spread out," approximating a uniform distribution, whereas if σ is small, ρ is very "concentrated" and there is little uncertainty present.

3.3 The Maximum Entropy Principle

In the last section, we developed enough motivational material to, hopefully, support the belief that the definitions of entropy given in 3.2.11 and 3.2.12 provide us with a measure of the ignorance, uncertainty, or disorder associated with a probability density. For purposes of discussion it will be convenient for us to introduce the dual notion, namely, that of information.

3.3.1. Definition.

(a) Let S be the standard $n-1$ simplex in R^n. By the *information function* on S, we mean the function

$$I : S \to \mathrm{R}$$

defined by

$$I(x_1, \ldots, x_n) = c\,\ell\mathrm{n}\, n - H_n(x_1, \ldots, x_n).$$

(b) If f is a p.d.f. for which $H(f)$ is defined, we define

$$I(f) = -H(f).$$

I is called the *information function.*

(c) If $\{p_1, \ldots, p_n\}$ is a discrete probability distribution, then $I(p_1, \ldots, p_n)$ is called the information content of $\{p_1, \ldots, p_n\}$. Likewise, if ρ is a p.d.f. on R^n, then $I(\rho)$ is called the *information* content of ρ.

Note that in the discrete case, the uniform distribution has zero information while the distribution that is concentrated on a single point has maximum information. In the continuous case, since H is unbounded in general, no such scaling is possible.

We are now ready to discuss the principal problem of this chapter. The discussion will be in terms of probability density functions on R^n since this is the topic of specific interest to us; the comparable notion for discrete spaces should be clear to the reader.

Suppose that we have a random vector $\mathbf{X} : \Omega \to \mathrm{R}^n$. Further, suppose that the probability measure $P_{\mathbf{X}}$ is unknown to us, but we do have some information about it (information in the Webster sense). In particular, we suppose the following.

(1) $P_{\mathbf{X}}$ is a continuous distribution, hence is generated by a p.d.f. ρ.

(2) ρ is known to satisfy a condition $c(\rho)$.

Condition (1) is clear; let us say a bit about (2). Our intention here is for the reader to interpret c as a predicate (see Definition C7 in Appendix C). Typically c would represent an equality or inequality that ρ is required to satisfy. In general, then, the information that ρ satisfies $c(\rho)$ can be thought of as a constraint on ρ in terms of the following problem.

Problem. Find the p.d.f. ρ such that $c(\rho)$ is true.

Obviously this is not a well-posed problem unless c contains sufficient information to uniquely characterize ρ, and in general this will not be the case. In other words, out of the many, many p.d.f.'s ρ such that $c(\rho)$ is true, how can we select a p.d.f. ρ_0 that is more "reasonable" than the others? The answer is the following generalization of Laplace's Principle of Insufficient Reason, namely, we choose ρ_0 so that it contains no more information than is inherent in the condition c. From our earlier work, we can easily quantify this idea, namely, we choose ρ_0 so that $c(\rho_0)$ is true and $I(\rho_0)$ is minimized. Why minimized? Two reasons. First of all it is clear that any p.d.f. ρ satisfying $c(\rho)$ must contain at least as much information as ρ_0, that is, $I(\rho) \geq I(\rho_0)$, so that ρ_0 utilizes all of the information available in c. The only way one could select ρ with $I(\rho) < I(\rho_0)$ would be to relax the constraint c, that is, to throw away some of the information in c. Second, if ρ_1 is chosen so that $I(\rho_1) \geq I(\rho_0)$ and $c(\rho_1)$ is true, then it is possible to select a p.d.f. ρ satisfying $I(\rho_1) > I(\rho)$ (namely, take ρ_0) *without* relaxing the conditions c. In other words, ρ_1 contains more information than is inherent in c.

It should be clear from Definition 3.3.1 that I is minimized exactly when H is maximized. Thus, the procedure outlined in the preceding paragraph can be formally stated as follows.

Maximum Entropy Principle. Among all of the p.d.f.'s ρ satisfying $c(\rho)$, choose the one that maximizes $H(\rho)$.

Note that this is *not* a theorem, it is a belief. One cannot prove it or disprove it, one can only use it and see how well it works. It is a metaprinciple and has the same status as the metaprinciple concerning independence in Section 1.4. In both cases, there are plausible reasons for belief in the principle, but each is still a belief! In the next section, we will see an amazing consequence of this belief.

3.4 The Prior Covariance Problem

We are now ready to apply the maximum entropy principle to a particular problem, namely, that of selecting the maximum entropy density ρ_0 on R^n when the mean $\boldsymbol{\mu}$ and covariance Σ are known. The choice of this particular problem is not an arbitrary one; it is quite germane to the Kalman filtering problem. The set of constraints is as follows.

$$\int_{\mathrm{R}^n} \rho \, d\mu_n = 1$$

$$\int_{\mathrm{R}^n} \mathbf{x}\rho(\mathbf{x}) \, d\mu_n = \boldsymbol{\mu} \tag{3.4-1}$$

$$\int_{\mathrm{R}^n} (\mathbf{x} - \boldsymbol{\mu})(\mathbf{x} - \boldsymbol{\mu})^T \rho(\mathbf{x}) \, d\mu_n = \Sigma,$$

where the vector and matrix integrations are element by element. Using the inner product generated by the trace operator on the $n \times n$ matrices, that is, $\langle A, B\rangle = \mathrm{tr}(AB^T)$, we can write the Lagrangian for this problem as

$$\begin{aligned}\mathcal{L}(\rho) &= -\int_{\mathbf{R}^n} \rho\,\ell\mathrm{n}\,\rho\,d\mu_n + \lambda_1\left[1 - \int_{\mathbf{R}^n} \rho\,d\mu_n\right] \\ &+ \langle \boldsymbol{\lambda}_2, \boldsymbol{\mu} - \int_{\mathbf{R}^n} \mathbf{x}\rho(\mathbf{x})\,d\mu_n\rangle \\ &+ \mathrm{tr}\left[\Lambda_3\left(\Sigma - \int_{\mathbf{R}^n} (\mathbf{x}-\boldsymbol{\mu})(\mathbf{x}-\boldsymbol{\mu})^T\right)d\mu_n\right],\end{aligned}$$

where λ_1 is a scalar, $\boldsymbol{\lambda}_2$ is a vector, and Λ_3 is an $n \times n$ matrix. These are the Lagrange multipliers. Taking the Gateaux (directional, variational) derivative, we have

$$\begin{aligned}\delta\mathcal{L}(\rho)(h) &= -\int_{\mathbf{R}^n} (h\,\ell\mathrm{n}\,\rho + h)\,d\mu_n - \lambda_1 \int_{\mathbf{R}^n} h\,d\mu_n \\ &- \langle \boldsymbol{\lambda}_2, \int_{\mathbf{R}^n} \mathbf{x}h(\mathbf{x})\,d\mu_n\rangle \\ &- \mathrm{tr}\left(\Lambda_3 \int_{\mathbf{R}^n} (\mathbf{x}-\boldsymbol{\mu})(\mathbf{x}-\boldsymbol{\mu})^T\right) h\,d\mu_n.\end{aligned}$$

Since the integral preserves linear combinations, this can be written as

$$\delta\mathcal{L}(\rho)(h) = -\int_{\mathbf{R}^n} \left[\ell\mathrm{n}\,\rho + 1 + \lambda_1 + \langle \boldsymbol{\lambda}_2, \mathbf{x}\rangle + \mathrm{tr}(\Lambda_3(\mathbf{x}-\boldsymbol{\mu})(\mathbf{x}-\boldsymbol{\mu})^T\right] h\,d\mu_n.$$

At ρ_0, the Lagrangian is stationary, that is, for every h, $\delta\mathcal{L}(\rho_0)(h) = 0$. This implies that

$$\ell\mathrm{n}\,\rho_0 + 1 + \lambda_1 + \langle \boldsymbol{\lambda}_2, \mathbf{x}\rangle + \langle \Lambda_3(\mathbf{x}-\boldsymbol{\mu}), \mathbf{x}-\boldsymbol{\mu}\rangle = 0. \tag{3.4-2}$$

Let $B \triangleq [\Lambda_3 + \Lambda_3^T]^{-1}$ so that $B = B^T$ and define

$$\gamma = 1 + \lambda_1 + \langle \boldsymbol{\lambda}_2, \boldsymbol{\mu}\rangle - \frac{1}{2}\langle B_3\boldsymbol{\lambda}_2, \boldsymbol{\lambda}_2\rangle.$$

It is then a simple matter of matrix algebra to show that (3.4-2) can be written as

$$\ell\mathrm{n}\,\rho_0 + \gamma + \frac{1}{2}\langle B^{-1}(\mathbf{x}-\boldsymbol{\mu}+B\boldsymbol{\lambda}_2), \mathbf{x}-\boldsymbol{\mu}+B\boldsymbol{\lambda}_2\rangle = 0$$

or

$$\rho_0 = e^{-\gamma}e^{-\langle B^{-1}(\mathbf{x}-\boldsymbol{\mu}+B\boldsymbol{\lambda}_2), \mathbf{x}-\boldsymbol{\mu}+B\lambda_2\rangle/2}. \tag{3.4-3}$$

Replacing the constant $e^{-\gamma}$ by A, we see that (3.4-3) is

$$\rho_0(\mathbf{x}) = Ae^{-\langle B^{-1}(\mathbf{x}-\boldsymbol{\mu}+B\boldsymbol{\lambda}_2, \mathbf{x}-\boldsymbol{\mu}+B\boldsymbol{\lambda}_2\rangle/2} \tag{3.4-4}$$

with

$$A \geq 0.$$

Now by Equation (1.8-14) in Section 1.8, we know that

$$\int_{\mathrm{R}^n} \rho_0(\mathbf{x})\, d\mu_n = A \cdot (2\pi)^{n/2}\sqrt{\det(B)}$$

and by the first of conditions (3.4-1), that

$$A = \frac{1}{(2\pi)^{n/2}\sqrt{\det(B)}}.$$

From this and Theorem 1.8.1, the second of conditions (3.4-1) implies that

$$\boldsymbol{\mu} = \boldsymbol{\mu} - B\boldsymbol{\lambda}_2$$

and so $\boldsymbol{\lambda}_2 = 0$. Thus, (3.4-4) has the form

$$\rho_0(\mathbf{x}) = \frac{1}{(2\pi)^{n/2}\sqrt{\det(B)}} e^{-\langle B^{-1}(\mathbf{x}-\boldsymbol{\mu}),\mathbf{x}-\boldsymbol{\mu}\rangle/2}. \tag{3.4-5}$$

Again, using Theorem 1.8.1, Equation (3.4-5), and the third condition in (3.4-1), we obtain the result that $B = \Sigma$. Hence, (3.4-5) becomes

$$\rho_0(\mathbf{x}) = \frac{1}{(2\pi)^{n/2}\sqrt{\det(B)}} e^{-\langle \Sigma^{-1}(\mathbf{x}-\boldsymbol{\mu}),\mathbf{x}-\boldsymbol{\mu}\rangle/2}. \tag{3.4-6}$$

Thus, if we are given the mean $\boldsymbol{\mu}$ and covariance matrix Σ as prior information, the maximum entropy density function ρ_0 is given by (3.4-6). In the next section, we will apply this result to our minimum variance estimation problem of the last chapter.

3.5 Minimum Variance Estimation with Prior Covariance

Returning to the notation in Chapter 2, let X be a random variable and $\mathbf{Y}$ a random vector defined on the probability space $(\Omega, \mathcal{E}, P)$. Recall that we showed that the minimum variance estimator of X given $\mathbf{Y}$ is the conditional expectation $E(X \mid \mathbf{Y})$. We now wish to address the problem of finding this estimator given only the mean and covariance matrix of the random vector $(X, \mathbf{Y})$. Had we not developed the maximum entropy principle, such a problem would seem ill defined and virtually impossible to solve. Now it is quite clear how to proceed. We simply take $P_{X\mathbf{Y}}$ to be the measure on R^{n+1} generated by the maximum entropy density ρ_0 and then use this density to calculate $E(X \mid \mathbf{Y})$.

For our convenience, we first state a useful corollary to Theorem 1.9.7.

3.5.1. Theorem. *Let* $\mathbf{Y}$ *and* $\mathbf{Z}$ *be random vectors on* Ω *such that* $B(\mathbf{Y}) = B(\mathbf{Z})$. *Then there is a* $B(\mathbf{Y})$ *measurable function* g *such that*

$$E(X \mid \mathbf{Y}) = g \circ Y; \qquad E(X \mid \mathbf{Z}) = g \circ Z.$$

This theorem is simply a way of emphasizing the fact that g depends only on X and $B(\mathbf{Y})$, not on the particular values that $\mathbf{Y}$ assumes. For us, the point is that since $B(\mathbf{Y}) = B(\mathbf{Y} - \boldsymbol{\mu}_y)$ and since $E(X \mid \mathbf{Y})$ is linear in the first variable (1.9.8(b)), we can calculate $E(X \mid \mathbf{Y})$ by calculating $E(X - \mu_x \mid \mathbf{Y} - \boldsymbol{\mu}_y)$ instead.

For convenience, let us introduce the following notation

$$\begin{aligned}
Z_2 &= \mathbf{Y} - \boldsymbol{\mu}_y \\
Z_1 &= X - \mu_x \\
\Gamma_{11} &= \operatorname{cov}(X - \mu_x, X - \mu_x) = E(Z_1^2) \\
\Gamma_{12} &= \operatorname{cov}(X - \mu_x, \mathbf{Y} - \boldsymbol{\mu}_y) = E(Z_1 \mathbf{Z}_2^T) \\
\Gamma_{21} &= \Gamma_{12}^T \\
\Gamma_{22} &= \operatorname{cov}(\mathbf{Y} - \boldsymbol{\mu}_y, \mathbf{Y} - \boldsymbol{\mu}_y) = E(\mathbf{Z}_2 \mathbf{Z}_2^T).
\end{aligned}$$

Note that Z_1 and $\mathbf{Z}_2$ have zero means. Let $\mathbf{Z} = \begin{bmatrix} Z_1 \\ \mathbf{Z}_2 \end{bmatrix}$. Then the maximum entropy density for $\mathbf{Z}$ is given by

$$\rho_0(\mathbf{Z}) = \frac{1}{(2\pi)^{n+1/2}\sqrt{\det \Gamma}} e^{-\langle \Gamma^{-1}\mathbf{Z}, \mathbf{Z}\rangle/2}, \tag{3.5-1}$$

where

$$\Gamma = \begin{bmatrix} \Gamma_{11} & \Gamma_{12} \\ \Gamma_{21} & \Gamma_{22} \end{bmatrix}.$$

If we let

$$\Gamma^{-1} = \begin{bmatrix} F_{11} & F_{12} \\ F_{21} & F_{22} \end{bmatrix},$$

then according to Theorem 1.7.1,

$$\begin{aligned}
F_{11} &= (\Gamma_{11} - \Gamma_{12}\Gamma_{22}^{-1}\Gamma_{21})^{-1} \\
F_{12} &= -F_{11}\Gamma_{12}\Gamma_{22}^{-1} \\
F_{21} &= -\Gamma_{22}^{-1}\Gamma_{21}F_{11} \\
F_{22} &= \Gamma_{22}^{-1} + \Gamma_{22}^{-1}\Gamma_{21}F_{11}\Gamma_{12}\Gamma_{22}^{-1}.
\end{aligned}$$

Thus,

$$\begin{aligned}
\langle \Gamma^{-1}\mathbf{Z}, \mathbf{Z}\rangle &= \mathbf{Z}^T\Gamma^{-1}\mathbf{Z} \\
&= [Z_1, \mathbf{Z}_2^T]\begin{bmatrix} F_{11} & F_{12} \\ F_{21} & F_{22} \end{bmatrix}\begin{bmatrix} Z_1 \\ \mathbf{Z}_2 \end{bmatrix} \\
&= Z_1F_{11}Z_1 + \mathbf{Z}_2^TF_{21}Z_1 + Z_1F_{12}\mathbf{Z}_2 + \mathbf{Z}_2^TF_{22}\mathbf{Z}_2 \\
&= Z_1F_{11}Z_1 - \mathbf{Z}_2^T\Gamma_{22}^{-1}\Gamma_{21}F_{11}Z_1 - Z_1F_{11}\Gamma_{12}\Gamma_{22}^{-1}\mathbf{Z}_2 \\
&\quad + \mathbf{Z}_2^T\Gamma_{22}^{-1}\mathbf{Z}_2 + \mathbf{Z}_2^T\Gamma_{22}^{-1}\Gamma_{21}F_{11}\Gamma_{12}\Gamma_{22}^{-1}\mathbf{Z}_2.
\end{aligned}$$

But each term is scalar, as is F_{11}, so

$$\begin{aligned}
\langle \Gamma^{-1}\mathbf{Z}, \mathbf{Z}\rangle &= F_{11}[Z_1^2 - \mathbf{Z}_2^T\Gamma_{22}^{-1}\Gamma_{21}Z_1 - Z_1\Gamma_{12}\Gamma_{22}^{-1}\mathbf{Z}_2 \\
&\quad + \mathbf{Z}_2^T\Gamma_{22}^{-1}\Gamma_{21}\Gamma_{22}^{-1}\mathbf{Z}_2] + \mathbf{Z}_2^T\Gamma_{22}^{-1}\mathbf{Z}_2 \\
&= F_{11}[Z_1 - \Gamma_{12}\Gamma_{22}^{-1}\mathbf{Z}_2]^T[Z_1 - \Gamma_{12}\Gamma_{22}^{-1}\mathbf{Z}_2] + \mathbf{Z}_2^T\Gamma_{22}^{-1}\mathbf{Z}_2 \\
&= F_{11}[Z_1 - \Gamma_{12}\Gamma_{22}^{-1}\mathbf{Z}_2]^2 + \mathbf{Z}_2^T\Gamma_{22}^{-1}\mathbf{Z}_2.
\end{aligned}$$

Thus, we can rewrite ρ_0 in the form

$$\rho_0(Z_1, \mathbf{Z}_2) = \frac{1}{(2\pi)^{n+1/2}(F_{11}^{-1})^{1/2}\sqrt{\det \Gamma_{22}}} e^{-F_{11}(Z_1 - \Gamma_{12}\Gamma_{22}^{-1}\mathbf{Z}_2)^2/2} e^{-\mathbf{Z}_2^T\Gamma_{22}^{-1}\mathbf{Z}_2},$$

where the constant is obtained from (1.5-1) using Lemma 1.7.4. Rewriting this expression as

$$\begin{aligned}
\rho_0(Z_1, \mathbf{Z}_2) &= \frac{1}{\sqrt{2\pi}(F_{11}^{-1})^{1/2}} e^{-((Z_1 - \Gamma_{11}\Gamma_{22}^{-1}\mathbf{Z}_2)/2F_{11}^{-1})^2} \\
&\quad \cdot \frac{1}{(2\pi)^{n/2}\sqrt{\det \Gamma_{22}}} e^{-\mathbf{Z}_2^T\Gamma_{22}^{-1}\mathbf{Z}_2},
\end{aligned}$$

we see that the first term has the form of a normal p.d.f. with mean $\Gamma_{11}\Gamma_{22}^{-1}\mathbf{Z}_2$ and variance F_{11}^{-1}. Thus, from earlier results, the marginal density for $\mathbf{Z}_2$, which we write as $\rho_0(\mathbf{Z}_2)$, is given by

$$\rho_0(Z_2) = \int_{-\infty}^{\infty} \rho_0(Z_1, \mathbf{Z}_2)\, dZ_1 = \frac{1}{(2\pi)^{n/2}\sqrt{\det \Gamma_{22}}} e^{-\mathbf{Z}_2^T\Gamma_{22}^{-1}\mathbf{Z}_2}.$$

Thus,

$$\rho_0(Z_1 \mid \mathbf{Z}_2) = \frac{\rho_0(Z_1, \mathbf{Z}_2)}{\rho_0(\mathbf{Z}_2)} = \frac{1}{\sqrt{2\pi}(F_{11}^{-1})^{1/2}} e^{-((Z_1 - \Gamma_{11}\Gamma_{22}^{-1}\mathbf{Z}_2)/2F_{11}^{-1})^2}.$$

Therefore,

$$\begin{aligned}
g(Z_2) &= \int_{\mathbf{R}^n} Z_2\rho_0(Z_1 \mid \mathbf{z}_2)\, d\mathbf{z}_2 \\
&= \Gamma_{12}\Gamma_{22}^{-1}\mathbf{Z}_2, \qquad (3.5\text{-}2)
\end{aligned}$$

the last equality following from Example 1.6.15. Note that since $\rho_0(Z_1 \,|\, \mathbf{Z}_2)$ has the form of a p.d.f. representing a random variable whose variance is F_{11}^{-1}, we might guess that F_{11}^{-1} is the variance of the estimation error. Of course, such a conclusion is based on treating $\Gamma_{11}\Gamma_{22}^{-1}Z_2$ as a constant, which is not the case. The conclusion is true, however, and we can calculate it directly.

$$\begin{aligned}
E((Z_1 - \Gamma_{12}\Gamma_{22}^{-1}Z_2)^2) &= E((Z_1 - \Gamma_{12}\Gamma_{22}^{-1}Z_2)(Z_1 - \Gamma_{12}\Gamma_{22}^{-1}Z_2)^T) \\
&= E((Z_1 - \Gamma_{12}\Gamma_{22}^{-1}Z_2)(Z_1 - Z_2^T\Gamma_{22}^{-1}\Gamma_{21})) \\
&= E(Z_1^2 - \Gamma_{12}\Gamma_{22}^{-1}Z_2Z_1 - Z_1Z_2^T\Gamma_{22}^{-1}\Gamma_{21} \\
&\quad + \Gamma_{12}\Gamma_{22}^{-1}Z_2Z_2^T\Gamma_{22}^{-1}\Gamma_{21}) \\
&= \Gamma_{11} - \Gamma_{12}\Gamma_{22}^{-1}\Gamma_{21} - \Gamma_{12}\Gamma_{22}^{-1}\Gamma_{21} \\
&\quad + \Gamma_{12}\Gamma_{22}^{-1}\Gamma_{22}\Gamma_{22}^{-1}\Gamma_{21} \\
&= \Gamma_{11} - \Gamma_{12}\Gamma_{22}^{-1}\Gamma_{21} \\
&= F_{11}^{-1}.
\end{aligned}$$

All of the above calculations, together with the remarks following Theorem 3.5.1, can be summarized by the following theorem.

3.5.2. **Theorem.** *Let X be a random variable and $\mathbf{Y}$ a random vector such that $(X, \mathbf{Y})$ is multivariate normal. Then if*

$$\begin{aligned}
\mu_x &= E(X) \\
\boldsymbol{\mu}_y &= E(\mathbf{Y}) \\
\Gamma_{11} &= \mathrm{var}(X) \\
\Gamma_{22} &= \mathrm{var}(\mathbf{Y}) \\
\Gamma_{21} &= \mathrm{cov}(X, \mathbf{Y}) = \Gamma_{21}^T,
\end{aligned}$$

it follows that the minimum variance estimate of X based on $\mathbf{Y}$ is given by

$$\hat{X} = K\mathbf{Y} + \mathbf{b},$$

where

$$K = \Gamma_{12}\Gamma_{22}^{-1}$$

and

$$\mathbf{b} = \mu_x - K\boldsymbol{\mu}_y.$$

Moreover, the covariance of the estimation error, that is, $E((\hat{X} - X)^2)$, is given by

$$E((\hat{X} - X)^2) = \Gamma_{11} - \Gamma_{12}\Gamma_{22}^{-1}\Gamma_{21}.$$

Proof. From the calculations above, namely, (3.5-2),

$$E(X - \mu_x \,|\, \mathbf{Y} - \boldsymbol{\mu}_y) = \Gamma_{12}\Gamma_{22}^{-1}(\mathbf{Y} - \boldsymbol{\mu}_y)$$

or

$$E(X \mid \mathbf{Y} - \boldsymbol{\mu}_y) - \mu_x = \Gamma_{12}\Gamma_{22}^{-1} Y - \Gamma_{12}\Gamma_{22}^{-1}\boldsymbol{\mu}_y.$$

By Theorem 3.5.1,

$$E(X \mid \mathbf{Y} - \boldsymbol{\mu}_y = \mathbf{y} - \boldsymbol{\mu}_y) = E(X \mid \mathbf{Y} = \mathbf{y}),$$

so that the above becomes

$$\hat{X} = E(X \mid \mathbf{Y}) = \Gamma_{12}\Gamma_{22}^{-1}\mathbf{Y} + \mu_x - \Gamma_{12}\Gamma_{22}^{-1}\boldsymbol{\mu}_y.$$

For the covariance, note that

$$\begin{aligned} E((\hat{X} - X)^2) &= E((\hat{X} - \mu_x + \mu_x - X)^2) \\ &= E((\Gamma_{12}\Gamma_{22}^{-1} Y - \Gamma_{12}\Gamma_{22}^{-1}\boldsymbol{\mu}_y - (X - \mu_x))^2 \\ &= E((\Gamma_{12}\Gamma_{22}^{-1}(\mathbf{Y} - \boldsymbol{\mu}_y) - (X - \mu_x))^2) \\ &= E((\hat{Z}_1 - Z_1)^2) \\ &= \Gamma_{11} - \Gamma_{12}\Gamma_{22}^{-1}\Gamma_{21}. \quad \Box \end{aligned}$$

Theorem 3.5.2 combined with the maximum entropy principle has enabled us to find the minimum variance estimator of X given $\mathbf{Y}$ in the case where we only know the first- and second-order statistics of $(X, \mathbf{Y})$. To be sure, there are some assumptions, and some subjective judgements, but at least we have a rational, well-defined procedure for tackling a very tough estimation problem.

Before we leave the maximum entropy principle, we feel obliged to point out an important connection that it has with physics. If one considers an ideal gas to be point masses that move without collisions, then the classical state space S is a $6N$ dimensional space, N being the number of particles, that depicts the position and momentum of each particle. Since the exact state of a gas is unobservable, that is, one can never know the exact position and momentum of each particle, one settles for a probabilistic description. Specifically, if one supposes that the total energy (the Hamiltonian) has an expected value E (determined by taking a temperature measurement for example), then one employs the maximum entropy principle with the expected value of energy equaling E as a constraint, and obtains a p.d.f. known as the Gibbs' Canonical ensemble. It turns out that the classical gas laws, thermodynamic entropy, and the like can be derived from this p.d.f., so it is consistent with experimental results. It thus seems that the probabilistic behavior of classical thermodynamics behaves in accordance with the maximum entropy principle, a fact that is somewhat comforting to know.

3.6 Some Criticisms and Conclusions

The derivation of the maximum entropy principle involves the assertion of beliefs that can never be completely justified on philosophical grounds. Likewise, the beliefs are not operational in nature so they cannot be checked by experimentation either. The best we can say is that the beliefs are quite plausible and the principle seems to work in many specific situations (such as classical thermodynamics). In spite of possible objections, the maximum entropy principle is a technique that enables us to attack a very hard problem when faced with a paucity of information, the alternative being no solution at all!

Aside from the above remarks, there are some other concerns and criticisms one can express. For one thing, there may be some "soft" information available that doesn't lend itself to a precisely formulated maximization problem but is nonetheless relevant. For example, suppose it is known that the p.d.f. should have bounded support although the bound is uncertain, or that the p.d.f. is skewed from the mean in an unknown fashion. This information tells us that the multivariate normal is not the correct choice, but it doesn't tell us how to proceed to obtain the correct p.d.f. For one thing, it is unlikely in most estimation problems that either of the means μ_x or μ_y are known. What is more likely (and this will be the case in the Kalman filter) is that one has some prior estimate $\hat{x}_1$ of X, some data $\mathbf{y}$, and a prior error covariance $P_1 = E((X - \hat{X}_1)^2)$, and from this, one wishes to deduce a new estimate $\hat{x}_2$ from a new estimator $\hat{X}_2$ and readjust the error covariance to $P_2 = E((X - \hat{X}_2)^2)$. Even worse, it can be the case that P_1 is unknown, $\hat{x}_1$ is unknown, and only $\mathbf{y}$ and some partial information about the statistical relation of x to $\mathbf{y}$ are known. Where does all this leave us? Are the problems we face simply too tough? Are there alternatives? The problems are indeed tough, but there are alternatives.

The solution to the minimum variance estimation problem given in 3.5.2 has a serendipitous consequence. Out of all the many possible Baire functions that might have been the appropriate estimating function, the one that turned out to be the correct choice was linear.[3] In view of the generality of the problem we faced, this result suggests that linear estimating functions are not as restrictive as they appear to be at first glance. Moreover, the final results in 3.5.2 make no mention of ρ_0. If this were true of linear estimators in general, that is, if the linear minimum variance estimation problem could be solved without specifically knowing the underlying p.d.f., then at least some of the above concerns would vanish. Unfortunately, 3.5.2 does not help us in this regard since that result, though linear in nature, was based on the properties of conditional expectation and the fact that it solves the general minimum variance estimation problem; the

[3] Technically speaking, *affine*.

general linear problem could look quite different. Thus, we have to reformulate our problem in a linear setting and start all over again. Chapter 5 will be devoted to this task. Before we do this, however, we will have to develop some additional mathematical facts, and that is the purpose of Chapter 4.

3.7 Exercises

1. (a) Prove that

$$H_{k+1}\left(\frac{1}{k+1},\ldots,\frac{1}{k+1}\right) \geq H_k\left(\frac{1}{k},\ldots,\frac{1}{k}\right)$$

for all k. (This is very easy.)

(b) Using part (a), explain how the conclusion to Theorem 3.2.9 is used within the proof to establish the conclusion. Sound strange? The argument is inductive. (Hint: Draw a tetrahedran to represent a 3-simplex in 4-space and see how the argument goes.)

2. Show that the matrix in (3.2-16) is positive definite by showing all of its principal minors are positive. (Hint: All of the minors have the same form, so induction works well.)

3. Let $(\Omega, \mathcal{E}, P)$ be a probability space. Define $\mathcal{L}_2^n(\Omega, P)$ to be the set

$$\mathcal{L}_2^n(\Omega, P) = \{(f_1, \ldots, f_n) \,|\, f_i \in \mathcal{L}_2(\Omega, P) \quad \text{each } i\}$$

equipped with coordinate-wise addition and scalar multiplication, and inner product

$$[f, g] = \sum_{i=1}^{n} \langle f_i, g_i \rangle.$$

(a) Show that $[f, g]$ is indeed an inner product.

(b) If $\mathbf{Y}$ is a random vector, show that

$$\mathcal{M}_i(\mathbf{Y}) \triangleq \{f \circ \mathbf{Y} \,|\, f \quad \text{is Baire,} \quad f \circ \mathbf{Y} \in \mathcal{L}_2^n(\Omega, P)\}.$$

Then each component of f is a Baire function (word for word, the form of proof in 1.1.6(a). Formulate the minimum variance estimation problem for $\mathbf{X}$ a random vector in R^n.

(c) Show that your answer to problem 1.10.7 solves the problem in (b).

(d) Using notation as in 3.5.1, give the maximum entropy solution to the prior covariance problem in this setting. Please use the results of Theorem 3.5.1 to obtain your answer (this is quite easy).

4. Suppose X is a random variable such that

   (1) $E(X) = \mu$, and

   (2) $X(\omega) \in [-a, a]$ for all $\omega \in \Omega$.

   (a) Find the maximum entropy density ρ_0 for X given (1) and (2) with $\mu \neq 0$.

   (b) What if $\mu = 0$?

   (c) What if condition (2) is removed?

5. Calculate $H(\rho)$ when ρ is the multivariate normal distribution.

4

Adjoints, Projections, Pseudoinverses

This chapter is concerned with developing some technical theorems in mathematics that will be used in the remainder of the text.

4.1 Adjoints

4.1.1. **Definitions.** Let $\mathcal{H}_1$ and $\mathcal{H}_2$ be Hilbert spaces and let A be a bounded operator from $\mathcal{H}_1$ to $\mathcal{H}_2$, that is, $A \in \mathcal{B}(\mathcal{H}_1, \mathcal{H}_2)$ (see Appendix E). By the *adjoint* of A, written A^*, we mean the (necessarily unique) operator satisfying

(a) $A^* \in \mathcal{L}(\mathcal{H}_2, \mathcal{H}_1)$,

(b) $\langle A\mathbf{x}, \mathbf{y}\rangle_2 = \langle \mathbf{x}, A^*\mathbf{y}\rangle_1$[1] for all $\mathbf{x} \in \mathcal{H}_1$ and $\mathbf{y} \in \mathcal{H}_2$.

4.1.2. **Lemma.** *A^* is well defined and continuous, that is, $A^* \in \mathcal{B}(\mathcal{H}_2, \mathcal{H}_1)$.*

Proof. For arbitrary $\mathbf{y} \in \mathcal{H}_2$, define a linear functional f on $\mathcal{H}_1$ via

$$f(\mathbf{x}) \triangleq \langle A\mathbf{x}, \mathbf{y}\rangle_2.$$

Since $A \in \mathcal{B}(\mathcal{H}_1, \mathcal{H}_2)$, we have

$$|f(\mathbf{x})| = |\langle A\mathbf{x}, \mathbf{y}\rangle| \leq ||A\mathbf{x}||\,||\mathbf{y}|| \leq ||A||\,||\mathbf{x}||\,||\mathbf{y}||, \tag{4.1-1}$$

so $f \in H_1^*$. By the Riesz representation theorem (Theorem E31), there exists a unique vector $\mathbf{f} \in \mathcal{H}_1$ such that

$$f(\mathbf{x}) = \langle \mathbf{x}, \mathbf{f}\rangle_1,$$

that is,

$$\langle A\mathbf{x}, \mathbf{x}\rangle_2 = \langle \mathbf{x}, \mathbf{f}\rangle_1$$

for all $\mathbf{x} \in \mathcal{H}_1$. Moreover, $||f|| = ||\mathbf{f}||$. Noting that $\mathbf{f}$ is unique and depends only on the choice of $\mathbf{y}$, we can define a function

$$A^* : \mathcal{H}_2 \to \mathcal{H}_1$$

---

[1]The subscripts indicate the space in which the inner product is calculated. We will drop this subscript unless clarity is sacrificed.

by

$$A^*(\mathbf{y}) = \mathbf{f},$$

and so A^* is a well-defined function. This established, it is clear that condition (b) of Definition 4.1.1 uniquely characterizes A^*.

The proof that A^* is linear is very easy and we omit it.

Finally, to show A^* is bounded, note that by (4.1-1)

$$\|f\| \leq \|A\| \, \|\mathbf{y}\|,$$

and since

$$\|f\| = \|\mathbf{f}\| = \|A^*\mathbf{y}\|,$$

it follows that

$$\|A^*\mathbf{y}\| \leq \|A\| \, \|\mathbf{y}\|. \tag{4.1-2}$$

This implies that $A^* \in \mathcal{B}(\mathcal{H}_2, \mathcal{H}_1)$. □

4.1.3. **Corollary.** $\|A^*\| \leq \|A\|$.

Proof. This follows directly from the definition of the norm of an operator and relation (4.1-2). □

4.1.4. **Theorem.** *Let* $A \in \mathcal{B}(\mathcal{H}_1, \mathcal{H}_2)$, $B \in \mathcal{B}(\mathcal{H}_2, \mathcal{H}_1)$.

(a) $\langle \mathbf{x}, B\mathbf{y}\rangle_1 = \langle B^*\mathbf{x}, \mathbf{y}\rangle_2$ *for all* $\mathbf{x} \in \mathcal{H}_1$, $\mathbf{y} \in \mathcal{H}_2$.

(b) $A^{**} = A$.

(c) $\|A\| = \|A^*\|$.

Proof.

(a) $\langle \mathbf{x}, B\mathbf{y}\rangle_1 = \langle B\mathbf{y}, \mathbf{x}\rangle_1^* = \langle \mathbf{y}, B^*\mathbf{x}\rangle_2^* = \langle B^*\mathbf{x}, \mathbf{y}\rangle_2$.

(b) $\langle A^{**}\mathbf{x}, \mathbf{y}\rangle = \langle \mathbf{x}, A^*\mathbf{y}\rangle$ *form* (a), *so* $\langle A^{**}\mathbf{x}, \mathbf{y}\rangle = \langle \mathbf{x}, \mathbf{y}\rangle$ *from the definition of* A^*.

(c) *Replace* A *in* 4.1.3 *by* A^* *and use part* (a). □

4.1.5. **Theorem.** *Let* $A \in \mathcal{B}(\mathcal{H}_1, \mathcal{H}_2)$ *and let* $\mathcal{R}(A)$ *denote the range of* A. *Then*

(a) $\mathcal{R}(A)^\perp = \ker(A^*)$

(b) $\overline{\mathcal{R}(A)} = \ker(A^*)^\perp$

(c) $\mathcal{R}(A^*)^\perp = \ker(A)$

(d) $\overline{\mathcal{R}(A^*)} = \ker(A)^\perp$.

Proof.

(a) Let $\mathbf{x} \in \ker(A^*)$. We will show that for every $\mathbf{y} \in \mathcal{R}(A)$, $\langle \mathbf{x}, \mathbf{y} \rangle_2 = 0$. If $\mathbf{y} \in \mathcal{R}(A)$, then there is a vector $\mathbf{z} \in \mathcal{H}_1$ such that $\mathbf{y} = A\mathbf{z}$. Hence,

$$\langle \mathbf{y}, \mathbf{x} \rangle_2 = \langle A\mathbf{z}, \mathbf{x} \rangle_2 = \langle \mathbf{z}, A^*\mathbf{x} \rangle_1 = \langle \mathbf{z}, \mathbf{0} \rangle = 0.$$

We have thus shown

$$\ker(A^*) \subset \mathcal{R}(A)^\perp.$$

Conversely, suppose $\mathbf{y} \in \mathcal{R}(A)^\perp$. Then, for every $\mathbf{x} \in \mathcal{H}_1$, $\langle \mathbf{y}, A\mathbf{x} \rangle_2 = 0$, that is, for every $\mathbf{x} \in \mathcal{H}_1$, $\langle A^*\mathbf{y}, \mathbf{x} \rangle_1 = 0$. Thus, we have shown that $A^*\mathbf{y} \perp \mathcal{H}_1$, that is, $A^*\mathbf{y} = 0$. But this means that $\mathbf{y} \in \ker(A^*)$, so

$$\mathcal{R}(A)^\perp \subset \ker(A^*).$$

(b) Applied $\perp$ to both sides of (a) and use Theorem E19 from Appendix E.

(c) and (d) Replace A by A^* in (a) and (b) and use 4.1.4. □

4.1.6. **Theorem.** *Let A and B be bounded operators (domain and range taken from context). Then*

(a) $(\alpha A)^* = \alpha^* A^*$;

(b) $(A + B)^* = A^* + B^*$;

(c) $(AB)^* = B^* A^*$;

(d) *If A has a bounded inverse, then so does A^*, and $(A^*)^{-1} = (A^{-1})^*$.*

Proof. These are all "one liners" and left to the reader. □

4.1.7. **Corollary.** *Let $A \in \mathcal{B}(\mathcal{H}_1, \mathcal{H}_2)$. Then A has a bounded inverse if and only if A^* does.*

In infinite dimensional spaces, it is conceivable that an operator $A \in \mathcal{B}(\mathcal{H}_1, \mathcal{H}_2)$ is one to one and onto, but A^{-1}, though existing and linear, is not bounded. This possibility is suggested by examples such as the following.

4.1.8. **Example.** In $\mathcal{L}_2[0,1]$ define the operator

$$A : \mathcal{L}_2[0,1] \to \mathcal{L}_2[0,1]$$

via

$$A(f)(t) = \int_0^t f(t)\, dt.$$

Since $A(f)$ is a continuous function on $[0,1]$, certainly $A(f) \in \mathcal{L}_2[0,1]$, so the above operator A is well defined. Now, if $A(f) = A(g)$, it follows that

$f = g$ [a.e.] (use 1.6.8) and so are equal in the sense of $\mathcal{L}_2[0,1]$. Thus, A is one to one.

If we let

$$\mathcal{D} = \{f \in \mathcal{L}_2[0,1], \mid f' \in \mathcal{L}_2[0,1]\}$$

(meaning that there is at least one function $g \in \mathcal{L}_2'[0,1]$ such that $f' = g$) then $\mathcal{D}$ is a subspace of $\mathcal{L}_2[0,1]$ (though not closed). Hence,

$$A^{-1} : \mathcal{D} \to \mathcal{L}_2[0,1]$$

is a well-defined linear transformation. However, if $f_n(t) = t^n$, then in $\mathcal{L}_2[0,1]$, we have $\lim_{n\to\infty} f_n(t) = 0$. (How do you reconcile this with the fact that $f_n(1) = 1$ for all n?) However,

$$\begin{aligned} \|A^{-1}(t^n)\|^2 &= \int_0^1 (nt^{n-1})^2 \\ &= n^2 \left. \frac{t^{2n-1}}{2n-1} \right|_0^1 \\ &= \frac{n^2}{2n-1}, \end{aligned}$$

which implies that

$$\lim_{n\to\infty} \|A^{-1}(f_n)\| = \infty.$$

Thus, A^{-1} is not bounded.

Now, the above example only suggests a possible problem. The issue we originally raised was concerned with one-to-one and onto functions, and the above example is not onto. We could restrict our attention to the inner product space $\mathcal{D}$, but this is not a Hilbert space (why?).

It turns out that there is no way we could have constructed a one-to-one onto bounded operator A without a bounded inverse. One of the happy surprises in operator theory is the following theorem.

4.1.9. **Theorem.** *If $A \in \mathcal{B}(\mathcal{H}_1, \mathcal{H}_2)$, A one-to-one and onto, then $A^{-1} \in \mathcal{B}(\mathcal{H}_2, \mathcal{H}_1)$.*

Proof. See Appendix F. □

Because of the above theorem, we will use A^{-1} to mean a bounded operator.

4.1.10. **Theorem.** *Let $A \in \mathcal{B}(\mathcal{H}_1, \mathcal{H}_2)$ and suppose that $\mathcal{R}(A)$ is closed. Let*

$$W \triangleq A\,|_{(\ker A)^\perp}$$

(A with its domain restricted to $\ker(A)^\perp$). Then W is one to one and onto, and so W^{-1} is continuous and linear.

Proof. Since $\ker(A)$ is closed,

$$\begin{aligned}\mathcal{R}(A) &= A(\mathcal{H}_1)\\ &= A((\ker A) + (\ker(A))^{\perp})\\ &= A(\ker A) + A((\ker(A))^{\perp})\\ &= 0 + W((\ker A)^{\perp})\\ &= \mathcal{R}(W).\end{aligned}$$

But $\mathcal{H}_2 = \mathcal{R}(A)$, and so W is onto.

Next, suppose $W(\mathbf{x}_1) = W(\mathbf{x}_2)$, where (necessarily) $\mathbf{x}_1, \mathbf{x}_2 \in (\ker A)^{\perp}$. Then

$$A(\mathbf{x}_1) = A(\mathbf{x}_2),$$

and so

$$A(\mathbf{x}_1 - \mathbf{x}_2) = \mathbf{0}.$$

But this means that

$$\mathbf{x}_1 - \mathbf{x}_2 \in (\ker A) \cap (\ker A)^{\perp} = \{\mathbf{0}\}$$

and so $\mathbf{x}_1 = \mathbf{x}_2$. □

Since $\mathcal{R}(A)$ and $(\ker A)^{\perp}$ are both closed subspaces of $\mathcal{H}_2$ and $\mathcal{H}_1$, respectively, they are Hilbert spaces in their own right (with the induced inner product). Thus, since A is continuous, W is continuous, and so W^{-1} is continuous by 4.1.9.

If we let $\mathcal{H}_1 = \mathrm{C}^n$ and $\mathcal{H}_2 = \mathrm{C}^m$ (see Appendix G) and identify the operator A with its standard matrix, then the range of A is the same as its column space. Thus the dimension of the range of A^* is the same as the column rank of the adjoint matrix (the transposed conjugate), also denoted A^*, which is the same as the row rank of the conjugate of A. But the row rank of conjugate A is the same as the row rank of A. Now, by 4.1.10, $\mathcal{R}(A)$ and $\ker(A)^{\perp}$ have the same dimension, and by 4.1.5(d), $\mathcal{R}(A^*) = \ker(A)^{\perp}$, so that $\mathcal{R}(A)$ and $\mathcal{R}(A^*)$ have the same dimension. Hence, we have a proof that the row rank of A is equal to the column rank of A.

4.2 Projections

4.2.1. **Definition.** Let $\mathcal{N}$ be a closed subspace of $\mathcal{H}$. Then, by Theorem E20, Appendix E, we know that if $\mathbf{x}$ is any element of $\mathcal{H}$, then $\mathbf{x}$ has the representation

$$\mathbf{x} = \mathbf{x}_1 + \mathbf{x}_2,$$

where $\mathbf{x}_1 \in \mathcal{N}$ and $\mathbf{x}_2 \in \mathcal{N}^{\perp}$, and this representation is unique. From the uniqueness of the representation, we have that

$$P(\mathbf{x}) = \mathbf{x}_1$$

is a well-defined function from $\mathcal{H}$ to itself. This function is called the *perpendicular projection* onto $\mathcal{N}$.

4.2.2. Theorem. *P is a linear operator and is bounded.*

4.2.3. Theorem. *P is a perpendicular projection if and only if $P \in \mathcal{B}(\mathcal{H})$ and $P = P^2 = P^*$.*

Proof. Let P be a perpendicular projection onto the subspace $\mathcal{N}$. Let $\mathbf{x} = \mathbf{x}_1 + \mathbf{x}_2$ be the unique decomposition of an arbitrary vector $\mathbf{x}$, where $\mathbf{x}_1 \in \mathcal{N}$ and $\mathbf{x}_2 \in \mathcal{N}^\perp$. Then

$$P^2(\mathbf{x}) = P(P(\mathbf{x})) = P(\mathbf{x}_1) = P(\mathbf{x}_1 + \mathbf{0}) = \mathbf{x}_1 = P(\mathbf{x}).$$

Thus, $P^2 = P$.

Next, let $\mathbf{x}, \mathbf{y} \in \mathcal{H}$, $\mathbf{x} = \mathbf{x}_1 + \mathbf{x}_2$, $\mathbf{y} = \mathbf{y}_1 + \mathbf{y}_2$ (with the convention of 4.2.1). Then

$$\begin{aligned}
\langle P\mathbf{x}, \mathbf{y} \rangle &= \langle \mathbf{x}_1, \mathbf{y} \rangle \\
&= \langle \mathbf{x}_1, \mathbf{y}_1 + \mathbf{y}_2 \rangle \\
&= \langle \mathbf{x}_1, \mathbf{y}_1 \rangle + \langle \mathbf{x}_1, \mathbf{y}_2 \rangle \\
&= \langle \mathbf{x}_1, \mathbf{y}_1 \rangle + 0 \\
&= \langle \mathbf{x}_1, \mathbf{y}_1 \rangle + \langle \mathbf{x}_2, \mathbf{y}_1 \rangle \\
&= \langle \mathbf{x}_1 + \mathbf{x}_2, \mathbf{y}_1 \rangle \\
&= \langle \mathbf{x}, \mathbf{y}_1 \rangle \\
&= \langle \mathbf{x}, P(\mathbf{y}) \rangle.
\end{aligned}$$

Thus,

$$\langle P\mathbf{x}, \mathbf{y} \rangle = \langle \mathbf{x}, P\mathbf{y} \rangle$$

for all $\mathbf{x}, \mathbf{y} \in \mathcal{H}$, which implies that $P = P^*$.

Conversely, suppose that $P = P^2 = P^*$. Define

$$\mathcal{N} \triangleq \{\mathbf{x} \mid P(\mathbf{x}) = \mathbf{x}\}.$$

Clearly $\mathcal{N}$ is a subspace and $\mathcal{N} \subset \mathcal{R}(P)$. But, if $\mathbf{y} \in \mathcal{R}(P)$, then $\mathbf{y} = P(\mathbf{z})$ for some $\mathbf{z} \in \mathcal{H}$. It follows that

$$\mathbf{y} = P(\mathbf{z}) = P^2(\mathbf{z}) = P(P(\mathbf{z})) = P(\mathbf{y}), \tag{4.2-1}$$

and this implies that $\mathbf{y} \in \mathcal{N}$. Hence,

$$\mathcal{N} = \mathcal{R}(P). \tag{4.2-2}$$

From this result, it follows from 4.1.5(c) and the fact that $P = P^*$ that

$$\mathcal{N}^\perp = \ker(P). \tag{4.2-3}$$

We next must show that $\mathcal{N}$ is closed. Hence, we suppose that $\mathbf{x}_n \in \mathcal{N}$ with $\mathbf{x}_n \to \mathbf{x}_0$. Since $P \in \mathcal{B}(\mathcal{H})$, P is continuous. Thus,

$$\begin{aligned} P(\mathbf{x}_0) &= P(\lim_{n\to\infty} x_n) \\ &= \lim_{n\to\infty} P(x_n) \\ &= \lim_{n\to\infty} x_n \\ &= x_0. \end{aligned}$$

Since $P(\mathbf{x}_0) = \mathbf{x}_0$, it follows that $\mathbf{x}_0 \in \mathcal{N}$, as required.

Since $\mathcal{N}$ is closed, the decomposition of any vector $\mathbf{x}$ as

$$\mathbf{x} = \mathbf{x}_1 + \mathbf{x}_2, \qquad x_1 \in \mathcal{N}, \qquad x_2 \in \mathcal{N}^\perp$$

works. From (4.2-1) and (4.2-3), we then have

$$P(\mathbf{x}) = P(\mathbf{x}_1) + P(\mathbf{x}_2) = \mathbf{x}_1,$$

so by Definition 4.2.1, P is indeed the perpendicular projection of $\mathcal{H}$ onto $\mathcal{N}$. □

4.2.4. **Definition.** Let $A \in \mathcal{B}(\mathcal{H}_1, \mathcal{H}_2)$. By A' we will mean the perpendicular projection onto $\mathcal{R}(A)^\perp$. If we define A'' to mean $\underline{(A')'}$, then A'' is the perpendicular projection onto $\mathcal{R}(A')^\perp = \mathcal{R}(A)^{\perp\perp} = \overline{\mathcal{R}(A)}$ (see Appendix E, E19, part (e)).

4.2.5. **Theorem.** *Let* $A \in \mathcal{B}(\mathcal{H}_1, \mathcal{H}_2)$, $B \in \mathcal{B}(\mathcal{H}_0, \mathcal{H}_1)$. *Then*

(a) $A''' = A'$;

(b) $(A^*)' =$ *perpendicular projection onto* $\ker(A)$;

(c) $(A \circ A^*)'' = A''$;[2]

(d) $(A^* \circ A)'' = (A^*)''$;

(e) $(A \circ B)'' = (A \circ B'')''$.

Proof. (a) First, note that

$$\overline{(\mathcal{R}(A))}^\perp = \mathcal{R}(A)^{\perp\perp\perp} = \mathcal{R}(A)^\perp.$$

Hence,

$$A''' = \text{perpendicular projection onto } \mathcal{R}(A'')^\perp.$$

However,

$$\mathcal{R}(A'')^\perp = \overline{\mathcal{R}(A)}^\perp = \mathcal{R}(A)^\perp$$

[2]This is the infinite dimensional analog of the theorem that rank $(A \circ A^*) =$ rank(A).

so

$$A''' = A'.$$

(b) $(A^*)' =$ perpendicular projection onto $\mathcal{R}(A^*)^\perp$. But by 4.1.5(c), $\mathcal{R}(A^*)^\perp = \ker(A)$, hence the result.

(c)

$$\begin{aligned}
\overline{\mathcal{R}(A \circ A^*)} &= [\ker(A \circ A^*)^*]^\perp \\
&= [\ker(A \circ A^*)]^\perp \\
&= [(A \circ A^*)^{-1}(\mathbf{0})]^\perp \\
&= [(A^*)^{-1}(A^{-1}(\mathbf{0}))]^\perp \\
&= [(A^*)^{-1}(\ker(A))]^\perp \\
&= [(A^*)^{-1}(\mathcal{R}(A^*)^\perp)]^\perp \\
&= [(A^*)^{-1}(\mathcal{R}(A^*)^\perp \cap \mathcal{R}(A^*))]^\perp \\
&\quad \text{(since } A^* \text{ is onto } \mathcal{R}(A^*)) \\
&= [(A^*)^{-1}(\mathbf{0})]^\perp \\
&= \ker(A^*)^\perp \\
&= \overline{\mathcal{R}(A)}.
\end{aligned}$$

The result follows from this.

(d) In (c), replace A by A^*;

(e)

$$\begin{aligned}
\mathcal{R}(A \circ B) &= A(\mathcal{R}(B)) \\
&\subset A(\overline{\mathcal{R}(B)}) \\
&= A(\mathcal{R}(B'')) \\
&= \mathcal{R}(A \circ B'').
\end{aligned}$$

Thus,

$$\mathcal{R}(A \circ B) \subset \mathcal{R}(A \circ B''). \tag{4.2-4}$$

Conversely,

$$\begin{aligned}
\mathcal{R}(A \circ B'') &= A(\mathcal{R}(B'') \\
&= A(\overline{\mathcal{R}(B)}) \\
&\subset \overline{A(\mathcal{R}(B))} \quad \text{this is continuity of } A \\
&= \overline{\mathcal{R}(A \circ B)}.
\end{aligned}$$

Hence,

$$\mathcal{R}(A \circ B'') \subset \mathcal{R}(A \circ B).$$

This, together with (4.2-4) implies the result. □

4.3 Pseudoinverses

In this section, we are going to study and "solve" the so-called general linear problem. Specifically, suppose we are given a vector $\mathbf{b} \in \mathcal{H}_2$ and a bounded operator $A \in \mathcal{B}(\mathcal{H}_1, \mathcal{H}_2)$. We then wish to find a vector $\mathbf{x} \in \mathcal{H}_1$ such that the equation

$$A\mathbf{x} = \mathbf{b}$$

is satisfied. Certainly if A is one to one and onto, hence invertible, the problem is easy. If A is just onto, the problem is solvable but (possibly) not uniquely. If A is not onto, the problem may have no solution. As we now see, however, by relaxing the equality requirement, we can obtain a solution to the general linear problem that is both meaningful and useful. We will first develop the necessary mathematics and then return to this problem.

Recall that in Theorem 4.1.10 we showed that when $\mathcal{R}(A)$ was closed the mapping W obtained by restricting A to $\ker(A)^\perp$ was one to one, onto, and continuous, whence

$$W^{-1} : \mathcal{R}(A) \to (\ker A)^\perp$$

was a continuous linear mapping. With these facts in mind, we make the following definition.

4.3.1. **Definition.** Let $A \in \mathcal{B}(\mathcal{H}_1, \mathcal{H}_2)$, $\mathcal{R}(A)$ be closed, and W be as in 4.1.10. We define the *pseudoinverse* of A, A^+, as

$$A^+ : \mathcal{H}_2 \to \mathcal{H}_1$$

$$A^+ \triangleq W^{-1} \circ A''.$$

The next theorem gives the algebraic properties of the pseudoinverse.

4.3.2. **Theorem.** *Let $A \in \mathcal{B}(\mathcal{H}_1, \mathcal{H}_2)$ with $\mathcal{R}(A)$ closed. Then*

(a) $A^+ \in \mathcal{B}(\mathcal{H}_2, \mathcal{H}_1)$;

(b) $A \circ A^+ = A''$;

(c) $A^+ \circ A'' = A^+$;

(d) $A \circ A^+ \circ A = A$;

(e) $A^+ \circ A \circ A^+ = A^+$;

(f) *if A^{-1} exists, then $A^{-1} = A^+$;*

(g) $A^{++} = A$;

(h) $A^+ \circ A = (A^*)''$;

(i) $(A^*)'' \circ A^+ = A^+$;

(j) $(A^*)^+$ *exists and* $(A^*)^+ = (A^+)^*$.

Proof. (a) That A^+ is linear is clear. Moreover, both W^{-1} and A'' are continuous, hence, so is A^+.

(b) Since $A(\mathbf{x}) = W(\mathbf{x})$ for $x \in (\ker A)^\perp = \mathcal{R}(W^{-1})$, it follows that $A \circ W^{-1} = W \circ W^{-1} = I_{\mathcal{R}(A)}$ (the identity operator on $\mathcal{R}(A)$). Hence,

$$A \circ A^+ = A \circ W^{-1} \circ A'' = I_{\mathcal{R}(A)} \circ A'' = A''.$$

(c) $A^+ \circ A'' = W^{-1} \circ A'' \circ A'' = W^{-1} \circ (A'')^2 = W^{-1} \circ A'' = A^+$.

(d) $A \circ A^+ \circ A = A'' \circ A = A$, the first equality following from part (b), and the second by noting that A'' is the identity on $\mathcal{R}(A)$.

(e) $A^+ \circ A \circ A^+ = A^+ \circ A'' = A^+$, the first equality follows from (b) and the second from (c).

(f) If A^{-1} exists, then $A'' = I$ and $W = A$, since $\ker(A) = \{\mathbf{0}\}$. Thus,

$$A^+ = A^{-1} \circ I = A^{-1}.$$

(g) First note that

$$\begin{aligned} \ker(A^+) &= (W^{-1} \circ A'')^{-1}(\mathbf{0}) \\ &= (A'')^{-1}(W(\mathbf{0})) \\ &= (A'')^{-1}(\mathbf{0}) \\ &= \ker(A'') \\ &= \mathcal{R}(A'')^\perp \\ &= \mathcal{R}(A)^\perp. \end{aligned}$$

By this calculation,

$$\ker(A^+)^\perp = \overline{\mathcal{R}(A)} = \mathcal{R}(A). \tag{4.3-1}$$

Thus, to construct $(A^+)^+$, we first restrict A^+ to $\ker(A^+)^\perp$ and call this mapping S, that is,

$$S : \ker(A^+)^\perp \to \mathcal{R}(A^+)$$

and

$$S(\mathbf{x}) = A^+(\mathbf{x}) \quad \text{for} \quad \mathbf{x} \in \ker(A^+)^\perp.$$

But by (4.3-1), this is the same as

$$S : \mathcal{R}(A) \to \mathcal{R}(A^+),$$

where

$$S(\mathbf{x}) = A^+(\mathbf{x}) \quad \text{for} \quad \mathbf{x} \in \mathcal{R}(A).$$

Note that S is playing the same role for A^+ that W plays for A. Hence,

$$A^{++} = S^{-1} \circ (A^+)''.$$

However, note that for $\mathbf{x} \in \mathcal{R}(A)$,

$$S(\mathbf{x}) = A^+(\mathbf{x}) = W^{-1} \circ A''(\mathbf{x}) = W^{-1}(\mathbf{x}),$$

and since S and W^{-1} are both defined on $\mathcal{R}(A)$,

$$S = W^{-1}.$$

It follows that

$$A^{++} = W \circ (A^+)''.$$

From 4.3.1, $(A^+)''$ is the perpendicular projection onto $\ker(A)^\perp$ (the range of W^{-1} is $\ker(A)^\perp$). Thus, for any $\mathbf{x} \in \mathcal{H}_1$, we can decompose $\mathbf{x}$ as

$$\mathbf{x} = \mathbf{x}_1 + \mathbf{x}_2, \qquad \mathbf{x}_1 \in \ker(A), \qquad \mathbf{x}_2 \in \ker(A)^\perp$$

and note that

$$\begin{aligned}
A^{++}(\mathbf{x}) &= W(A^+)''(\mathbf{x}_1 + \mathbf{x}_2) \\
&= W((A^+)''(\mathbf{x}_1) + (A^+)''(\mathbf{x}_2)) \\
&= W(\mathbf{0} + \mathbf{x}_2) \\
&= W(\mathbf{x}_2) \\
&= A(\mathbf{x}_2) \\
&= A(\mathbf{x}_1) + A(\mathbf{x}_2) \quad (A(\mathbf{x}_1) = \mathbf{0}) \\
&= A(\mathbf{x}_1 + \mathbf{x}_2) \\
&= A(\mathbf{x}).
\end{aligned}$$

This is the desired result.

(h) $A^+ \circ A = W^{-1} \circ A'' \circ A = W^{-1} \circ A$. If $\mathbf{x} \in \ker(A)$, then $A^+ \circ A(\mathbf{x}) = \mathbf{0}$. If $\mathbf{x} \in (\ker A)^\perp$, then $A(\mathbf{x}) = W(\mathbf{x})$, so

$$A^+ \circ A(\mathbf{x}) = W^{-1}W(\mathbf{x}) = \mathbf{x}.$$

Thus, $A^+ \circ A$ is the perpendicular projection onto $\ker(A)^\perp$. But $\ker(A)^\perp = \overline{\mathcal{R}(A^*)}$ so $A^+ \circ A = (A^*)''$.

(i) $(A^*)'' \circ A^+ = A^+ \circ A \circ A^+ = A^+$, the first equality following from (h), the second from (e).

(j) Let

$$T : \ker(A^*)^\perp \to \overline{\mathcal{R}(A^*)}$$

be obtained by restricting A^* to $\ker(A^*)^\perp$, that is,

$$T(\mathbf{x}) = A^*(\mathbf{x}) \quad \text{for} \quad x \in \ker(A^*)^\perp.$$

[The reason we use $\overline{\mathcal{R}(A^*)}$ rather than $\mathcal{R}(A^*)$ is that we want T to map into a Hilbert space. Of course, this means that T may not be onto.] If $\mathbf{x} \in \ker(A^*)^\perp = \mathcal{R}(A)$ and $\mathbf{y} \in \overline{\mathcal{R}(A^*)} = \ker(A)^\perp$, then

$$\langle T\mathbf{x}, \mathbf{y}\rangle_1 = \langle A^*\mathbf{x}, \mathbf{y}\rangle_1 = \langle \mathbf{x}, A\mathbf{y}\rangle_2 = \langle \mathbf{x}, W\mathbf{y}\rangle_2.$$

Denoting the adjoint with respect to the Hilbert subspaces $\overline{\mathcal{R}(A^*)}$ and $\mathcal{R}(A)$ by #, the above calculation shows that

$$W^\# = T.$$

Since W^{-1} exists, it follows from 4.16(d) that

$$(W^{-1})^\# = (W^\#)^{-1} = T^{-1} \tag{4.3-2}$$

so that T^{-1} exists. It follows that T is onto and so

$$\mathcal{R}(A^*) = \mathcal{R}(T) = \overline{\mathcal{R}(A^*)}. \tag{4.3-3}$$

We have therefore shown that A^* is range closed. At this point, we can then assert that $(A^*)^+$ exists. Then using part (i), it follows that for all $\mathbf{x} \in \mathcal{H}_1$ and all $\mathbf{y} \in \mathcal{H}_2$

$$\begin{aligned}
\langle (A^*)^+\mathbf{x}, \mathbf{y}\rangle_2 &= \langle A''(A^*)^+\mathbf{x}, \mathbf{y}\rangle_2 \quad \text{replace } A \text{ by } A^* \text{ in (i))} \\
&= \langle (A^*)^+(\mathbf{x}, A''\mathbf{y}\rangle_2 \\
&= \langle T^{-1}(A^*)''\mathbf{x}, A''\mathbf{y}\rangle_2 \\
&= \langle (W^{-1})^\#(A^*)''\mathbf{x}, A''\mathbf{y}\rangle_2 \quad \text{(by (4.3-2))} \\
&= \langle (A^*)''\mathbf{x}, W^{-1}A''\mathbf{y}\rangle_1 \\
&= \langle (A^*)''\mathbf{x}, A^+\mathbf{y}\rangle_1 \\
&= \langle \mathbf{x}, (A^*)''A^+\mathbf{y}\rangle_1 \\
&= \langle \mathbf{x}, A^+\mathbf{y}\rangle_1 \quad \text{(by part (i))} \\
&= \langle (A^+)^*\mathbf{x}, \mathbf{y}\rangle_1.
\end{aligned}$$

Since this calculation holds for all $\mathbf{x}$ and $\mathbf{y}$, the result follows. □

4.3.3. **Corollary** (to proof of j). *If $\mathcal{R}(A)$ is closed then $\mathcal{R}(A^*)$ is closed.*

4.3.4. **Theorem.** *Let $A \in \mathcal{B}(\mathcal{H}_1, \mathcal{H}_2)$. Suppose there exists an $R \in \mathcal{B}(\mathcal{H}_2, \mathcal{H}_1)$ such that*

(a) $A \circ R = A''$.

(b) $(A^*)'' \circ R = R$.

Then A^+ exists and equals $\mathcal{R}$.

Proof. Assume R satisfies the hypotheses. If $\mathbf{y}_n \in \mathcal{R}(A)$ for each n and $\mathbf{y}_n \to \mathbf{y}_0$, then we will show that $\mathbf{y}_0 \in \mathcal{R}(A)$. Let $\mathbf{x}_n \triangleq R\mathbf{y}_n$. Since

R is uniformly continuous and $\mathbf{y}_n$ converges (hence is Cauchy), it follows that $\{\mathbf{x}_n\}$ is a Cauchy sequence. By the completeness of $\mathcal{H}_1$, there exists an $\mathbf{x}_0 \in \mathcal{H}_1$ with $\mathbf{x}_n \to \mathbf{x}_0$, and by continuity of R, $\mathbf{x}_0 = R\mathbf{y}_0$. Thus, by assumption (a) and the fact that $\mathbf{y}_0 \in \overline{\mathcal{R}(A)}$, we have

$$A\mathbf{x}_0 = AR\mathbf{y}_0 = A''\mathbf{y}_0 = \mathbf{y}_0.$$

Hence, $\mathbf{y}_0 \in \mathcal{R}(A)$ and so $\mathcal{R}(A)$ is closed. We therefore know that A^+ exists. Using parts (e) and (h) of 4.3.2 together with assumptions (a) and (b), we then have

$$R = (A^*)'' \circ R = A^+ \circ A \circ R = A^+ \circ A'' = A^+$$

as required. □

4.3.5. **Theorem.** *Let $A \in \mathcal{B}(\mathcal{H}_1, \mathcal{H}_2)$. Suppose there exists an $R \in \mathcal{B}(\mathcal{H}_2, \mathcal{H}_1)$ such that*

(a) $R \circ A = (A^*)''$;

(b) $R \circ A'' = R$.

Then, A^+ exists and equals R.

The proof of this is very similar to 4.3.4 and so we omit it.

4.3.6. **Corollary.** *If A^+ exists, then so do $(A \circ A^*)^+$ and $(A^* \circ A)^+$. Moreover,*

(a) $A^+ = A^* \circ (A \circ A^*)^+$;

(b) $A^+ = (A^* \circ A)^+ \circ A^*$.

Proof. We will do (a). First, note that

$$\begin{aligned}
\mathcal{R}(A) &= A(\mathcal{H}_1) \\
&= A(\ker(A) + (\ker(A))^\perp) \\
&= A(\ker(A)^\perp) \\
&= A\overline{\mathcal{R}(A^*)}) \\
&= A(\mathcal{R}(A^*)) \quad \text{(Corollary 4.3.3)} \\
&= \mathcal{R}(A \circ A^*).
\end{aligned}$$

Since $\mathcal{R}(A)$ is closed, so is $\mathcal{R}(A \circ A^*)$; thus, $(AA^*)^+$ exists. We therefore define

$$R \triangleq A^* \circ (A \circ A^*)^+$$

and show that R so defined satisfies (a) and (b) of Theorem 4.3.4. First, (a):

$$A \circ R = A \circ A^* \circ (A \circ A^*)^+ = (A \circ A^*)'' = A'',$$

the last equality being 4.2.5(c). For (b),

$$(A^*)'' \circ R = (A^*)'' \circ A^* \circ (A \circ A^*)^+ = A^* \circ (A \circ A^*)^+ = R.$$

The conclusion now follows from Theorem 4.3.4. $\square$

4.3.7. **Lemma.** *If* $\mathbf{x} \in \mathcal{H}_1$, $\mathbf{b} \in \mathcal{H}_2$, *then*

$$\langle A\mathbf{x} - A''\mathbf{b}, A''\mathbf{b} - \mathbf{b}\rangle_2 = 0.$$

Proof. Clearly, $\mathbf{b} = A'(\mathbf{b}) + A''(\mathbf{b})$. Thus,

$$\begin{aligned} \langle A\mathbf{x} - A''\mathbf{b}, A''\mathbf{b} - \mathbf{b}\rangle_2 &= \langle A\mathbf{x} - A''\mathbf{b}, -A'\mathbf{b}\rangle \\ &= -\langle A\mathbf{x}, A'\mathbf{b}\rangle + \langle A''\mathbf{b}, A'\mathbf{b}\rangle \\ &= -0 + 0 \\ &= 0. \quad \square \end{aligned}$$

4.3.8. **Theorem.** *If* A^+ *exists, let* $\mathbf{x}_0 = A^+\mathbf{b}$, *where* $\mathbf{b}$ *is any vector in* $\mathcal{H}_2$. *Then for any* $\mathbf{x} \in \mathcal{H}$,

$$||A\mathbf{x}_0 - \mathbf{b}\}_2 \leq ||A\mathbf{x} - \mathbf{b}||_2.$$

Proof. Choose any $\mathbf{x} \in \mathcal{H}_1$. By Lemma 4.3.7 and the Pythagorean theorem,

$$\begin{aligned} ||A\mathbf{x} - \mathbf{b}||^2 &= ||A\mathbf{x} - A''\mathbf{b} + A''\mathbf{b} - \mathbf{b}||^2 \\ &= ||A\mathbf{x} - A''\mathbf{b}||^2 + ||A''\mathbf{b} - \mathbf{b}||^2 \\ &\geq ||A''\mathbf{b} - \mathbf{b}||^2 \\ &= ||A \circ A^+\mathbf{b} - \mathbf{b}||^2 \\ &= ||A\mathbf{x}_0 - \mathbf{b}||^2. \quad \square \end{aligned}$$

4.3.9. **Corollary.** *If* $\mathbf{b} \in \mathcal{R}(A)$, *then* $\mathbf{x}_0$ *as defined in Theorem* 4.3.8 *satisfies* $A\mathbf{x}_0 = \mathbf{b}$.

We now see the sense in which the general linear problem can be solved. Given any $\mathbf{b} \in \mathcal{H}_2$, we can always find an $\mathbf{x}_0 \in \mathcal{H}_1$, namely, $A^+\mathbf{b}$, such that $A\mathbf{x}_0$ is closer to $\mathbf{b}$ than $A\mathbf{x}$ for any other $\mathbf{x} \in \mathcal{H}_1$. If $\mathbf{b}$ happens to be in $\mathcal{R}(A)$, then $||A\mathbf{x}_0 - \mathbf{b}|| = 0$. Referring to the proof of Theorem 4.3.8, we see that $A\mathbf{x}_0 = A''\mathbf{b}$, the projection of $\mathbf{b}$ onto the range of A.

4.3.10. **Theorem.** *If* A *is range closed, then*

$$A'' = A \circ (A^* \circ A)^+ \circ A^*.$$

Proof. Use Corollary 4.3.6, part (b). $\square$

4.4 Calculating the Pseudoinverse in Finite Dimensions

In this section, we will give two methods for calculating the pseudoinverse of a matrix, that is, for operators on finite dimensional spaces. The first result is suitable for pencil and paper calculations, the second for machine calculations.

From Corollary 4.3.6, it follows that to calculate A^+ it is sufficient to describe an algorithm for calculating a self-adjoint matrix, either $A \circ A^*$ or $A^* \circ A$. Hence, the following.

4.4.1. **Theorem** (Foulis). *Let $A \in \mathcal{B}(\mathrm{R}^n)$, $A = A^*$, and suppose that*

$$\pm\chi_A(\lambda) = \lambda^n + \alpha_{n-1}\lambda^{n-1} + \cdots + \alpha_k\lambda^k, \qquad \alpha_k \neq 0$$

is the characteristic polynomial for A (or its negative). Then

$$A'' = -\frac{1}{\alpha_k}[A^{n-k} + \alpha_{n-1}A^{n-k-1} + \cdots + \alpha_{k+1}A]$$

and

$$A^+ = -\frac{1}{\alpha_k}[A^{n-k-1} + \alpha_{n-1}A^{n-k-2} + \cdots + \alpha_{k+2}A + \alpha_{k+1}A''].$$

Proof. Since $A = A^*$, $(\ker A)^\perp = \mathcal{R}(A)$. Thus,

$$\begin{aligned}
\mathcal{R}(A^2) &= A(\mathcal{R}(A)) \\
&= A((\ker A)^\perp) \\
&= A((\ker A)^\perp + \ker A) \\
&= A(\mathrm{R}^n) \\
&= \mathcal{R}(A).
\end{aligned}$$

But if $\mathcal{R}(A^2) = \mathcal{R}(A)$, then

$$\begin{aligned}
\mathcal{R}(A^3) &= A(\mathcal{R}(A^2)) \\
&= A(\mathcal{R}(A)) \\
&= \mathcal{R}(A^2) \\
&= \mathcal{R}(A).
\end{aligned}$$

A simple inductive argument establishes that

$$\mathcal{R}(A^i) = \mathcal{R}(A) \quad \text{for all} \quad i.$$

It follows at once that

$$\ker(A^i) = \ker(A) \quad \text{for all} \quad i.$$

Now, let

$$R \triangleq -\frac{1}{\alpha_k}[A^{n-k} + \alpha_{n-1}A^{n-k-1} + \cdots + \alpha_{k+1}A].$$

Clearly, if $\mathbf{x} \in \ker(A)$, then $R(\mathbf{x}) = \mathbf{0}$. Thus,

$$\ker(A) \subset \ker(R),$$

and so

$$\ker(R)^{\perp} \subset \ker(A)^{\perp}.$$

Since R and A are self-adjoint, this last inclusion implies that

$$\mathcal{R}(R) \subset \mathcal{R}(A). \tag{4.4-1}$$

However, since $\mathcal{R}(A) = \mathcal{R}(A^k)$, it follows that if $\mathbf{y} \in \mathcal{R}(A)$, there exists some vector $\mathbf{z}$ such that $\mathbf{y} = A^k\mathbf{z}$. Hence,

$$\begin{aligned} R\mathbf{y} &= R \circ A^k\mathbf{z} \\ &= -\frac{1}{\alpha_k}[A^n + \alpha_{n-1}A^{n-1} + \cdots + \alpha_{k+1}A^{k+1}](\mathbf{z}) \\ &= -\frac{1}{\alpha_k}[A^n + \alpha_{n-1}A^{n-1} + \cdots + \alpha_{k+1}A^{k+1} + \alpha_k A^k - \alpha_k A^k](\mathbf{z}) \\ &= -\frac{1}{\alpha_k}[\pm\chi_A(A) - \alpha_k A^k](\mathbf{z}) \\ &= A^k(\mathbf{z}) \end{aligned}$$

since $\chi_A(A) = 0$ by the Caley–Hamilton theorem. Thus,

$$R\mathbf{y} = \mathbf{y}, \tag{4.4-2}$$

so, among other things $\mathbf{y} \in \mathcal{R}(R)$. Hence,

$$\mathcal{R}(A) \subset \mathcal{R}(R),$$

which combined with (4.4-1) implies

$$\mathcal{R}(A) = \mathcal{R}(R).$$

It follows that

$$\ker(A) = \ker(R). \tag{4.4-3}$$

Moreover,

$$R|_{\mathcal{R}(A)} = I_{\mathcal{R}(A)} \quad \text{(by (4.4-2))}$$

and

$$R|_{\mathcal{R}(A)^{\perp}} = R|_{\ker(A)} = 0 \quad \text{(by (4.4-3))},$$

so that

$$R = A''.$$

Next define

$$W \triangleq -\frac{1}{\alpha_k}[A^{n-k-1} + \alpha_{n-1}A^{n-k-2} + \cdots + \alpha_{k+2}A + \alpha_{k+1}A''].$$

Since $(A^*)'' = A''$ and $A \circ A'' = A'' \circ A = A$, we have

$$(A^*)''W = A''W = W$$

and

$$AW = -\frac{1}{\alpha_k}[A^{n-k} + \alpha_{n-1}A^{n-k-1} + \cdots + \alpha_{k+2}A^2 + \alpha_{k+1}A] = A''.$$

By Theorem 4.3.4, $W = A^+$. □

Theorem 4.4.1 settles the fact that in finite dimensional spaces, A^+ is computable. Although this theorem is fine for "pencil and paper" calculations, the large number of multiplications involved in the matrix products degrade machine calculations because of roundoff error. A better technique uses the spectral theorem for self-adjoint operators (Appendix G) and is done below.[3] A complete discussion of the numerical techniques used in effecting a spectral decomposition (power methods, deflation schemes, Hausholder transformations, etc.) would take us too far afield. For these, and other numerical techniques involved with Kalman filtering, the reader is referred to Reference [26].

We begin with a simple lemma and then prove the main theorem concerning the calculation of A^+.

4.4.2. **Lemma.** *Let D be an operator on $\mathbb{R}^n$ whose matrix is (using D to also represent the matrix)*

$$D = \begin{bmatrix} \lambda_1 & & & \\ & \lambda_2 & & 0 \\ & & \ddots & \\ 0 & & & \lambda_n \end{bmatrix} \qquad \lambda_1 \geq \lambda_2 \geq \cdots \geq \lambda_n \geq 0.$$

Then if $\lambda_p > 0$, $\lambda_{p+1} = 0$, we have

$$D'' = \left[\begin{array}{cccc|c} 1 & & & 0 & \\ & 1 & & & \\ & & \ddots & & 0 \\ & 0 & & 1 & \\ \hline & & 0 & & 0 \end{array}\right] \quad \} \ p \text{ rows},$$

[3] A related technique, called singular value decomposition, is also frequently employed.

and

$$D^+ = \left[\begin{array}{cccc|c} \lambda_1^{-1} & & & & \\ & \lambda_2^{-1} & & 0 & \\ & & \ddots & & 0 \\ & 0 & & \lambda_p^{-1} & \\ \hline & & 0 & & 0 \end{array}\right].$$

Proof. D'' is obviously a projection. Since the column space of D'' and D are identical (first p coordinates), D'' is clearly the projection onto the range of D. D^+ can be checked using Theorem 4.3.4. □

4.4.3. **Theorem.** *Let Λ be an $n \times n$ matrix, $A = A^*$, and let*

$$D = U^*AU$$

be its spectral decomposition, U unitary. Then

$$A^+ = UD^+U^*.$$

Proof. By parts (c) and (e) of Theorem 4.2.5, we have

$$\begin{aligned} A'' &= (AA^*)'' \\ &= (UDU^*UDU^*)'' \\ &= (UDD(U^*)'')'' \\ &= (UDD)'' \quad \text{(since } (U^*)'' = I) \\ &= (UDD'')'' \\ &= (UD)'' \quad \text{(since } D = D^* \Rightarrow DD'' = D) \\ &= (UD'')'' \\ &= (UD''(U^*)'')'' \end{aligned}$$

$$A'' = (UD''U^*)''. \tag{4.4-4}$$

But $(UD''U^*)(UD''U^*) = UD''U^*$ and $UD''U^*$ is clearly self-adjoint, so by Theorem 4.2.3, it is a projection. This observation combined with the fact that $UD''U^*$ and A have the same range (Equation (4.4-4)) implies that

$$A'' = UD''U^*. \tag{4.4-5}$$

Hence, defining

$$R = UD^+U^*, \tag{4.4-6}$$

we have

$$AR = UDU^*UD^+U^* = UDD^+U^+ = UD''U^+ = A''$$

and

$$(A^*)''R = A''R = UD''U^*UD^+U^* = UD''D^+U^* = UD^+U^*,$$

so

$$(A^*)''R = R.$$

By Theorem 4.3.4, $R = A^+$. □

4.5 The Grammian

4.5.1. **Theorem.** *Let $\mathcal{H}$ be a Hilbert space, $\{\mathbf{z}_1, \mathbf{z}_2, \ldots, \mathbf{z}_n\} \subset \mathcal{H}$ (not necessarily independent) and let $\mathbf{x} \in \mathcal{H}$. Let*

$$\hat{\mathbf{x}} = \sum_{i=1}^{n} \alpha_i \mathbf{z}_i$$

denote the projection of $\mathbf{x}$ onto the linear span of $\mathbf{z}_1, \mathbf{z}_2, \ldots, \mathbf{z}_n$. Then

$$\begin{bmatrix} \alpha_1 \\ \alpha_2 \\ \vdots \\ \alpha_n \end{bmatrix} = \begin{bmatrix} \langle \mathbf{z}_1, \mathbf{z}_2 \rangle & \ldots & \langle \mathbf{z}_1, \mathbf{z}_n \rangle \\ \vdots & & \vdots \\ \langle \mathbf{z}_n, \mathbf{z}_1 \rangle & \ldots & \langle \mathbf{z}_n, \mathbf{z}_n \rangle \end{bmatrix}^+ \begin{bmatrix} \langle \mathbf{z}_1, \mathbf{x} \rangle \\ \vdots \\ \langle \mathbf{z}_n, \mathbf{x} \rangle \end{bmatrix}.$$

Proof. Let

$$A : \mathrm{R}^n \to \mathcal{H}$$

be defined by

$$A \begin{bmatrix} \beta_1 \\ \vdots \\ \beta_n \end{bmatrix} = \sum_{i=1}^{n} \beta_i \mathbf{z}_i.$$

Then $\mathcal{R}(A) = \text{span } \{\mathbf{z}_1, \ldots, \mathbf{z}_n\}$, so

$$\hat{\mathbf{x}} = A''\mathbf{x} = AA^+\mathbf{x} = A(A^* \circ A)^+ A^*\mathbf{x}.$$

Note that

$$A^* \circ A : \mathrm{R}^n \to \mathrm{R}^n,$$

so we can compute its standard matrix in the usual way. Letting $\{\mathbf{e}_1, \ldots, \mathbf{e}_n\}$ denote the standard basis for R^n, we have

$$A^* \circ A(\mathbf{e}_i) = \sum_{j=1}^{n} \beta_{ji} \mathbf{e}_j,$$

so that

$$\langle A^* \circ A\mathbf{e}_i, \mathbf{e}_k \rangle = \langle \sum_{j=1}^{n} \beta_{ji} \mathbf{e}_j, \mathbf{e}_k \rangle = \beta_{ki}.$$

However,

$$\langle A^* \circ A\mathbf{e}_i, \mathbf{e}_k \rangle = \langle A\mathbf{e}_i, A\mathbf{e}_k \rangle = \langle \mathbf{z}_i, \mathbf{z}_k \rangle,$$

so that

$$\beta_{ki} = \langle \mathbf{z}_i, \mathbf{z}_k \rangle. \tag{4.5-1}$$

We next claim that

$$A^*\mathbf{x} = \sum_{j=1}^{n} \langle \mathbf{x}, \mathbf{z}_j \rangle \mathbf{e}_j. \tag{4.5-2}$$

For, if

$$A^*\mathbf{x} = \sum_{j=1}^{n} \gamma_j \mathbf{e}_j,$$

then

$$\gamma_i = \langle A^*\mathbf{x}, \mathbf{e}_i \rangle = \langle \mathbf{x}, A\mathbf{e}_i \rangle = \langle \mathbf{x}, \mathbf{z}_i \rangle,$$

which proves our claim.

Next define

$$P \triangleq \begin{bmatrix} \rho_{11} & \cdots & \rho_{1n} \\ \vdots & & \vdots \\ \rho_{n1} & \cdots & \rho_{nn} \end{bmatrix} \triangleq \begin{bmatrix} \langle \mathbf{z}_1, \mathbf{z}_1 \rangle & \cdots & \langle \mathbf{z}_1, \mathbf{z}_n \rangle \\ \vdots & & \vdots \\ \langle \mathbf{z}_n, \mathbf{z}_1 \rangle & \cdots & \langle \mathbf{z}_n, \mathbf{z}_n \rangle \end{bmatrix}^+,$$

so that from (4.5-1), P is the matrix of $(A^* \circ A)^+$. Recall that

$$\hat{\mathbf{x}} = A \circ (A^* \circ A)^+ \circ A^*\mathbf{x}. \tag{4.5-3}$$

From (4.5-2), the components of $A^*\mathbf{x}$ with respect to the standard basis are $\langle \mathbf{z}_1, \mathbf{x} \rangle, \ldots, \langle \mathbf{z}_n, \mathbf{x} \rangle$, so the components of $(A^* \circ A)^+ A^*\mathbf{x}$ are $c_1, \ldots, c_n$ as given by

$$\begin{bmatrix} c_1 \\ c_2 \\ \vdots \\ c_n \end{bmatrix} = \begin{bmatrix} \rho_{11} & \cdots & \rho_{1n} \\ \vdots & & \vdots \\ \rho_{n1} & \cdots & \rho_{nn} \end{bmatrix} \begin{bmatrix} \langle \mathbf{z}_1, \mathbf{x} \rangle \\ \vdots \\ \langle \mathbf{z}_n, \mathbf{x} \rangle \end{bmatrix}. \tag{4.5-4}$$

Hence,

$$\hat{\mathbf{x}} = A \begin{bmatrix} c_1 \\ \vdots \\ c_n \end{bmatrix} = \sum_{j=1}^{n} c_j \mathbf{z}_j,$$

which implies that $\alpha_i = c_i$. Replacing c_i by α_i in (4.5-4), we obtain the theorem. □

4.5.2. **Definition.** The matrix

$$G = \begin{bmatrix} \langle \mathbf{z}_1, \mathbf{z}_1 \rangle & \cdots & \langle \mathbf{z}_1, \mathbf{z}_n \rangle \\ \vdots & & \vdots \\ \langle \mathbf{z}_n, \mathbf{z}_1 \rangle & \cdots & \langle \mathbf{z}_n, \mathbf{z}_n \rangle \end{bmatrix}$$

is called the *Grammian* and $\det(G)$ is called the *Gram determinant.*

Note that if the inner products are in the space $\mathcal{L}_2(\Omega, P)$, then G has the form of a correlation matrix. This observation together with Theorem 4.5.1 will be central to our study.

4.6 Exercises

1. (a) Show that $(A \circ B)^* = B^* \circ A^*$.

   (b) Show that $A'' \circ A = A$ and $A \circ (A^*)'' = A$.

2. Show that if $\mathcal{M}$ and $\mathcal{N}$ are subspaces with $\mathcal{M} \perp \mathcal{N}$, M the perpendicular projection onto $\mathcal{M}$, N the perpendicular projection onto $\mathcal{N}$, then the perpendicular projection onto $\mathcal{M} + \mathcal{N}$ is $N + M$. (Note: in general, the sum of two closed subspaces need not be closed. Hence, you first need to show that $\mathcal{M} + \mathcal{N}$ is closed.)

3. Let

$$A = \begin{bmatrix} 0 & 1 & 1 \\ -1 & 0 & 1 \\ -1 & 1 & 2 \end{bmatrix}$$

   represent a transformation from R^3 to itself. Using the same symbol for the matrix and the transformation, determine the following:

   (a) Calculate the matrix representing A'';

   (b) verify that $(A'')^2 = A'' = (A'')^*$;

   (c) using your answer to (a), given an algebraic characterization of $\mathcal{R}(A)$;

   (d) repeat (c) using only the definition of $\mathcal{R}(A)$;

   (e) calculate A^+;

   (f) calculate $(A^*)''$ (easily done using (e)!);

   (g) using (f), a calculation as in (c), and the definition of $\ker(A)$, verify that $\mathcal{R}(A^*)^\perp = \ker(A)$.

4. Let $\mathcal{L}(\mathrm{C}^n, \mathrm{C}^m)$ represent all linear transformations from C^n to C^m. In the usual fashion, this set can be identified with the set of $m \times n$ complex matrices. Using addition and scalar multiplication of operators (matrices) this has a vector space structure. Using the adjoint operator, $\mathcal{L}(\mathrm{C}^n, \mathrm{C}^m)$ can be made into an inner product space via

$$\langle A, B \rangle \triangleq \mathrm{tr}(AB^*)$$

   where tr is the trace operator.

(a) Show that if operator A has the matrix representation M with respect to an orthonormal basis, then A^* has the matrix representation M^* (same basis where M^* represents the transpose conjugate of M).

(b) Verify that $\langle A, B\rangle = \mathrm{tr}(AB^*)$ is an inner product on $\mathcal{L}(\mathrm{C}^n, \mathrm{C}^m)$.

(c) Let A denote the usual operator norm for A (see Appendix E). Prove that $\|A\| = \sqrt{\lambda_p}$, where λ_p is the largest eigenvalue of AA^* (or A^*A). (Note: you need to show $\lambda_p \geq 0$.)

(d) Let $\|A\|_{\mathrm{tr}} = \sqrt{\langle A, A^*\rangle}$ denote the trace norm. Show that

$$\|A\| \leq \|A\|_{\mathrm{tr}} \leq \sqrt{m}\,\|A\|.$$

5

Linear Minimum Variance Estimation

5.1 Reformulation

After wading through all the technical details of Chapter 4, it is probably wise to refresh ourselves in terms of minimum variance estimation. At the end of Chapter 3, our status could be described as follows.

We agreed to model our processes as random vectors (variables) defined on a probability space (Ω, ϵ, P), where P is, at best, partially described.

We defined a minimum variance estimator $\hat{X}$ of a random variable X based on random vector $\mathbf{Y}$ to be a random variable of the form

$$\hat{X} = g \circ \mathbf{Y}, \qquad g \text{ a Baire function,}$$

where the norms are taken in $\mathcal{L}_2(\Omega, P)$.

The function g was called the estimating function. If $\mathbf{Y}(\omega) = (y_1, \ldots, y_n)$, then the corresponding estimate $\hat{\mathbf{x}}$ of X was given by

$$\hat{\mathbf{x}} = g(y_1, \ldots, y_n).$$

The minimum variance estimator was given by

$$\hat{X} = E(X \mid \mathbf{Y}).$$

We developed the notion of entropy and stated the maximum entropy principle.

We showed that if the only description of the probabilistic behavior of a random vector X is knowledge of its mean vector and covariance matrix, then the maximum entropy choice of a p.d.f. for X is the multivariate normal distribution.

We showed that if X is a random variable, $\mathbf{Y}$ a random vector, and the means and covariance of X and $\mathbf{Y}$ are known (including $\text{cov}(X, \mathbf{Y})$), then applying the maximum entropy principle, the minimum variance estimator of X given $\mathbf{Y}$ was an affine function of $\mathbf{Y}$.

> This last result suggested to us that perhaps affine estimators are not as restrictive as one might think, and hence deserve study.

In order to pursue the study of affine estimators, we feel it is prudent to first give the general classification scheme, as we see it, for minimum variance estimators.

5.1.1. **Definition.** Let $\mathbf{Y}$ be a random vector on (Ω, ϵ, P) and let $\mathcal{A}$ be a set of real valued functions on R^n such that the following hold.

(a) $\mathcal{A}$ is algebraically closed, that is, closed under addition and scalar multiplication.

(b) If $f \in \mathcal{A}$, then f is a Baire function.

(c) If $f \in \mathcal{A}$, then $f \circ \mathbf{Y} \in \mathcal{L}^2(\Omega, P)$.

(d) $\mathcal{M}(\mathcal{A}, Y) = \{f \circ \mathbf{Y} \mid f \in \mathcal{A}\}$ is topologically closed in $\mathcal{L}^2(\Omega, P)$, that is, the limit of a sequence of elements in $\mathcal{M}(\mathcal{A}, \mathbf{Y})$ is again in $\mathcal{M}(\mathcal{A}, \mathbf{Y})$.

Then, $\mathcal{M}(\mathcal{A}, \mathbf{Y})$ is called the set of $\mathcal{A}$ *estimators based on* $\mathbf{Y}$.

5.1.2. **Examples.**

(a) If $\mathcal{A}$ is equal to the set of all Baire functions on R^n, then $\mathcal{M}(\mathcal{A}, \mathbf{Y})$ is the set $\mathcal{M}(\mathbf{Y})$ that as introduced in Definition 2.3.3.

(b) Let $\mathcal{A} = \{g : \mathrm{R}^n \rightarrow \mathrm{R} \mid g \text{ is a linear transformation}\}$. Clearly, $\mathcal{A}$ is closed algebraically and since $\mathcal{M}(\mathcal{A}, \mathbf{Y})$ is finite dimensional, of dimension at most n, it is also a closed subspace of $\mathcal{L}_2(\Omega, P)$. In this case, $\mathcal{M}(\mathcal{A}, \mathbf{Y})$ is called the set of *linear estimators* based on $\mathbf{Y}$.

(c) If $\mathcal{A}$ is the set of affine transformations from R^n to R, that is, functions g of the form
$$g(\mathbf{x}) = A\mathbf{x} + b,$$
A a $1 \times n$ matrix, and b a constant, then again $\mathcal{M}(\mathcal{A}, \mathbf{Y})$ is a closed subspace of $\mathcal{L}_2(\Omega, P)$ and is called the set of *affine estimators* based on $\mathbf{Y}$.

Note that the estimator we calculated in Section 3.4 was an affine estimator. It is common for people to refer to both of the estimators in (b) and (c) as being "linear," and in fact we did this back in Chapter 3. For the remainder of these lectures, however, we are going to distinguish between linear and affine estimators. The reason for doing this will be made clear as we proceed.

5.1.3. **Definition.** Let X be a random variable, $\mathbf{Y}$ a random vector, and $\mathcal{M}(\mathcal{A}, \mathbf{Y})$ a set of $\mathcal{A}$ estimators. By the best $\mathcal{A}$-minimum variance estimator of X based on $\mathbf{Y}$, we simply mean the perpendicular projection $\hat{X}$ of X onto $\mathcal{M}(\mathcal{A}, \mathbf{Y})$.

Note that by the projection theorem, $\hat{X}$ always exists and is a square-summable, $\mathcal{B}(\mathbf{Y})$-measurable function. Again, we remind the reader that a necessary and sufficient condition for X to be the best $\mathcal{A}$-minimum variance estimator is that $(X - \hat{X}) \perp \mathcal{M}(\mathcal{A}, \mathbf{Y})$.

The notion in Definition 5.1.3 can be easily extended to the case where X is a random vector rather than a random variable. The idea has already been introduced for the general minimum variance estimation problem, namely, by Exercise 3.6.3. We now give the details for the specific case of $\mathcal{M}(\mathcal{A}, \mathbf{Y})$, leaving most proofs to the interested reader.

5.1.4. **Definition.**

(a) On $\mathcal{L}_2^n(\Omega, P)$, we define the function

$$[\mathbf{f}, \mathbf{g}] = \sum_{i=1}^{n} \langle f_i, g_i \rangle,$$

where $\mathbf{f} = (f_1, \ldots, f_n)$, $\mathbf{g} = (g_1, \ldots, g_n)$, and $\langle \cdot, \cdot \rangle$ is the usual inner product on $\mathcal{L}_2(\Omega, P)$.

(b) $\mathcal{M}_n(\mathcal{A}, \mathbf{Y}) \triangleq \mathcal{M}(\mathcal{A}, \mathbf{Y})^n$. In other words, an element of $\mathcal{M}_n(\mathcal{A}, \mathbf{Y})$ is a random vector of the form $(f_1 \circ \mathbf{Y}, \ldots, f_n \circ \mathbf{Y})$, where each $f_i \in \mathcal{A}$.

The salient facts are contained in the following theorem.

5.1.5. **Theorem.**

(a) $[\cdot, \cdot]$ *is an inner product on* $\mathcal{L}_2^n(\Omega, P)$ *and* $\mathcal{L}_2^n(\Omega, P)$ *so equipped is a Hilbert space.*

(b) *If* M *is a closed subspace of* $\mathcal{L}_2(\Omega, P)$ *then* M^n *is a closed subspace of* $\mathcal{L}_2^n(\Omega, P)$.

(c) $\hat{\mathbf{X}} = (\hat{X}_1, \ldots, \hat{X}_n)$ *is the projection of* $\mathbf{X} = (X_1, \ldots, X_n)$ *onto* M^n *if and only if* $\hat{X}_i$ *is the projection of* X_i *onto* M *for each* i.

Proof. Items (a) and (b) are left to the reader. Item (c) follows at once from the observation that $(\mathbf{X} - \hat{\mathbf{X}}) \perp M^n$ if and only if $(X_i - \hat{X}_i) \perp M$ for each i. Details are Exercise 5.4.1. □

5.1.6. **Corollary.** $\mathcal{M}_n(\mathcal{A}, \mathbf{Y})$ *is a closed subspace of* $\mathcal{L}_2^n(\Omega, P)$. *Moreover,* $\hat{\mathbf{X}}$ *is the best* $\mathcal{A}$*-minimum variance estimate of random vector* $\mathbf{X}$ *if and only if* $\hat{X}_i$ *is the best* $\mathcal{A}$*-minimum variance estimate of* X_i.

5.2 Linear Minimum Variance Estimation

5.2.1. **Definition.** Let $\mathbf{X} \in \mathcal{L}_2^n(\Omega, P)$, $\mathbf{Y} \in \mathcal{L}_2^m(\Omega, P)$. By the best linear minimum variance estimator (BLMVE) of $\mathbf{X}$ based on $\mathbf{Y}$, we mean the random vector $\hat{\mathbf{X}} \in \mathcal{L}_2^n(\Omega, P)$ satisfying

(a) $\hat{\mathbf{X}} = K\mathbf{Y}$, where K is an $n \times m$ matrix of constants, and

(b) $\|\hat{\mathbf{X}} - \mathbf{X}\|$ is minimal with respect to the constraint in (a).

5.2.2. **Lemma.**

(a) *If* $\mathbf{Y} = (Y_1, \ldots, Y_m)$, *then* $\mathcal{M}(\mathcal{A}, \mathbf{Y})$, $\mathcal{A}$ *the set of linear operators, is given by*

$$\mathcal{M}(\mathcal{A}, \mathbf{Y}) = \text{span}\{Y_1, Y_2, \ldots, Y_m\}.$$

(b) *For convenience let* $M \triangleq \mathcal{M}(\mathcal{A}, Y)$. M^n *is a subspace of* $\mathcal{L}_2^n(\Omega, P)$ *of dimension at most* $m \cdot n$.

(c) *A vector* $\mathbf{Z} \in \mathcal{L}_2^n(\Omega, P)$ *has the property* $\mathbf{Z} \perp M^n$ *if and only if* $E(\mathbf{Z}\mathbf{Y}^T) = 0$.

Proof. Part (a) is clear; (b) follows from the fact that M is at most dimension m; and for (c), using Exercise 5.4.1, $\mathbf{Z} \perp M^n$ if and only if $Z_i \perp M$ for each i, that is, $Z_i \perp Y_j$ for each i and j. But this is equivalent to $E(\mathbf{Z}\mathbf{Y}^T) = 0$. □

5.2.3. **Theorem** (Gauss–Markov). *Let* $\mathbf{X}$ *and* $\mathbf{Y}$ *be random vectors. The* BLMVE *of* $\mathbf{X}$ *based on* $\mathbf{Y}$ *is given by*

$$\hat{\mathbf{X}} = K\mathbf{Y},$$

where

$$K = E(XY^T)E(YY^T)^+.$$

Moreover,

$$E((\hat{\mathbf{X}} - \mathbf{X})(\hat{\mathbf{X}} - \mathbf{X})^T) = E(\mathbf{X}\mathbf{X}^T) - KE(\mathbf{Y}\mathbf{X}^T).$$

Proof. Let $K = (k_{ij})$. Then (using the notation in 5.1.5) $\hat{X}_i$ is the projection of X_i onto M and has the form

$$\hat{X}_i = \sum_{j=1}^{m} k_{ij} Y_j.$$

By Theorem 4.5.1, these k_{ij}'s are given by

$$\begin{bmatrix} k_{i1} \\ k_{i2} \\ \vdots \\ k_{im} \end{bmatrix} = \begin{bmatrix} \langle Y_1, Y_1 \rangle & \cdots & \langle Y_1, Y_m \rangle \\ \vdots & & \vdots \\ \langle Y_m, Y_1 \rangle & \cdots & \langle Y_m, Y_m \rangle \end{bmatrix}^+ \begin{bmatrix} \langle Y_1, X_i \rangle \\ \vdots \\ \langle Y_m, X_i \rangle \end{bmatrix}.$$

If we denote the ith row of K by K_i, then the above expression is just

$$K_i^T = E(\mathbf{Y}\mathbf{Y}^T)^+ E(\mathbf{Y}X_i^T),$$

so that

$$K_i = E(X_i \mathbf{Y}^T)E(\mathbf{YY}^T)^+.$$

Since

$$E(\mathbf{XY}^T) = \begin{bmatrix} E(X_1\mathbf{Y}^T) \\ E(X_2\mathbf{Y}^T) \\ \vdots \\ E(X_n\mathbf{Y}^T) \end{bmatrix},$$

it follows that

$$K = E(\mathbf{XY}^T)E(\mathbf{YY}^T)^+.$$

To calculate the error covariance, we note that

$$E((\hat{\mathbf{X}} - \mathbf{X})(\hat{\mathbf{X}} - \mathbf{X})^T) = E((\hat{\mathbf{X}} - \mathbf{X})\hat{\mathbf{X}}^T) - E((\hat{\mathbf{X}} - \mathbf{X})\mathbf{X}^T).$$

But $(\hat{\mathbf{X}} - \mathbf{X}) \perp M^n$, so by 5.2.2(c),

$$E((\hat{\mathbf{X}} - \mathbf{X})\mathbf{Y}^T) = 0. \tag{5.2-1}$$

Hence,

$$\begin{aligned} E((\hat{\mathbf{X}} - \mathbf{X})\hat{\mathbf{X}}^T) &= E((\hat{\mathbf{X}} - \mathbf{X})(K\mathbf{Y})^T) \\ &= E((\hat{\mathbf{X}} - \mathbf{X})\mathbf{Y}^T) \cdot K^T \\ &= 0 \cdot K^T \qquad \text{(using (5.2-1))} \\ &= 0. \end{aligned}$$

Thus,

$$\begin{aligned} E((\hat{\mathbf{X}} - \mathbf{X})(\hat{\mathbf{X}} - \mathbf{X})^T) &= -E((\hat{\mathbf{X}} - \mathbf{X})\mathbf{X}^T) \\ &= E(\mathbf{XX}^T) - E(\hat{\mathbf{X}}\mathbf{X}^T) \\ &= E(\mathbf{XX}^T) - E(K\mathbf{YX}^T) \\ &= E(\mathbf{XX}^T) - KE(\mathbf{YX}^T). \qquad \square \end{aligned}$$

5.3 Unbiased Estimators, Affine Estimators

The main purpose of this section is to clear up some subtle misunderstandings about the connections between linear estimators and unbiased linear estimators; they are often thought to be the same. It turns out that, in general, they are quite different, unless all of the means are zero. Fortunately, the "natural" way to handle estimators in the case of nonzero means turns out to be the correct way, and the above misconception seldom causes problems. Here are the facts.

5.3.1. **Definition.** An estimator $\hat{\mathbf{X}}$ of a random vector $\mathbf{X}$ is said to be *unbiased* if and only if

$$E(\hat{\mathbf{X}}) = E(\mathbf{X}). \tag{5.3-1}$$

Suppose that $\hat{\mathbf{X}}$ is a best linear minimum variance estimator of $\mathbf{X}$ and is also unbiased. Then $\hat{\mathbf{X}} = K\mathbf{Y}$ for some suitable K and relation (5.3-1) holds. Thus,

$$\mu_{\mathbf{x}} = E(\mathbf{X}) = E(\hat{\mathbf{X}}) = E(K\mathbf{Y}) = KE(\mathbf{Y}) = K\mu_{\mathbf{y}},$$

and so

$$\hat{\mathbf{X}} - \mu_{\mathbf{x}} = K\mathbf{Y} - K\mu_{\mathbf{y}}$$

or

$$\hat{\mathbf{X}} - \mu_{\mathbf{x}} = K(\mathbf{Y} - \mu_{\mathbf{y}}).$$

It would appear, therefore, that to obtain the BLMVE unbiased estimate of $\mathbf{X}$ based on $\mathbf{Y}$, one should simply use the Gauss–Markov theorem to calculate the appropriate K to obtain the BLMVE of $\mathbf{X} - \mu_{\mathbf{x}}$ based on $\mathbf{Y} - \mu_{\mathbf{y}}$, and this K gives the BLMVE unbiased estimate of X based on $\mathbf{Y}$. Not so! The problem is that using the Gauss–Markov theorem we do indeed have a linear minimum variance estimate $\widehat{\mathbf{X} - \mu_{\mathbf{x}}}$ of $\mathbf{X} - \mu_{\mathbf{x}}$ and that $E(\widehat{\mathbf{X} - \mu_{\mathbf{x}}}) = 0$. There is no assurance, however, that $\widehat{\mathbf{X} - \mu_{\mathbf{x}}} = \hat{\mathbf{X}} - \mu_x$, where $\hat{\mathbf{X}} = K\mathbf{Y}$. It turns out that, in general, they are not the same, and that in turn raises the question as to what $\widehat{\mathbf{X} - \mu_{\mathbf{x}}} = K(\mathbf{Y} - \mu_{\mathbf{y}})$ is telling us.

5.3.2. **Theorem.** *Let X be a random variable, $\mathbf{Y}^T = (Y_1, \ldots, Y_m)$ a random vector. Then the best linear minimum variance unbiased estimator of X based on $\mathbf{Y}$ is given by*

$$\hat{X} = E(X\mathbf{Y}^T)E(\mathbf{Y}\mathbf{Y}^T)^+\mathbf{Y} + [\mu_{\mathbf{x}} - E(X\mathbf{Y}^T)E(\mathbf{Y}\mathbf{Y}^T)^+\mu_{\mathbf{y}}]\frac{\mu_{\mathbf{y}}^T E(\mathbf{Y}\mathbf{Y}^T)^+\mathbf{Y}}{\mu_{\mathbf{y}}^T E(\mathbf{Y}\mathbf{Y}^T)^+\mu_{\mathbf{y}}}.$$

Proof. Exercise 5.4.2. □

Note that $E(\hat{X}) = \mu_{\mathbf{x}}$ so that the above estimate is indeed unbiased. But also note that the appropriate matrix K is *not* the matrix one obtains by estimating $X - \mu_{\mathbf{x}}$ based on $\mathbf{Y} - \mu_{\mathbf{y}}$. The above estimate is the sum of the BLMVE of X based on $\mathbf{Y}$ plus a correction term that moves the mean.

As we will see next, the above result is more of a curiosity than a useful theorem, and is a consequence of our stubborn insistence on using linear estimators when we know the means. The next theorem provides us with a more reasonable way to proceed in this situation. It's what people generally do anyway, even though they may not be using the correct language to describe their actions. First, a definition is presented.

5.3.3. **Definition.** Let $\mathbf{X}$ and $\mathbf{Y}$ be random vectors. By the *best affine minimum variance estimator* $\hat{\mathbf{X}}$ of $\mathbf{X}$ based on $\mathbf{Y}$, we mean an estimator of the form

$$\hat{\mathbf{X}} = K\mathbf{Y} + \mathbf{b}$$

such that

$$\|\hat{\mathbf{X}} - \mathbf{X}\|$$

is minimized in $\mathcal{L}_2^n(\Omega, P)$. $\hat{\mathbf{X}}$ is abbreviated as BAMVE.

5.3.4. **Lemma.** *Let* $\mathbf{Z}$ *be any random vector. Then*

$$\mathbf{Z} = E(\mathbf{Z}\mathbf{Z}^T)E(\mathbf{Z}\mathbf{Z}^T)^+\mathbf{Z}.$$

Proof. From Theorem 5.2.3, the right-hand side of the above expression is the BLMVE of $\mathbf{Z}$ based on itself, and that is obviously $\mathbf{Z}$. (See also Exercise 5.4.3.) □

5.3.5. **Theorem.** *Let* $\mathbf{X}$ *and* $\mathbf{Y}$ *be random vectors with means* $\boldsymbol{\mu}_x$ and $\boldsymbol{\mu}_y$, *respectively, and with covariances given by*

$$\Gamma_{11} = \text{cov}(\mathbf{X}, \mathbf{X}), \quad \Gamma_{12} = \Gamma_{21}^T = \text{cov}(\mathbf{X}, \mathbf{Y}), \quad \Gamma_{22} = \text{cov}(\mathbf{Y}, \mathbf{Y}).$$

Then the BAMVE *of* $\mathbf{X}$ *based on* $\mathbf{Y}$ *is given by*

$$\hat{\mathbf{X}} = K\mathbf{Y} + \mathbf{b}, \tag{5.3-2}$$

where

$$K = \Gamma_{12}\Gamma_{22}^+ \tag{5.3-3}$$

and

$$\mathbf{b} = \boldsymbol{\mu}_x - K\boldsymbol{\mu}_y. \tag{5.3-4}$$

Moreover, the covariance of the estimation error is given by

$$P \triangleq E((\mathbf{X} - \hat{\mathbf{X}})(\mathbf{X} - \hat{\mathbf{X}})^T) = \Gamma_{11} - K\Gamma_{21}. \tag{5.3-5}$$

Proof. First, note that we can write the relation

$$\hat{\mathbf{X}} = K\mathbf{Y} + \mathbf{b}$$

as

$$\hat{\mathbf{X}} = [K, \mathbf{b}]\begin{bmatrix} \mathbf{Y} \\ 1 \end{bmatrix}$$

so the estimate $\hat{\mathbf{X}}$ can be considered as the BLMVE of $\mathbf{X}$ based on $\begin{bmatrix} \mathbf{Y} \\ 1 \end{bmatrix}$.

From Theorem 5.2.3, therefore,

$$\begin{aligned}
[K, \mathbf{b}] &= E\left(\mathbf{X}\begin{bmatrix} \mathbf{Y} \\ 1 \end{bmatrix}^T\right) E\left(\begin{bmatrix} \mathbf{Y} \\ 1 \end{bmatrix}\begin{bmatrix} \mathbf{Y} \\ 1 \end{bmatrix}^T\right)^+ \\
&= E(\mathbf{X}(\mathbf{Y}^T, 1))E\begin{bmatrix} \mathbf{Y}\mathbf{Y}^T & \mathbf{Y} \\ \mathbf{Y}^T & 1 \end{bmatrix}^+
\end{aligned}$$

or

$$[K, b] = [E(\mathbf{X}\mathbf{Y}^T), \boldsymbol{\mu}_x] \begin{bmatrix} E(\mathbf{Y}\mathbf{Y}^T) & \boldsymbol{\mu}_y \\ \boldsymbol{\mu}_y^T & 1 \end{bmatrix}^+ . \tag{5.3-6}$$

Let us define A by

$$A \triangleq \begin{bmatrix} E(YY^T) & \boldsymbol{\mu}_y \\ \boldsymbol{\mu}_y^T & 1 \end{bmatrix}.$$

Then the range of A can be calculated as follows. For $\mathbf{x}_1 \in \mathrm{R}^n$, $x_2 \in \mathrm{R}$,

$$\begin{bmatrix} \mathbf{z}_1 \\ z_2 \end{bmatrix} \triangleq A \begin{bmatrix} \mathbf{x}_1 \\ x_2 \end{bmatrix},$$

so that

$$\begin{aligned} \mathbf{z}_1 &= E(\mathbf{Y}\mathbf{Y}^T)\mathbf{x}_1 + \boldsymbol{\mu}_y x_2 \\ z_2 &= \boldsymbol{\mu}_y^T \mathbf{x}_1 + x_2. \end{aligned} \tag{5.3-7}$$

From Lemma 5.3.4, we have that

$$\boldsymbol{\mu}_y = E(\mathbf{Y}\mathbf{Y}^T)E(\mathbf{Y}\mathbf{Y}^T)^+ \boldsymbol{\mu}_y$$

(just take the expected value of both sides), so if we apply $E(\mathbf{Y}\mathbf{Y}^T)$ to the vector $\mathbf{x}_1 + E(\mathbf{Y}\mathbf{Y}^T)^+ \boldsymbol{\mu}_y x_2$, we obtain

$$E(\mathbf{Y}\mathbf{Y}^T)(\mathbf{x}_1 + E(\mathbf{Y}\mathbf{Y}^T)^+ \boldsymbol{\mu}_y x_2) = E(\mathbf{Y}\mathbf{Y}^T)\mathbf{x}_1 + \boldsymbol{\mu}_y x_2.$$

In other words, for any $\mathbf{z}_1$ satisfying the first of Equation (5.3-7), $\mathbf{z}_1$ is in the range of $E(YY^T)$. The second of Equations (5.3-7) is a scalar equation, so we have shown that the range of A is

$$\mathcal{R}(A) = \mathcal{R}(E(\mathbf{Y}\mathbf{Y}^T)) \times \mathrm{R}.$$

From this it follows that

$$A'' = \begin{bmatrix} E(\mathbf{Y}\mathbf{Y}^T)'' & 0 \\ 0 & 1 \end{bmatrix}.$$

Multiplying (5.3-6) through on the right by A, we obtain

$$[K, \mathbf{b}]A = [E(\mathbf{X}\mathbf{Y}^T), \boldsymbol{\mu}_x]A^+ A$$

or

$$[K, \mathbf{b}]A = [E(\mathbf{X}\mathbf{Y}^T), \boldsymbol{\mu}_x](A^*)''.$$

But A is self-adjoint, so we have $(A^*)'' = A''$, and the above becomes

$$[K, \mathbf{b}] \begin{bmatrix} E(\mathbf{Y}\mathbf{Y}^T) & \boldsymbol{\mu}_y \\ \boldsymbol{\mu}_y^T & 1 \end{bmatrix} = [E(\mathbf{X}\mathbf{Y}^T), \boldsymbol{\mu}_x] \begin{bmatrix} E(\mathbf{Y}\mathbf{Y}^T)'' & 0 \\ 0 & 1 \end{bmatrix},$$

which in turn is

$$[KE(\mathbf{Y}\mathbf{Y}^T) + \mathbf{b}\,\boldsymbol{\mu}_y^T, K\boldsymbol{\mu}_y + \mathbf{b}] = [E(\mathbf{X}\mathbf{Y}^T)E(\mathbf{Y}\mathbf{Y}^T)'', \boldsymbol{\mu}_x].$$

Comparing components, we have

$$KE(\mathbf{Y}\mathbf{Y}^T) + \mathbf{b}\boldsymbol{\mu}_y^T = E(\mathbf{X}\mathbf{Y}^T)E(\mathbf{Y}\mathbf{Y}^T)'' \tag{5.3-8}$$

and

$$K\boldsymbol{\mu}_y + \mathbf{b} = \boldsymbol{\mu}_x. \tag{5.3-9}$$

Equation (5.3-9) is one of our desired results. Multiplying (5.3-9) on the right by $\boldsymbol{\mu}_y^T$, we obtain

$$K\boldsymbol{\mu}_y\boldsymbol{\mu}_y^T + \mathbf{b}\boldsymbol{\mu}_y^T = \boldsymbol{\mu}_x\boldsymbol{\mu}_y^T,$$

so substituting this into (5.3-8), we obtain

$$K[E(\mathbf{Y}\mathbf{Y}^T) - \boldsymbol{\mu}_y\boldsymbol{\mu}_y^T] + \boldsymbol{\mu}_x\boldsymbol{\mu}_y^T = E(\mathbf{X}\mathbf{Y}^T)E(\mathbf{Y}\mathbf{Y}^T)''.$$

Now, $\mathbf{Y} = E(\mathbf{Y}\mathbf{Y}^T)''\mathbf{Y}$ (essentially Lemma 5.3.4) so that $\mathbf{Y}^T = \mathbf{Y}^T E(\mathbf{Y}\mathbf{Y}^T)''$. Thus,

$$E(\mathbf{X}\mathbf{Y}^T)E(\mathbf{Y}\mathbf{Y}^T)'' = E(\mathbf{X}\mathbf{Y}^T E(\mathbf{Y}\mathbf{Y}^T)'') = E(\mathbf{X}\mathbf{Y}^T),$$

so the above expression becomes

$$K[E(\mathbf{Y}\mathbf{Y}^T) - \boldsymbol{\mu}_y\boldsymbol{\mu}_y^T] = E(\mathbf{X}\mathbf{Y}^T) - \boldsymbol{\mu}_x\boldsymbol{\mu}_y^T. \tag{5.3-10}$$

Now, one can easily check that

$$\Gamma_{22} = E((\mathbf{Y} - \boldsymbol{\mu}_y)(\mathbf{Y} - \boldsymbol{\mu}_y)^T) = E(\mathbf{Y}\mathbf{Y}^T) - \boldsymbol{\mu}_y\boldsymbol{\mu}_y^T$$

and

$$\Gamma_{12} = E((\mathbf{X} - \boldsymbol{\mu}_x)(\mathbf{Y} - \boldsymbol{\mu}_y)^T = E(\mathbf{X}\mathbf{Y}^T) - \boldsymbol{\mu}_x\boldsymbol{\mu}_y^T,$$

so (5.3-10) can be rewritten as

$$K\Gamma_{22} = \Gamma_{12}. \tag{5.3-11}$$

From Lemma 5.3.4 (with $\mathbf{Z} = \mathbf{Y} - \boldsymbol{\mu}_y$), we have

$$\mathbf{Y} - \boldsymbol{\mu}_y = \Gamma_{22}\Gamma_{22}^{+}(\mathbf{Y} - \boldsymbol{\mu}_y). \tag{5.3-12}$$

We thus have the result that the BAMVE of $\mathbf{X}$ based on $\mathbf{Y}$ is given by

$$\hat{\mathbf{X}} = K\mathbf{Y} + \mathbf{b}, \tag{5.3-13}$$

where K and $\mathbf{b}$ must satisfy (5.3-9) and (5.3-11). Multiplying (5.3-11) through on the right by $\Gamma_{22}^{+}(\mathbf{Y} - \boldsymbol{\mu}_y)$ and noting (5.3-12), we obtain

$$K(\mathbf{Y} - \boldsymbol{\mu}_y) = \Gamma_{12}\Gamma_{22}^{+}(\mathbf{Y} - \boldsymbol{\mu}_y)$$

and from (5.3-9),

$$K\mathbf{Y} + \mathbf{b} - \boldsymbol{\mu}_x = \Gamma_{12}\Gamma_{22}^{+}(\mathbf{Y} - \boldsymbol{\mu}_y).$$

Thus, using (5.3-13), this expression becomes

$$\hat{\mathbf{X}} - \boldsymbol{\mu}_x = \Gamma_{12}\Gamma_{22}^{+}(\mathbf{Y} - \boldsymbol{\mu}_y) \tag{5.3-14}$$

or

$$\hat{\mathbf{X}} = \Gamma_{12}\Gamma_{22}^{+}\mathbf{Y} + (\boldsymbol{\mu}_x - \Gamma_{12}\Gamma_{22}^{+}\boldsymbol{\mu}_y). \tag{5.3-15}$$

Clearly, the choices

$$K = \Gamma_{12}\Gamma_{22}^{+}$$

and

$$\mathbf{b} = \boldsymbol{\mu}_x - \Gamma_{12}\Gamma_{22}^{+}\boldsymbol{\mu}_y$$

give us the desired result. □

The expression for P is obtained in exactly the same fashion as the error covariance was obtained in the proof of Theorem 3.5.2. The argument is word for word the same as it was there and depends only on the result in (5.3-14) and the relation $\Gamma_{22}^{+} = \Gamma_{22}^{+}\Gamma_{22}\Gamma_{22}^{+}$ (Theorem 4.3.2(e)). Note that by (5.3-14), we have

$$E(\hat{\mathbf{X}}) = \boldsymbol{\mu}_x,$$

so that the BAMVE is unbiased. Moreover, as we have seen, it can be calculated by calculating the BLMVE of $\mathbf{X} - \boldsymbol{\mu}_x$ based on $\mathbf{Y} - \boldsymbol{\mu}_y$ and adding $\boldsymbol{\mu}_x$ to the result. In other words, the BLMVE of $\mathbf{X} - \boldsymbol{\mu}_x$ based on $\mathbf{Y} - \boldsymbol{\mu}_y$ is $\hat{\mathbf{X}} - \boldsymbol{\mu}_x$, where $\hat{\mathbf{X}}$ is the BAVME. Of course, if the means $\boldsymbol{\mu}_x$ and $\boldsymbol{\mu}_y$ are unknown (which is not uncommon), the best one can do is calculate the BLMVE of $\mathbf{X}$ given $\mathbf{Y}$, and this may be a biased estimate.

One word about the method of proof used in Theorem 5.3.5: the amount of work we did could have been greatly reduced had we assumed the existence of inverses everywhere, for then we could have used the matrix inversion lemma from Chapter 1. The result in that case would, of course, be weaker. In case Γ_{22} is invertible, the result we obtained in 5.3.5 is identical with that we obtained in Theorem 3.5.2. Although this is not surprising, our result in 3.5.2 did not preclude our proving Theorem 5.3.5. For although the result in 3.5.2 was a minimum variance estimator and was also affine, it was obtained using conditional expectations. Conceivably, the BAMVE could have turned out to be an affine transformation different from that we obtained in 3.5.2. Happily, such is not the case.

Before we end this chapter we should like to point out that the various estimates we have calculated, although making use of the first and second order statistics of the random variables involved, make no mention of the underlying densities: the estimate is the same no matter what the density! Of course, we pay a price. Unless the random variables involved are multivariate normal, the BLMVE (or BAMVE) may not, in fact, be the best minimum variance estimate. There truly is no such thing as a free lunch!

5.4 Exercises

1. Prove that if M is a subspace of $\mathcal{L}_2(\Omega, P)$, then $\mathbf{Z} \perp M^n$ if and only if $Z_i \perp M$ for each i ($\mathbf{Z} = (z_1, \mathbf{z}_2 \dots, z_n)$).

2. Prove Theorem 5.3.2. Hints: Let $\alpha^T = (\alpha_1, \alpha_2, \dots, \alpha_n)$ so that X is of the form
$$\hat{\mathbf{X}} = \boldsymbol{\alpha}^T Y = \mathbf{Y}^T \boldsymbol{\alpha}.$$
If the estimate is to be unbiased
$$\boldsymbol{\mu}_x = \boldsymbol{\alpha}^T \boldsymbol{\mu}_y.$$
(1) Show that $J(\boldsymbol{\alpha}) = \|\mathbf{X} - \hat{\mathbf{X}}\|^2$ is given by
$$J(\alpha) = E(\mathbf{X}^2) - 2E(\mathbf{X}\mathbf{Y}^T)\boldsymbol{\alpha} + \boldsymbol{\alpha}^T E(\mathbf{Y}\mathbf{Y}^T)\boldsymbol{\alpha}.$$
(2) To minimize $J(\alpha)$ subject to $\boldsymbol{\mu}_x = \alpha^T \mu_y$, form the Lagrangian
$$L(\lambda, \boldsymbol{\alpha}) = J(\boldsymbol{\alpha}) + \lambda(\boldsymbol{\mu}_x - \boldsymbol{\mu}_y^T \boldsymbol{\alpha}).$$
If $\boldsymbol{\alpha}_0$ represents the optimal $\boldsymbol{\alpha}$ and λ_0 is the corresponding Lagrange multiplier, show that
$$\boldsymbol{\alpha}_0^T E(\mathbf{Y}\mathbf{Y}^T) = E(\mathbf{X}\mathbf{Y}^T) + \frac{\lambda_0}{2} \boldsymbol{\mu}_y^T.$$
(3) Multiply this last expression by $E(\mathbf{Y}\mathbf{Y}^T)^+$ on the right and use 5.3.4.

3. Prove Lemma 5.3.4 by showing that
$$E((\mathbf{Z} - E(\mathbf{Z}\mathbf{Z}^T)E(\mathbf{Z}\mathbf{Z}^T)^+\mathbf{Z})(\mathbf{Z} - E(\mathbf{Z}\mathbf{Z}^T)E(\mathbf{Z}\mathbf{Z}^T)^+\mathbf{Z})^T) = 0.$$

4. Using the notation in Theorem 5.3.5, define the *correlation matrix* ρ by
$$\rho(X, Y) = \Gamma_{11}^+ \Gamma_{12} \Gamma_{22}^+ \Gamma_{21}.$$
(a) Prove that $\Gamma_{11}'' \Gamma_{12} = \Gamma_{12}$ (see 5.3.4 and the proof of Equation (5.3-9)).
(b) Show that if $\rho(X, Y) = I$, then
$$\mathbf{X} - \boldsymbol{\mu}_x = \Gamma_{12} \Gamma_{22}^+ (\mathbf{Y} - \boldsymbol{\mu}_y).$$
Hint: Expand $\|\mathbf{X} - \boldsymbol{\mu}_x - \Gamma_{12}\Gamma_{22}^{-1}(\mathbf{Y} - \boldsymbol{\mu}_y\|^2$ and make use of the fact that
$$[\mathbf{X}, \mathbf{Y}] = \operatorname{tr}(E(\mathbf{X}\mathbf{Y}^T))$$
for any $\mathbf{X}$, $\mathbf{Y}$.

6

Recursive Linear Estimation (Bayesian Estimation)

6.1 Introduction

In the last chapter, we developed formulas enabling us to calculate the linear minimum variance estimator of $\mathbf{X}$ based on knowledge of a random vector $\mathbf{Y}$. Moreover, we also calculated an expression for the so-called covariance of the estimation error. Specifically, then, our output was a random variable $\hat{\mathbf{X}}$ representing an estimator of $\mathbf{X}$, and a matrix P defined by

$$P \triangleq E((\mathbf{X} - \hat{\mathbf{X}})(\mathbf{X} - \hat{\mathbf{X}})^T). \tag{6.1-1}$$

Both of these outputs were based on knowledge of the means $\boldsymbol{\mu}_x$ and $\boldsymbol{\mu}_y$, the covariances of $\mathbf{X}$ and $\mathbf{Y}$, and their combined covariance, $\text{cov}(\mathbf{X}, \mathbf{Y})$. Of course, as discussed in Chapter 2, one generally would not have knowledge of $\mathbf{Y}$. Instead, one would have knowledge of some realization of $\mathbf{Y}$, that is, $\mathbf{Y}(\omega)$ for some ω. The real "output" is, therefore, an estimating function (the g of Chapter 2) and a covariance matrix.

In this chapter, we are going to discuss a slightly different estimation problem. Specifically, suppose that somehow or other (perhaps using results from the last chapter—how is unimportant), we have obtained a linear minimum variance estimator $\hat{\mathbf{X}}_1$ of a random vector $\mathbf{X}$ based on a random vector $\mathbf{Y}_1$ and an error covariance matrix P_1 of the form given in (6.1-1) (again, how this was secured is unimportant). Further, suppose that we now have some new information about $\mathbf{X}$, call it $\mathbf{Y}_2$, that is obtained by taking a linear "measurement" of $\mathbf{X}$ that is corrupted by "additive measurement noise" $\mathbf{W}$, specifically,

$$\mathbf{Y}_2 = H\mathbf{X} + \mathbf{W}. \tag{6.1-2}$$

For the time being, the word "noise" associated with the random vector $\mathbf{W}$ means that

$$E(\mathbf{X}\mathbf{W}^T) = 0 \tag{6.1-3}$$

and

$$E(\mathbf{Y}_1\mathbf{W}^T) = 0 \tag{6.1-4}$$

$$\boldsymbol{\mu}_W = \mathbf{0}. \tag{6.1-5}$$

Equation (6.1-5) expresses the belief that noise should be unbiased—if there is a known bias then $\mathbf{W}$ is not strictly noise since it has some structure.

Equations (6.1-3) and (6.1-4) represent the belief that true noise is not causally related to either the quantity being measured nor to any past information about it (recall independence from Chapter 1).

From the last chapter, we know that, however it was obtained, $\hat{\mathbf{X}}_1$ is the projection of $\mathbf{X}$ onto M^n, where M is the span of the components of $\mathbf{Y}_1$. If $\hat{\mathbf{X}}_2$ is to represent the linear minimum variance estimate based on

$$\mathbf{Y} = \begin{bmatrix} \mathbf{Y}_1 \\ \mathbf{Y}_2 \end{bmatrix}, \tag{6.1-6}$$

then $\hat{\mathbf{X}}_2$ would be the projection of $\mathbf{X}$ onto the space N^n, where N is the span of the components of $\mathbf{Y}$. Note that if M_2 represents the span of the components of $\mathbf{Y}_2$, then

$$N = M_1 + M_2, \tag{6.1-7}$$

whence

$$N^n = (M_1 + M_2)^n. \tag{6.1-8}$$

We now pose the problem to be solved. Suppose that we know $\hat{\mathbf{X}}_1$, P_1, $\mathbf{Y}_2$, H, and R, where

$$R \triangleq E(\mathbf{W}\mathbf{W}^T). \tag{6.1-9}$$

Moreover, we suppose that (6.1-2), (6.1-3), and (6.1-4) all hold. We do not know $\mathbf{Y}_1$, $\mathbf{X}$, or any means or covariances of any of the random variables, other than what we have stated above. Under these circumstances, can we calculate $\hat{\mathbf{X}}_2$? Surprisingly, the answer is yes and the details are given in the next section.

6.2 The Recursive Linear Estimator

We begin with a technical lemma.

6.2.1. **Lemma.** *If*

(a) $\mathbf{X} \in \mathcal{L}_2^n(\Omega, P)$, $\mathbf{Y}_2 \in \mathcal{L}_2^m(\Omega, P)$,

(b) $M_1 \subset \mathcal{L}_2(\Omega, P)$ *is a closed subspace,*

(c) M_2 $=$ *span of components of* $\mathbf{Y}_2$,

(d) $\hat{\mathbf{X}}_1$ $=$ *projection of* $\mathbf{X}$ *onto* M_1^n,

(e) $\hat{\mathbf{Y}}_2$ $=$ *projection of* $\mathbf{Y}_2$ *onto* M_1^n, *and*

(f) $\tilde{\mathbf{Y}}_2 \triangleq \mathbf{Y}_2 - \hat{\mathbf{Y}}_2$, *then the projection of* $\mathbf{X}$ *onto* $(M_1 + M_2)^n$ *is given by*

$$\hat{\mathbf{X}}_2 = \hat{\mathbf{X}}_1 + E(\mathbf{X}\tilde{\mathbf{Y}}_2^T)E(\tilde{\mathbf{Y}}_2\tilde{\mathbf{Y}}_2^T)^+\tilde{\mathbf{Y}}_2. \tag{6.2-1}$$

Proof. Since $\tilde{\mathbf{Y}}_2 = \mathbf{Y}_2 - \hat{\mathbf{Y}}_2$, it follows from the projection theorem that $\tilde{\mathbf{Y}}_2 \perp M_1^n$. From Exercise 5.4.1, $(\tilde{\mathbf{Y}}_2)_i \perp M_1$, where $(\tilde{\mathbf{Y}}_2)_i$ represents the ith component of $\tilde{\mathbf{Y}}_2$. Letting $\tilde{M}_2 \triangleq$ span of the components of $\tilde{\mathbf{Y}}_2$, we then have that $\tilde{M}_2 \perp M_1$ in $\mathcal{L}_2(\Omega, P)$. From Exercise 5.4.1, we then have that $\tilde{M}_2^n \perp M_1^n$ in $\mathcal{L}_2^n(\Omega, P)$.

Next note that since $\tilde{\mathbf{Y}}_2 = \mathbf{Y}_2 - \hat{\mathbf{Y}}_2$, each component of $\tilde{\mathbf{Y}}_2$ is the sum of a vector in M_2 and a vector in M_1; hence, each vector in $\tilde{M}_2$ is the sum of vectors in M_2 and in M_1. Thus,

$$\tilde{M}_2 \subset M_1 + M_2. \tag{6.2-2}$$

Also, since $\mathbf{Y}_2 = \tilde{\mathbf{Y}}_2 + \hat{\mathbf{Y}}$, a similar argument shows that

$$M_2 \subset \tilde{M}_2 + M_1. \tag{6.2-3}$$

Adding M_1 to each side of (6.2-2) and (6.2-3), we obtain

$$M_1 + \tilde{M}_2 \subset M_1 + M_2$$

and

$$M_1 + M_2 \subset M_1 + \tilde{M}_2$$

from which it follows that

$$M_1 + M_2 = M_1 + \tilde{M}_2.$$

We then have

$$(M_1 + M_2)^n = (M_1 + \tilde{M}_2)^n = M_1^n + \tilde{M}_2^n, \tag{6.2-4}$$

the last equality following trivially from the definition of addition in $\mathcal{L}_2^n(\Omega, P)$, that is, from the fact that addition is coordinate-wise. We showed above that $\tilde{M}_2^n \perp M_1^n$, so from Exercise 4.6.2, it follows that the projection of $\mathbf{X}$ onto $M_1^n + \tilde{M}_2^n$, which is $\hat{\mathbf{X}}_2$ by (6.2-4), is the sum of the projection of $\mathbf{X}$ onto M_1, namely, $\hat{\mathbf{X}}_1$, plus the projection of $\mathbf{X}$ onto $\tilde{M}_2^n$. However, this latter estimate is simply the BLMVE of $\mathbf{X}$ based on $\tilde{\mathbf{Y}}_2$, and by 5.2.3 it is

$$E(\mathbf{X}\tilde{\mathbf{Y}}_2^T)E(\tilde{\mathbf{Y}}_2\tilde{\mathbf{Y}}_2^T)^+\tilde{\mathbf{Y}}_2$$

so that

$$\hat{\mathbf{X}}_2 = \hat{\mathbf{X}}_1 + E(\mathbf{X}\tilde{\mathbf{Y}}_2^T)E(\tilde{\mathbf{Y}}_2\tilde{\mathbf{Y}}_2^T)^+\tilde{\mathbf{Y}}_2. \quad \square$$

A picture in R^3 that is analogous to the setup in Lemma 6.2.1 is given in Figure 6.2-1.

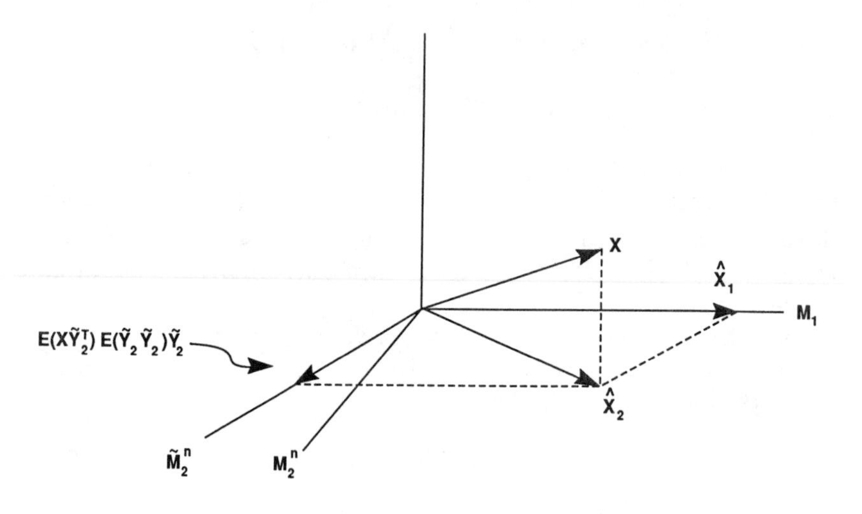

FIGURE 6.1. A geometric interpretation of Lemma 6.2.1.

6.2.2. **Theorem** (Static Updating Theorem—Bayesian Estimation). *Suppose that the random vectors* $\mathbf{X}$ *and* $\mathbf{Y}_2$ *are related by*

$$\mathbf{Y}_2 = H\mathbf{X} + \mathbf{W},$$

where H *is a known* $m \times n$ *matrix and* $\mathbf{W}$ *is a random vector such that*

$$R = E(\mathbf{W}\mathbf{W}^T)$$

is known. Further suppose that $\hat{\mathbf{X}}_1$ *is the* BLMVE *of* $\mathbf{X}$ *based on a random vector* $\mathbf{Y}_1$ *and that both* $\hat{\mathbf{X}}_1$ *and*

$$P_1 \triangleq E((\mathbf{X} - \hat{\mathbf{X}}_1)(\mathbf{X} - \hat{\mathbf{X}}_1)^T)$$

are known. Finally, suppose that

$$E(\mathbf{X}\mathbf{W}^T) = 0$$

and

$$E(\mathbf{Y}_1\mathbf{W}^T) = 0.$$

If

$$\mathbf{Y} = \begin{bmatrix} \mathbf{Y}_1 \\ \mathbf{Y}_2 \end{bmatrix},$$

then the BLMVE $\hat{\mathbf{X}}_2$ *of* $\mathbf{X}$ *based on* $\mathbf{Y}$ *is given by*

(a) $\hat{\mathbf{X}}_2 = \hat{\mathbf{X}}_1 + P_1 H^T [H P_1 H^T + R]^+ [\mathbf{Y}_2 - H\hat{\mathbf{X}}_1]$, *and the new error covariance*

$$P_2 \triangleq E((\mathbf{X} - \hat{\mathbf{X}}_2)(\mathbf{X} - \hat{\mathbf{X}}_2)^T)$$

is given by

(b) $P_2 = P_1 - P_1H^T[HP_1H^T + R]^+HP_1$.

Proof. Using the notation of Lemma 6.2.1 with M_1 = span of $\mathbf{Y}_1$, we have by 5.2.3 that the projection $\hat{\mathbf{Y}}_2$ of $\mathbf{Y}_2$ onto M_1^n is given by

$$\hat{\mathbf{Y}}_2 = E(\mathbf{Y}_2\mathbf{Y}_1^T)E(\mathbf{Y}_1\mathbf{Y}_1^T)^+\mathbf{Y}_1.$$

Since $\mathbf{Y}_2 = H\mathbf{X} + \mathbf{W}$, it follows that

$$\begin{aligned}\hat{\mathbf{Y}}_2 &= E((H\mathbf{X}+\mathbf{W})\mathbf{Y}_1^T)E(\mathbf{Y}_1\mathbf{Y}_1^T)^+\mathbf{Y}_1\\ &= [HE(\mathbf{X}\mathbf{Y}_1^T) + E(\mathbf{W}Y_1^T)]E(\mathbf{Y}_1\mathbf{Y}_1^T)^+\mathbf{Y}_1\\ &= HE(\mathbf{X}\mathbf{Y}_1^T)E(\mathbf{Y}_1\mathbf{Y}_1^T)^+\mathbf{Y}_1\end{aligned}$$

$$\hat{\mathbf{Y}}_2 = H\hat{\mathbf{X}}_1. \tag{6.2-5}$$

As in 6.2.1, let $\tilde{\mathbf{Y}}_2 = \mathbf{Y}_2 - \hat{\mathbf{Y}}_2$. Then, from Lemma 6.2.1, we have

$$\hat{\mathbf{X}}_2 = \hat{\mathbf{X}}_1 + E(\mathbf{X}\tilde{\mathbf{Y}}_2^T)E(\tilde{\mathbf{Y}}_2\tilde{\mathbf{Y}}_2^T)^+\tilde{\mathbf{Y}}_2. \tag{6.2-6}$$

From (6.2-5), we see at once that

$$\tilde{\mathbf{Y}}_2 = \mathbf{Y}_2 - H\hat{\mathbf{X}}_1, \tag{6.2-7}$$

and so

$$\tilde{\mathbf{Y}}_2 = \mathbf{Y}_2 - H\hat{\mathbf{X}}_1 = H\mathbf{X} + \mathbf{W} - H\hat{\mathbf{X}}_1$$

or

$$\tilde{\mathbf{Y}}_2 = H(\mathbf{X} - \hat{\mathbf{X}}_1) + \mathbf{W}. \tag{6.2-8}$$

We then have

$$\begin{aligned}E(\mathbf{X}\tilde{\mathbf{Y}}_2^T) &= E(\mathbf{X}(H(\mathbf{X}-\hat{\mathbf{X}}_1)+\mathbf{W})^T)\\ &= E(\mathbf{X}(\mathbf{X}-\hat{\mathbf{X}}_1)^TH^T) + E(\mathbf{X}\mathbf{W}^T)\\ &= E(\mathbf{X}(\mathbf{X}-\hat{\mathbf{X}}_1)^T)H^T.\end{aligned} \tag{6.2-9}$$

Now, since $(\mathbf{X} - \hat{\mathbf{X}}_1) \perp M_1^n$, it follows from Exercise 5.4.1 that $\mathbf{X}_i - (\hat{\mathbf{X}}_1)_i \perp M_1$ for each i. Since $(\hat{\mathbf{X}}_1)_j \in M_1$ for each j, it then follows that for all i, j, $(\mathbf{X}_i - (\hat{\mathbf{X}}_1)_i) \perp (\hat{\mathbf{X}}_1)_j$. Hence,

$$E(\hat{\mathbf{X}}_1(\mathbf{X} - \hat{\mathbf{X}}_1)^T = 0. \tag{6.2-10}$$

It follows from this and (6.2-9) that

$$E(\mathbf{X}\tilde{\mathbf{Y}}_2^T) = E((\mathbf{X}-\hat{\mathbf{X}}_1)(\mathbf{X}-\hat{\mathbf{X}}_1)^T)H^T = P_1H^T. \tag{6.2-11}$$

Similarly

$$\begin{aligned}E(\tilde{\mathbf{Y}}_2\tilde{\mathbf{Y}}_2^T) &= E((H(\mathbf{X}-\hat{\mathbf{X}}_1)+\mathbf{W})(H(\mathbf{X}-\hat{\mathbf{X}}_1)+\mathbf{W})^T)\\ &= HE((\mathbf{X}-\hat{\mathbf{X}}_1)(\mathbf{X}-\hat{\mathbf{X}}_1)^T)H^T + HE((\mathbf{X}-\hat{\mathbf{X}}_1)\mathbf{W}^T)\end{aligned}$$

$$+ E(W(\mathbf{X} - \hat{\mathbf{X}}_1)^T)H^T + E(\mathbf{W}\mathbf{W}^T). \tag{6.2-12}$$

Now $\hat{\mathbf{X}}_1 = K\mathbf{Y}_1$ for K the constant matrix given in Theorem 5.2.3. Hence,

$$0 = K \cdot 0 = KE(\mathbf{Y}_1\mathbf{W}^T) = E(K\mathbf{Y}_1\mathbf{W}^T) = E(\hat{\mathbf{X}}_1\mathbf{W}^T).$$

By assumption, $E(\mathbf{X}\mathbf{W}^T) = 0$, so with this and the previous calculation, (6.2-12) becomes

$$E(\tilde{\mathbf{Y}}_2\tilde{\mathbf{Y}}_2^T) = HP_1 + R. \tag{6.2-13}$$

From (6.2-6), (6.2-7), (6.2-11), and (6.2-13), it follows that

$$\hat{\mathbf{X}}_2 = \hat{\mathbf{X}}_1 + P_1H^T[HP_1H^T + R]^+[\mathbf{Y}_2 - H\mathbf{X}_1], \tag{6.2-14}$$

which is the first part of the theorem.

Next, temporarily set

$$M \triangleq [HP_1H^T + R]^+$$

and note that $M = M^T$. From (6.2-14),

$$\hat{\mathbf{X}}_2 = \hat{\mathbf{X}}_1 + P_1H^TM[\mathbf{Y}_2 - H\hat{\mathbf{X}}_1].$$

Thus,

$$\begin{aligned}
P_2 &= E((\mathbf{X} - \hat{\mathbf{X}}_2)(\mathbf{X} - \hat{\mathbf{X}}_2)^T) \\
&= E((\mathbf{X} - \hat{\mathbf{X}}_1 - P_1H^TM(\mathbf{Y}_2 - H\hat{\mathbf{X}}_1)) \\
&\quad \times (\mathbf{X} - \hat{\mathbf{X}}_1 - P_1H^TM(\mathbf{Y}_2 - H\hat{\mathbf{X}}_1))^T) \\
&= E((\mathbf{X} - \hat{\mathbf{X}}_1)(\mathbf{X} - \hat{\mathbf{X}}_1)^T) - P_1H^TME((\mathbf{Y}_2 - H\hat{\mathbf{X}}_1)(\mathbf{X} - \hat{\mathbf{X}}_1)) \\
&\quad - E((\mathbf{X} - \hat{\mathbf{X}}_1)(P_1H^TM(\mathbf{Y}_2 - H\hat{\mathbf{X}}_1))^T) \\
&\quad + P_1H^TME((\mathbf{Y}_2 - M\hat{\mathbf{X}}_1)(\mathbf{Y}_2 - H\hat{\mathbf{X}}_1)^T)MHP_1 \\
&= P_1 - P_1H^TME((H(\mathbf{X} - \hat{\mathbf{X}}_1) + W)(\mathbf{X} - \hat{\mathbf{X}}_1)^T) \\
&\quad - E((\mathbf{X} - \hat{\mathbf{X}}_1)(H(\mathbf{X} - \hat{\mathbf{X}}_1) + W)^T)MHP_1 \\
&\quad + PH^TME((H(\mathbf{X} - \hat{\mathbf{X}}_1) + W)(H(\mathbf{X} - \hat{\mathbf{X}}_1) + W)^T)MHP_1 \\
&= P_1 - P_1H^TMHP_1 - P_1H^TMHP_1 + PH^TM[HP_1H^T + R]MHP_1 \\
&= P_1 - 2P_1H^TMHP_1 + P_1H^TMM^+MHP_1 \\
&= P_1 - 2P_1H^TMHP_1 + P_1H^TMHP_1 \\
&= P_1 - P_1H^TMHP_1.
\end{aligned}$$

Substituting for M, we obtain

$$P_2 = P_1 - P_1H^T[HP_1H^T + R]^+HP_1, \tag{6.2-15}$$

and we are done. □

It should be clear that 6.2.2 implicitly describes a recursive estimation scheme. If $\hat{\mathbf{X}}_1$, P_1 are known and $\mathbf{Y}_2$ is secured, then we can calculate $\hat{\mathbf{X}}_2$ and P_2. If $\mathbf{Y}_3$ is then secured, we can calculate $\hat{\mathbf{X}}_3$ and P_3, and so on. Of course, in practice one doesn't really have $\hat{\mathbf{X}}_1$ and $\mathbf{Y}_2$. Rather one has a vector $\hat{\mathbf{x}}_1 \in \mathrm{R}^n$ representing a realization of $\hat{\mathbf{X}}_1$, an estimate of $\mathbf{X}$ as described in Chapter 2, and a vector $\mathbf{y}_2 \in \mathrm{R}^m$ representing a measurement, that is, a realization of $\mathbf{Y}_2$. One then calculates an estimate $\hat{\mathbf{x}}_2 \in \mathrm{R}^n$ based on $\mathbf{y}_2$ using (6.2-14) with $\hat{\mathbf{X}}_1$ replaced by $\hat{\mathbf{x}}_1$ and $\mathbf{Y}_2$ replaced by $\mathbf{y}_2$. P_2 is calculated as before using (6.2-15).

There is one feature of such a recursive estimation that, at first glance, seems a bit curious. Referring to Equation (6.2-15), it is apparent that we can calculate the covariance of the error in the estimator $\hat{\mathbf{X}}_2$ before any data is collected or $\hat{\mathbf{X}}_2$ is even calculated. The reason this sounds strange has to do with our language and the discussion in the preceding paragraph. Specifically, P_2 is a statement about the properties of the estimator $\hat{\mathbf{X}}_2$ based on the properties of the random vector $\mathbf{Y}_2$. It is not a statement about the estimate $\hat{\mathbf{x}}_2$ based on the measurement $\mathbf{y}_2$. If one made many, many estimates of $\mathbf{X}$, using measurements having the statistical properties stated in the theorem, and calculated a sample covariance of $\mathbf{X} - \hat{\mathbf{X}}_2$ using $\mathbf{x} - \hat{\mathbf{x}}_2$ (we would have to know $\mathbf{x}$, of course), this sample covariance would be approximately equal to P_2. In other words, P_2 is an assessment of the statistical features of the algorithm in part (a) used to process data and not an assessment of a specific estimate.

There is another issue we must settle. Nowhere in the above arguments did we use assumption (6.1-5) (other than to make (6.1-3) and (6.1-4) plausible). The reason is that we used linear estimates of $\mathbf{X}$ and these require no information about the various means. The next theorem, which is an exercise, settles the recursive estimation problem for affine estimates, and the result is pleasantly surprising.

6.2.3. **Theorem.** *Let* $\hat{\mathbf{X}}_1$ *be the* BAMVE *of* $\mathbf{X}$ *based on* $\mathbf{Y}_1$ *and* $\hat{\mathbf{X}}_2$ *the* BAMVE *of* $\mathbf{X}$ *based on*

$$Y = \begin{bmatrix} Y_1 \\ Y_2 \end{bmatrix}.$$

If $\mathbf{W}$ *satisfies* (6.1-5) *as well as all of the other conditions in Theorem* 6.2.2, *then* $\hat{\mathbf{X}}_1$, $\hat{\mathbf{X}}_2$, P_1, *and* P_2 *satisfy* (a) *and* (b) *of Theorem* 6.2.2.

Proof. Exercise 6.3.1. □

Let us emphasize the conclusion. If we start with linear estimates and apply the algorithms in 6.2.2, we end up with linear estimates of $\mathbf{X}$. If we start with affine estimates and apply the algorithms in 6.2.2, we end up with affine estimates. The algorithms are exactly the same in either case.

6.3 Exercises

1. Prove Theorem 6.2.3 using the following hints.

   (a) Establish that the BLMVE of $\mathbf{X} - \boldsymbol{\mu}_x$ based on $\mathbf{Y} - \boldsymbol{\mu}_y$ (any $\mathbf{Y}$) is of the form $\hat{\mathbf{X}}_1 - \boldsymbol{\mu}_x$, where $\hat{\mathbf{X}}_1$ is the BAMVE of $\mathbf{X}$. Moreover $E(\hat{\mathbf{X}}_1) = \boldsymbol{\mu}_x$.

   (b) Show $\boldsymbol{\mu}_{y_2} - H\boldsymbol{\mu}_x$.

   (c) Show that the projection of $\mathbf{Y}_2 - \boldsymbol{\mu}_{y_2}$ onto the subspace spanned by the components of $\mathbf{Y}_1 - \boldsymbol{\mu}_{y_1}$, call it $\widehat{\mathbf{Y}_2 - \boldsymbol{\mu}_{y_2}}$, is given by

   $$\widehat{\mathbf{Y}_2 - \boldsymbol{\mu}_{y_2}} = H(\hat{\mathbf{X}}_1 - \boldsymbol{\mu}_x).$$

   (d) Define $\tilde{\mathbf{Y}}_2 = (\mathbf{Y}_2 - \boldsymbol{\mu}_{y_2}) - (\widehat{\mathbf{Y}_2 - \boldsymbol{\mu}_{y_2}})$ and show that

   $$\tilde{\mathbf{Y}}_2 = \mathbf{Y}_2 - H\hat{\mathbf{X}}_1 = H(\mathbf{X} - \hat{\mathbf{X}}_1) + \mathbf{W}.$$

   (e) By Lemma 6.2.1 and part (a)

   $$\hat{\mathbf{X}}_2 - \boldsymbol{\mu}_x = \hat{\mathbf{X}}_1 - \boldsymbol{\mu}_x + E((\mathbf{X} - \boldsymbol{\mu}_x)\tilde{Y}_2^T)E(\tilde{\mathbf{Y}}_2\tilde{\mathbf{Y}}_2^T)^+\tilde{\mathbf{Y}}_2.$$

   Now use part (d) and the proof of 6.2.2.

2. Suppose that P_1, P_2, R, and $HP_1H^T + R$ are all invertible. Then show that the conclusion to Theorem 6.2.2 can be written as

   $$P_2 = [P_1^{-1} + H^T R^{-1} H]^{-1}$$

   $$\hat{\mathbf{X}}_2 = P_2 H^T R^{-1}\mathbf{Y}_2 + P_2 P_1^{-1}\hat{\mathbf{X}}_1.$$

   Hint: Use the matrix inversion lemma from Chapter 1.

3. The matrices $F_1 = P_1^{-1}$, $F_2 = P_2^{-1}$ are called *Fisher information matrices.* (Recall problem 3.6.5 and the definition of information.) Show that if we define the *information vector* I_i by

   $$I_i = P_i^{-1}\hat{X}_i,$$

   then

   $$F_2 = F_1 + H^T R^{-1} H$$

   and

   $$I_2 = I_1 + H^T R^{-1} Y_2;$$

   $H^T R^{-1} Y_2$ is interpreted as the new information about X contained in Y_2.

7

The Discrete Kalman Filter

7.1 Discrete Linear Dynamical Systems

In Chapter 6, we discussed the problem of making recursive estimates of a random vector $\mathbf{X}$. The problem was static in the sense that every measurement was used to update or improve the estimate of the same random vector $\mathbf{X}$. We now consider the case where the random vector changes in time, between measurements, according to a specified statistical dynamic.

7.1.1. **Definition.** (a) By a *discrete random process* (or time series) we mean a sequence of random vectors. More precisely, a discrete random process $\mathbf{x}$ is a function

$$\mathbf{x} : \mathrm{N} \times \Omega \to \mathrm{R}^n,$$

where N denotes the natural numbers and $(\Omega, \mathcal{E}, P)$ is a probability space. Unless clarity is sacrificed, it is customary to suppress the second variable and write $\mathbf{x}(k)$ rather than $\mathbf{x}(k;\omega)$. Note that $\mathbf{x}(k)$ is a random vector. Although we have been used to using upper case letters for random vectors, the above notation is common when studying discrete processes and we will conform to tradition.

(b) A discrete random process $\mathbf{u}$ is called *white* or *uncorrelated* in case

$$\operatorname{cor}(\mathbf{u}(k), \mathbf{u}(j)) = Q(k)\delta_{kj}. \tag{7.1-1}$$

Obviously,

$$Q(k) = E(\mathbf{u}(k)\mathbf{u}(k)^T).$$

7.1.2. **Definition.** An *n-dimensional linear dynamic model* of a discrete random process consists of the following: (a) a vector difference equation

$$\mathbf{x}(k+1) = \phi(k)\mathbf{x}(k) + \mathbf{u}(k); \qquad k = 0, 1, 2, \ldots, \tag{7.1-2}$$

where $\mathbf{x}$ is a discrete random process, called the *state vector;* $\phi(k)$ is a known $n \times n$ matrix of constants (that can change with k); and $\mathbf{u}(k)$ is an n-dimensional white process called *process noise.* It is assumed that

$$Q(k) = E(\mathbf{u}(k)\mathbf{u}(k)^T) \tag{7.1-3}$$

is known for each k, and that

$$E(\mathbf{u}(k)\mathbf{x}(j)^T) = 0 \quad \text{for} \quad j \leq k, \text{ and} \tag{7.1-4}$$

(b) A *measurement equation* of the form

$$\mathbf{z}(k) = H(k)\mathbf{x}(k) + \mathbf{w}(k), \tag{7.1-5}$$

where $H(k)$ is a known $m \times n$ matrix, called the *measurement matrix*, $\mathbf{w}(k)$ is a discrete white process called *measurement noise*,

$$R(k) = E(\mathbf{w}(k)\mathbf{w}(k)^T) \tag{7.1-6}$$

is known, and the conditions

$$E(\mathbf{w}(k)\mathbf{u}(j)^T) = 0 \tag{7.1-7}$$

and

$$E(\mathbf{w}(k)\mathbf{x}(j)^T) = 0 \tag{7.1-8}$$

hold for all j and k.

7.1.3. **Examples.** (a) Suppose one has some dynamical system such that when operating in a noise-free environment satisfies

$$\dot{\mathbf{x}} = A\mathbf{x}, \tag{7.1-9}$$

A being a constant $n \times n$ matrix. Thus, there is a state transition matrix

$$\Phi(t) = e^{At} \tag{7.1-10}$$

such that

$$\mathbf{x}(t_2) = e^{A(t_2-t_1)}\mathbf{x}(t_1),$$

that is,

$$\mathbf{x}(t_2) = \Phi(t_2 - t_1)\mathbf{x}(t_1). \tag{7.1-11}$$

If the system is not stationary, that is, A is a function of t, then there still exists a state transition matrix (t_2, t_1) such that

$$\mathbf{x}(t_2) = \Phi(t_2, t_1)\mathbf{x}(t_1); \tag{7.1-12}$$

however, in this case, Φ is considerably more difficult to calculate than in the stationary case. If we let Δt represent some fixed time increment and (with the obvious abuse of notation) we define

$$\mathbf{x}(k) = \mathbf{x}(k\,\Delta t), \tag{7.1-13}$$

then

$$\mathbf{x}(k+1) = \phi(k)\mathbf{x}(k), \tag{7.1-14}$$

where

$$\phi(k) \triangleq \Phi((k+1)\,\Delta t, k\,\Delta t). \tag{7.1-15}$$

If the system now operates in a noisy environment, we "adjust" (7.1-14) by adding a noise term $q(k)$ to obtain

$$\mathbf{x}(k+1) = \phi(k)\mathbf{x}(k) + q(k).$$

Suppose that $\mathbf{x}$ is not completely observable, that is, physically there is no way that we can observe the value of each state. For example, if we use a pitot tube to measure velocity in a moving aircraft, what we really measure is the difference between the vehicle's ground speed and the speed of the wind relative to the ground. In general then, we suppose that

$$\mathbf{z}(k) = H(k)\mathbf{x}(k) + \mathbf{w}(k),$$

where $H(k)$ is a matrix. The term $\mathbf{w}(k)$ is added to account for the imperfections in the measurement process.

(b) Suppose that a certain process is described by the vector differential equation

$$\dot{\mathbf{x}} = f(\mathbf{x}, t) \tag{7.1-16}$$

when the system operates in a noise-free environment. If the process is initialized at some known vector $\mathbf{x}_0$, then the initial value theorem for vector differential equations assures us of a unique solution $\mathbf{x}_N$ satisfying

$$\dot{\mathbf{x}}_N = f(\mathbf{x}_N, t), \quad \mathbf{x}_N(0) = \mathbf{x}_0, \tag{7.1-17}$$

provided f satisfies certain smoothness conditions. We call $\mathbf{x}_N$ the nominal solution (or nominal path or nominal track). In principle, therefore, $\mathbf{x}_N$ is known. Another possibility is that f includes "controls" in the sense that if $\mathbf{x}$ is perfectly observed, then f adjusts itself so that the solution to (7.1-17) is a previously specified track given by $\mathbf{x}_N$ (think of a pilot/navigator as part of a navigation system that, with perfect information, adheres to a specified flight plan).

If the system is now forced to operate in a noisy environment, the differential equation in (7.1-16) is altered to

$$\dot{\mathbf{x}} = f(\mathbf{x}, t) + \mathbf{q}(t), \tag{7.1-18}$$

where $\mathbf{q}(t)$ represents the noise. For this text, we will have to be somewhat vague in describing $\mathbf{q}(t)$; for references purposes, we will only say that it is continuous white noise. If we assume that f is (Frechet) differentiable in $\mathbf{x}$, then defining

$$\epsilon(t) \triangleq \mathbf{x}(t) - \mathbf{x}_N(t), \tag{7.1-19}$$

we have that

$$f(\mathbf{x}(t), t) = f(\mathbf{x}_N(t) + \epsilon(t), t) = f(\mathbf{x}_N(t), t) + f'(\mathbf{x}_N(t), t)\epsilon(t) + \mathbf{r},$$

where the remainder $\mathbf{r}$ has the property that

$$\lim_{\|x\|\to 0} \frac{\mathbf{r}(\mathbf{x})}{\|\mathbf{x}\|} = 0.$$

Thus,

$$f(\mathbf{x}(t),t) - f(\mathbf{x}_N(t),t) \simeq f'(\mathbf{x}_N(t),t)\epsilon(t).$$

From (7.1-17) and (7.1-18), this can be written

$$\dot{\mathbf{x}}(t) - q(t) - \dot{\mathbf{x}}_N(t) \simeq f'(\mathbf{x}_N(t),t)\epsilon(t)$$

or, letting $\simeq$ be $=$,

$$\dot{\epsilon}(t) = f'(\mathbf{x}_N(t),t)\epsilon(t) + q(t). \tag{7.1-20}$$

The equation (7.1-20) is called a *dynamic error model.* If we write this as

$$\dot{\epsilon}(t) = F(t)\epsilon(t) + \mathbf{q}(t), \tag{7.1-21}$$

we see that this is a linear model. If $\Phi(t_2,t_1)$ is the state transition matrix for

$$\dot{\epsilon} = F\epsilon,$$

then (using the notation from part (a))

$$\epsilon(k+1) = \phi(k)\epsilon(k) + \mathbf{q}(k),$$

where $\mathbf{q}(k) = \int_0^{kt} \Phi(kT,t)q(t)\,dt$[1] turns out to be a discrete white process. This is called a discrete error model. Note that if

$$\mathbf{z}_N(k) = H(k)\mathbf{x}_N(k)$$

represents the noise-free measurements, and

$$\mathbf{z}(k) = H(k)\mathbf{x}(k) + \mathbf{w}(k)$$

represents the noisy measurements of the noise-driven random vector $\mathbf{x}(k)$, then

$$\mathbf{z}_\epsilon(k) \triangleq \mathbf{z}(k) - \mathbf{z}_N(k)$$

satisfies

$$\mathbf{z}_\epsilon(k) = H(k)\epsilon(k) + \mathbf{w}(k)$$

and is the error measurement equation.

The above examples are both instances of using discrete systems to represent continuous systems, some only approximately. It is sometimes the case

[1]This is not an ordinary integral, it is called a stochastic integral.

that a system's dynamical structure is inherently discrete and is known. For example, a system that is driven by digital inputs or is itself dynamically structured by use of digital devices would be an inherently discrete system. Sometimes, however, a system is nothing but a string of digital outputs (the $\mathbf{z}(k)$'s), and the nature of the dynamics is totally unknown. This is called a time series. The problem of finding a discrete dynamical system having the given time series as its output is called the plant (or system) identification problem, and the mathematical discipline devoted to solving this problem is known as time series analysis. We have neither time nor space to delve into this fascinating problem, but for reference purposes we mention that the most common analysis techniques are autoregressive processes, moving average processes, autoregressive moving averages, and canonical variates (see References [1], [16], and [26]).

7.2 The Kalman Filter

In this section, we present one of the principle results in this text and one of the most important theorems in estimation theory in the last thirty years (if not more). This theorem, attributable to R. E. Kalman, first appeared in March 1960 [14] and solves the recursive estimation problem for discrete dynamical systems. About a year later, Kalman and Bucy [15] published an analagous theorem for continuous systems. In this text, we are restricting our attention to discrete systems, so we will present the original discrete version of the theorem.

7.2.1. **Definition.**

(a) Let $\mathbf{x}$ be the state vector associated with the linear system of 7.1.2. By $\hat{\mathbf{x}}(k \mid j)$, we will mean the BLMVE of $\mathbf{x}(k)$ based on

$$\mathbf{y}_j = \begin{bmatrix} \mathbf{z}(0) \\ \mathbf{z}(1) \\ \vdots \\ \mathbf{z}(j) \end{bmatrix}.$$

Thus, $\hat{\mathbf{x}}(k \mid j)$ is the projection of $\mathbf{x}(k)$ onto the n-fold product of M_j with itself, M_j being the linear span of the components of $\mathbf{y}_j$.

(b) $P(k \mid j) = E((\hat{\mathbf{x}}(k \mid j) - \mathbf{x}(k))(\hat{\mathbf{x}}(k \mid j) - \mathbf{x}(k))^T)$.

(c) When $j = k$, $\hat{\mathbf{x}}(k \mid k)$ is called the *filtered* estimate of $\mathbf{x}(k)$; when $j < k$, $\hat{\mathbf{x}}(k \mid j)$ is called the *predicted* estimate of $\mathbf{x}(k)$; when $j > k$, $\hat{\mathbf{x}}(k \mid j)$ is called the *smoothed* estimate of $\mathbf{x}(k)$.

7.2.2. **Theorem** (Kalman, 1960). *The* BLMVE $\hat{\mathbf{x}}(k \mid k)$ *may be generated recursively by*

(a) $\hat{x}(k+1 \mid k+1) = \phi(k)\hat{\mathbf{x}}(k \mid k) + K(k+1)[\mathbf{z}(k+1) - H(k+1)\phi(k)\hat{\mathbf{x}}(k \mid k)]$, *where* $K(k+1)$, *the Kalman gain matrix, is given by*

(b) $K(k+1) = P(k+1 \mid k)H(k+1)^T[H(k+1)P(k+1 \mid k)H(k+1)^T + R(k+1)]^+$ *and* $P(k+1 \mid k)$ *is generated recursively by the equations*

(c) $P(k \mid k) = [I - K(k)H(k)]P(k \mid k-1)$ (covariance update) *and*

(d) $P(k+1 \mid k) = \phi(k)P(k \mid k)\phi(k)^T + Q(k)$ (the covariance extrapolations).

Proof. Suppose $\mathbf{z}(0), \mathbf{z}(1), \ldots, \mathbf{z}(k)$ have been secured and that estimate $\hat{\mathbf{x}}(k \mid k)$ and covariance $P(k \mid k)$ have been calculated. Using the notation in 7.2.1(a), we have, in particular, that

$$\hat{\mathbf{x}}(k \mid k) = E(\mathbf{x}(k)\mathbf{y}_k^T)E(\mathbf{y}_k\mathbf{y}_k^T)^+\mathbf{y}_k. \tag{7.2-1}$$

Note that

$$E(\mathbf{u}(k)\mathbf{y}_j^T) = (E(\mathbf{u}(k)[\mathbf{z}(0)^T, \mathbf{z}(1)^T, \ldots, \mathbf{z}(k)^T])$$

$$E(\mathbf{u}(k)\mathbf{y}_j^T) = [E(\mathbf{u}(k)\mathbf{z}(0)^T), \ldots, E(\mathbf{u}(k)\mathbf{z}(j)^T)]. \tag{7.2-2}$$

But by assumption, each $\mathbf{u}(k)$ is uncorrelated with $\mathbf{x}(j)$ for $j \le k$ and is uncorrelated with $w(j)$ for all j. Hence, for $i \le j$,

$$E(\mathbf{u}(k)\mathbf{z}(i)^T) = E(\mathbf{u}(k)\mathbf{x}(i)^T)H(i) + E(\mathbf{u}(k)\mathbf{w}(i)^T) = 0,$$

and so by (7.2-2)

$$E(\mathbf{u}(k)\mathbf{y}_j^T) = 0; \quad j \le k. \tag{7.2-3}$$

Now, from Definition 7.2.1(a), Equation (7.1-2), and Equation (7.2-3), we have

$$\begin{aligned}
\hat{\mathbf{x}}(k+1 \mid k) &= E(\mathbf{x}(k+1)\mathbf{y}_k^T)E(\mathbf{y}_k\mathbf{y}_k^T)^+\mathbf{y}_k \\
&= E((\phi(k)\mathbf{x}(k) + \mathbf{u}(k))\mathbf{y}_k^T)E(\mathbf{y}_k\mathbf{y}_k^T)^+\mathbf{y}_k \\
&= [\phi(k)E(\mathbf{x}(k)\mathbf{y}_k^T) + E(\mathbf{u}(k)\mathbf{y}_k^T)]E(\mathbf{y}_k\mathbf{y}_k^T)^+\mathbf{y}_k \\
&= \phi(k)E(\mathbf{x}(k)\mathbf{y}_k^T)E(\mathbf{y}_k\mathbf{y}_k^T)^+\mathbf{y}_k
\end{aligned}$$

$$\hat{x}(k+1 \mid k) = \phi(k)\hat{\mathbf{x}}(k \mid k). \tag{7.2-4}$$

Using (7.2-4), we can calculate $P(k+1 \mid k)$ as follows.

$$\begin{aligned}
P(k+1 \mid k) &= E((\hat{\mathbf{x}}(k+1 \mid k) - \mathbf{x}(k+1))(\hat{\mathbf{x}}(k+1 \mid k) - \mathbf{x}(k+1))^T) \\
&= E((\phi(k)(\hat{x}(k \mid k) - \mathbf{x}(k)) - \mathbf{u}(k))(\phi(k)(\hat{\mathbf{x}}(k \mid k) - \mathbf{x}(k)) \\
&\quad -\mathbf{u}(k))^T) \\
&= \phi(k)E((\hat{\mathbf{x}}(k \mid k) - \mathbf{x}(k))(\hat{\mathbf{x}}(k \mid k) - \mathbf{x}(k))^T)\phi(k)^T \\
&\quad - \phi(k)E((\hat{\mathbf{x}}(k \mid k) - \mathbf{x}(k))\mathbf{u}(k)^T) \\
&\quad - E(\mathbf{u}(k)((\hat{\mathbf{x}}(k \mid k) - \mathbf{x}(k))^T)\phi(k)^T + E(\mathbf{u}(k)\mathbf{u}(k)^T) \\
P(k+1 \mid k) &= \phi(k)P(k \mid k)\phi(k)^T + Q(k).
\end{aligned}$$

The two middle terms are zero by (7.2-3) and the fact that $\hat{\mathbf{x}}(k \mid k)$ is a linear combination of the elements in M_j^n (see Definition 7.2.1(a)). We have thus established the equation in part (d).

Next note that by part (a) of Theorem 6.2.2 we have

$$\hat{\mathbf{x}}(k+1 \mid k+1) = \hat{\mathbf{x}}(k+1 \mid k) + K(k+1)[\mathbf{z}(k+1) - H(k+1)\hat{\mathbf{x}}(k+1 \mid k)],$$

where $K(k+1)$ is as defined in (b) of the present theorem. The result in part (a) is thus obtained by substituting (7.2-4) into the above expression. Finally, from part (b) of 6.2.2, we have

$$P(k+1 \mid k+1) = P(k+1 \mid k) - K(k+1)H(k+1)P(k+1 \mid k).$$

Part (c) is obtained by replacing k by $k-1$ in this expression. □

Equations (c) and (d) taken together constitute a recursively solvable matrix difference equation known as the *discrete Riccati equation.* If the process and measurement noise covariances are known in advance, the set of Kalman gains can be calculated before any measurements are taken (see the discussion following Theorem 6.2.2). Even when this is not possible, real time estimates of K are possible since its calculation only requires knowledge of Q at the previous time step and knowledge of R at the current time step.

The quantity $\mathbf{z}(k+1) - H(k+1)\phi(k)\hat{\mathbf{x}}(k \mid k)$ is called a *residual.* It is the difference between the BLMVE of $\mathbf{z}(k+1)$ based on $\mathbf{z}(0), \ldots, \mathbf{z}(k)$ (see Equation (6.2-5)) and the actual measurement state $\mathbf{z}(k+1)$.

The dynamic model

$$\mathbf{x}(k+1) = \phi(k)\mathbf{x}(k) + \mathbf{u}(k)$$

is called the *filter reference model.* This model may or may not provide an accurate description of the actual physical process being observed. When it is believed that this model is correct, the filter is called a *matched filter.* This is most often an idealization. The quantification of the degradation in accuracy due to implementing a mismatched filter (often referred to as a suboptimal filter) is known as sensitivity analysis. The reader can find a discussion of this in Reference [10].

A linear predicted estimate is easily obtained from the filtered estimate (as is the corresponding error covariance), and this is Exercise 7.4.1. The smoothed estimate is quite another matter, and we address that in the last chapter.

Using the results in Exercise 6.3.2, we have the following corollary.

7.2.3. **Corollary.** *If* $R(K)^{-1}$ *and* $[H(k)P(k \mid k-1)H(k)^T + R(k)]^{-1}$ *both exist, then*

(a) $P(k \mid k)^{-1} = P(k \mid k-1)^{-1} + H(k)^T R(k)^{-1} H(k)$, *and*

(b) $K(k) = P(k \mid k)H(k)^T R(k)^{-1}$.

Part (b) of this corollary shows the sensitivity of $K(k)$ to the measurement noise. If the noise is large (as measured by the eigenvalues, for example), then the gain is small and the resulting estimate is influenced less by the residual than if the noise were small. This will also be discussed in Exercise 7.4.3.

One final result is the following extension of Theorem 6.2.3.

7.2.4. **Theorem.** *Let* $\hat{x}(k\,|\,j)$ *be the* BAMVE *of* $\mathbf{x}(k)$ *based on* $\mathbf{z}(0), \ldots, \mathbf{z}(u)$. *Then the exact same results* (a)–(d) *of Theorem* 7.2.2 *hold for affine estimates.*

Proof. This is Exercise 7.4.2. □

7.3 Initialization, Fisher Estimation

We now come to the issue that many regard as the most disturbing feature of Bayesian estimation, namely, how and where do we obtain priors? In the case of the Kalman filter, this is equivalent to the problem of obtaining $\hat{\mathbf{x}}(0\,|\,0)$ and $P(0\,|\,0)$.

In some cases, this initialization problem is quite easy. For example, suppose $\mathbf{x}(0)$ is completely known, perhaps even a constant vector. This situation would occur, for example, if a system was deterministically initialized prior to being put into operation in a noisy environment. Here, we would simply take

$$\hat{\mathbf{x}}(0\,|\,0) = \mathbf{x}(0)$$

and

$$P(0\,|\,0) = 0.$$

In this situation, there would be no reason to obtain $\mathbf{z}(0)$ since $\mathbf{x}(0)$ would be completely known. Thus, the first step in the filtering procedure would be

$$\hat{\mathbf{x}}(1\,|\,0) = \phi(0)\mathbf{x}(0\,|\,0)$$

and

$$P(1\,|\,0) = Q(0).$$

Another situation we might envisage is that wherein the mean and covariance matrices of $\mathbf{x}(0)$ are known. Here, we simply take Y in Theorem 5.3.5 to be $\mathbf{z}(0)$ and obtain $\hat{\mathbf{x}}(0\,|\,0)$ and $P(0\,|\,0)$ from (5.3-1) and (5.3-4), respectively. Unfortunately, in real situations, where $\mathbf{x}(0)$ is unknown, unless one has a multitude of past experience on which to make a reasonable guess, the mean and covariance is generally unavailable. What do we do in such a case?

We seem to be faced with the problem of estimating a random vector $\mathbf{x}$ from a single measurement $\mathbf{y}$ of the form

$$\mathbf{y} = H\mathbf{x} + \mathbf{w}, \tag{7.3-1}$$

where H and $R = E(\mathbf{w}\mathbf{w}^T)$ are known matrices and $\mathbf{x}$ is uncorrelated with $\mathbf{w}$, that is,

$$E(\mathbf{x}\mathbf{w}^T) = 0. \qquad (7.3\text{-}2)$$

Ideally, we should like to obtain an estimate $\mathbf{x}$ that is the BLMVE of $\mathbf{x}$ based on $\mathbf{y}$.[2] Let us use some of the properties that such estimates have and try to obtain $\hat{\mathbf{x}}$ in the present circumstances.

We know that

$$\hat{\mathbf{x}} = K\mathbf{y} \qquad (7.3\text{-}3)$$

for some suitable K. In the case of BLMVE's, we know that

$$E((\hat{\mathbf{x}} - \mathbf{x})\mathbf{y}^T = 0 \qquad (7.3\text{-}4(a))$$

and

$$E((\hat{\mathbf{x}} - \mathbf{x})\hat{\mathbf{x}}^T) = 0. \qquad (7.3\text{-}4(b))$$

Now

$$\begin{aligned} E(\hat{\mathbf{x}}\mathbf{w}^T) &= E(K\mathbf{y}\mathbf{w}^T) \quad \text{(use 7.3-3)} \\ &= KE(\mathbf{y}\mathbf{w}^T) \\ &= KE((H\mathbf{x} + \mathbf{w})\mathbf{w}^T) \quad \text{(use 7.3-1))} \\ &= KE(\mathbf{w}\mathbf{w}^T) \quad \text{(by 7.3-2)} \end{aligned}$$

$$E(\hat{\mathbf{x}}\mathbf{w}^T) = KR. \qquad (7.3\text{-}5)$$

Since $E(\mathbf{x}\mathbf{w}^T) = 0$ (Equation (7.3-2)), we can write (7.3-5) as

$$\begin{aligned} KR &= E(\hat{\mathbf{x}}\mathbf{w}^T) - E(\mathbf{x}\mathbf{w}^T) \\ &= E((\hat{\mathbf{x}} - \mathbf{x})\mathbf{w}^T) \\ &= E((\hat{\mathbf{x}} - \mathbf{x})(\mathbf{y}^T - \mathbf{x}^T H^T)) \quad \text{(use 7.3-1)} \\ KR &= -E((\hat{\mathbf{x}} - \mathbf{x})\mathbf{x}^T)H^T \quad \text{(by 7.3-4(a))}. \end{aligned}$$

Using the fact that $E((\hat{\mathbf{x}} - \mathbf{x})\hat{\mathbf{x}}^T) = 0$ (Equation (7.3-4(b)), this last expression can be written as

$$KR = E((\hat{\mathbf{x}} - \mathbf{x})\hat{\mathbf{x}})H^T - E((\hat{\mathbf{x}} - \mathbf{x})\mathbf{x}^T)H^T$$

or

$$KR = E((\hat{\mathbf{x}} - \mathbf{x})(\hat{\mathbf{x}} - \mathbf{x})^T)H^T.$$

It follows from this, assuming R^{-1} exists, that

$$K = PH^T R^{-1}.^3 \qquad (7.3\text{-}6)$$

[2] Without knowledge of means, the BAMVE is out of the question.

[3] Note that we have proved part (b) of Corollary 7.2.3 under the single assumption that $R(k)^{-1}$ exists.

Thus, one can specify K by specifying P, however, not just any P. Let us attempt to calculate P. From Equations (7.3-3) and (7.3-1), we have

$$\hat{\mathbf{x}} = K(H\mathbf{x} + \mathbf{w}),$$

so

$$\hat{\mathbf{x}} - \mathbf{x} = KM\mathbf{x} - \mathbf{x} + K\mathbf{v}. \tag{7.3-7}$$

Using Equations (7.3-4(a)) and (7.3-7), we have

$$\begin{aligned} P &= E((\hat{\mathbf{x}} - \mathbf{x})(\mathbf{x} - \mathbf{x})^T) \\ &= E(\hat{\mathbf{x}} - \mathbf{x})(-\mathbf{x})^T \\ &= E((I - KM)\mathbf{x}\mathbf{x}^T - K\mathbf{v}\mathbf{x}^T) \end{aligned}$$

$$P = (I - KH)E(\mathbf{x}\mathbf{x}^T), \tag{7.3-8}$$

the last equality following from (7.3-2). Thus, using Equation (7.3-6), Equation (7.3-8) becomes

$$P = [I - PH^T R^{-1} H]E(\mathbf{x}\mathbf{x}^T).$$

Assuming the indicated inverses, this equation can be solved for P to obtain

$$P = E[(\mathbf{x}\mathbf{x}^T)^{-1} + H^T R^{-1} H]^{-1}. \tag{7.3-9}$$

At this point it is clear that we are stuck, because the determination of P explicitly requires knowledge of the correlation matrix of $\mathbf{x}$, and $\mathbf{x}$ is completely unknown. However, rather than throw up our hands and quit, we settle for a less ambitious estimate, known as the Fisher estimate. There are four general procedures for obtaining this estimate, each exhibiting plausibility in its own particular way. They are, the classical Fisher approach, the improper prior approach, the variational approach, and the maximum likelihood approach. We will do the first three of these, the fourth being done in the exercises. The reader will note the lack of theorems and definitions throughout this section. This is simply because we are attempting to make a plausible estimate, of an estimate, without setting forth criteria before we begin. The criteria we do impose are, therefore, ad hoc in nature. The plan here is to chip away at the problem until we can obtain a solution that is both meaningful and precise in its criteria.

7.3.1 Classical Fisher Estimation

We chose to discuss this estimate first for a number of reasons. First, it is historically appropriate. Second, proponents of classical Fisher estimation have rather vociferous philosophical objections to the estimation problem as we have formulated it and therefore to the three solutions that follow this discussion. We should like to make these objections clear. Finally, we should like to be in a position of answering the aforementioned objections before we present the next three solutions.

In the discussion immediately following (7.3-9), we pointed out that the problem posed in this section is essentially ill posed. This is not at all surprising. If we look at Equation (7.3-1), we see that knowledge of the statistical behavior of $\mathbf{y}$ could be caused by the random behavior of either $\mathbf{x}$ or $\mathbf{w}$. Since $\mathbf{x}$ is completely unknown, this leaves us with the problem of trying to infer statistical information about $\mathbf{x}$ from statistical information about $\mathbf{y}$, without knowing the exact statistical relation of $\mathbf{x}$ to $\mathbf{y}$. The problem is ill posed indeed!

The proponents of classical Fisher estimation avoid addressing the above ill-posed problem by simply claiming that to do so is meaningless. Rather, they change the problem and suppose that $\mathbf{x}$ is a constant vector that is completely unknown, a so-called vector of unknown parameters. This assumption, of course, means that the entire statistical behavior of $\mathbf{y}$ is attributable to $\mathbf{w}$, a much different situation than that described above. In the analysis that follows we will suppose that

$$E(\mathbf{w}) = 0, \tag{7.3-10}$$

an assumption about noise that we are always willing to make if needed.

In the usual fashion, we wish to find an estimator $\hat{\mathbf{x}}$ for $\mathbf{x}$ of the form

$$\hat{\mathbf{x}} = K\mathbf{y}.$$

From Equation (7.3-1), it then follows that

$$\hat{\mathbf{x}} = KH\mathbf{x} + K\mathbf{w}. \tag{7.3-11}$$

Taking the expected value of both sides of (7.3-11) while imposing (7.3-10) and the assumption that $\mathbf{x}$ is constant, we obtain

$$E(\hat{\mathbf{x}}) = KH\mathbf{x}.$$

The next assumption made in the classical Fisher estimation scheme is that the estimator $\hat{\mathbf{x}}$ be unbiased. As we will see later when we develop the variational approach, this assumption can be avoided. Anyway, under such an assumption, the above equation becomes

$$\mathbf{x} = KH\mathbf{x}. \tag{7.3-12}$$

Next comes a rather troublesome assumption. Since $\mathbf{x}$ is completely unknown, it is argued that (7.3-12) must hold for all $\mathbf{x}$, and thus,

$$KH = I. \tag{7.3-13}$$

We will have more to say about (7.3-13) below. Continuing, we substitute (7.3-12) into (7.3-11) and obtain

$$\hat{\mathbf{x}} = \mathbf{x} + K\mathbf{w}$$

or

$$\hat{\mathbf{x}} - \mathbf{x} = K\mathbf{w}.$$

It follows that

$$\begin{aligned} \|\hat{\mathbf{x}} - \mathbf{x}\|^2 &= \langle K\mathbf{w}, K\mathbf{w} \rangle \\ &= \operatorname{tr}(E(K\mathbf{w}\mathbf{w}^T K^T)) \end{aligned}$$

$$\|\hat{\mathbf{x}} - \mathbf{x}\|^2 = \operatorname{tr}(KRK^T). \tag{7.3-14}$$

The classical Fisher estimation problem is, therefore, to minimize $\operatorname{tr}(KRK^T)$ subject to the constraint $KH = I$. The solution, assuming that the indicated inverses exist, is given by

$$K = PH^T R^{-1}, \tag{7.3-15}$$

where

$$P \triangleq E((\hat{\mathbf{x}} - \mathbf{x})(\hat{\mathbf{x}} - \mathbf{x})^T)$$

is given by

$$P = (H^T R^{-1} H)^{-1}. \tag{7.3-16}$$

We are not going to derive (7.3-15) and (7.3-16) here because a more general derivation, which includes these results, will be given when we discuss the nonclassical variational approach. Also, see References [18] and [24].

Before we leave the above analysis, we would like to make a few comments. First, as mentioned above, the unbiasedness assumption in (7.3-12) is not at all necessary. This is fortunate, for it is this assumption that led us to Equation (7.3-13), the requirement that H have a left inverse. However, there is no reason at all to suppose that H has a left inverse. For example, if H were a simple state selection matrix, a distinct possibility, then H would have no left inverse. Second, there is no a priori reason to suppose, as indicated in (7.3-16), that $H^T R^{-1} H$ is invertible. In fact, we will see that this can be weakened. Neither of these criticisms, however, is indiginous to the classical Fisher assumption that $\mathbf{x}$ is a constant vector, and it is this assumption that is our real concern. Specifically, in the context of the problem we face, namely, that of initializing a Kalman filter, we feel the assumption is unwarranted. For, if we attempt to initialize a Kalman filter by estimating the state vector for an on going dynamical system, then we are in fact trying to estimate a random vector and not an unknown constant vector.

7.3.2 The Improper Prior Approach

Here, we will show that the results obtained in 7.3.1, namely, (7.3-15) and (7.3-16), can be obtained by a rather compelling procedure that does not involve (7.3-13) nor the assumption that $\mathbf{x}$ is constant. It does involve a limiting procedure that can be carried out in a mathematically rigorous fashion, but whose philosophical underpinnings can be questioned. Since we are not going to "hang our hat" on this approach, we will present these objections but not feel obliged to respond to them.

Let us suppose that assumptions (7.3-1) and (7.3-2) hold. In addition, let us suppose, for the moment, that we also have a prior estimate $\hat{\mathbf{x}}_{-1}$ of $\mathbf{x}$ and a corresponding error covariance matrix P_{-1}. Then, from Exercise 6.3.2, assuming the necessary inverse exists, we have

$$P^{-1} = P_{-1}^{-1} + H^T R^{-1} H \tag{7.3-17}$$

and

$$\hat{\mathbf{x}} = P H^T R^{-1} \mathbf{y} + P P_{-1}^{-1} \hat{\mathbf{x}}_{-1}. \tag{7.3-18}$$

The "improper prior" approach takes the attitude that if we have no information about $\mathbf{x}$ at all, that is, if $\hat{\mathbf{x}}_1$ is completely unknown, then the random vector $\hat{\mathbf{x}}_1 - \mathbf{x}$ must be uniformly distributed and so have an "infinite covariance." Accordingly, we set $P_{-1}^{-1} = 0$ and obtain

$$P^{-1} = H^T R^{-1} H \tag{7.3-19}$$

from (7.3-17) and

$$\hat{\mathbf{x}} = P H^T R^{-1} \mathbf{y} \tag{7.3-20}$$

from (7.3-18). Equations (7.3-19) and (7.3-20) imply (7.3-15) and (7.3-16).

In spite of the strange looking equation $P_{-1}^{-1} = 0$, the above argument does at least suggest that Fisher estimates are limiting cases of Bayesian estimates when the prior information becomes sparse. One can make such a limit mathematically rigorous, but this really begs the philosophical issues.

Referring to problem 6.3.3, one can also carry out the above analysis by setting the initial information $I_1 = 0$ and setting the Fisher information matrix $F_1 = 0$. Again, this does not really address the philosophical issues at hand.

The reason the above is called the "improper prior" approach is as follows. On R^n we cannot have a uniformly distributed random vector, so the above "trick" interchanges the process of choosing the prior and then making the estimate, with the procedure of first making the estimate and then choosing the (improper) prior. It is this interchange of operations that is vulnerable to criticism. Nevertheless, this approch does suggest that some sort of Fisher-type estimate is a reasonable way to initialize a Kalman filter.

There is another method, the maximum likelihood method, that provides yet another plausible rationale for making Fisher estimates. This method is developed in Exercise 7.5.8.

7.3.3 The Variational Approach, Nonsingular R

If we reconsider the measurement equation (7.3-1), we see that if $\mathbf{x} \in \ker(H)$, then the measurement $\mathbf{y}$ contains no information about $\mathbf{x}$. Similarly, if $\mathbf{x}$ has a component $\mathbf{x}_1 \in \ker(H)$, this component will be indiscernible via any measurement of the form in (7.3-1). We thus invoke an informal sort of maximum entropy principle and require that any estimate $\hat{\mathbf{x}}$, of $\mathbf{x}$ should contain no more information than is inherent in the data $\mathbf{y}$, meaning that $\mathbf{x}$ should contain *no* component in $\ker(H)$. Thus, we make the requirement that

$$\hat{\mathbf{x}} \in \ker(H)^{\perp}. \tag{7.3-21}$$

From results in Chapter 4 (see Theorem 4.1.5), it follows that (7.3-21) is equivalent to

$$\hat{\mathbf{x}} \in \mathcal{R}(H^T). \tag{7.3-22}$$

If one had no measurement noise at all, then as in Chapter 4, one would estimate $\mathbf{x}$ subject to the constraint (7.3-22) by simply projecting $\mathbf{x}$ onto the range of H^T (using K as the pseudoinverse, for example). Our strategy here will be to require K to be of such form that even in the presence of noise, KH is the projection onto the range of H^T, that is,

$$KH = (H^T)''. \tag{7.3-23}$$

Clearly, (7.3-23) is a constraint that is always possible (unlike the Fisher requirement that $KH = I$). Moreover, consistent with our motivation, when we study the case of singular noise in the next section, we will be obliged to show that our choice of K reduces to the pseudoinverse H^+ when the noise is zero. For the present, however, we will study the case where the noise covariance R is nonsingular. Our reasons for starting with this case will be apparent when we reach the next section.

The problem we pose, therefore, is given by the following definition.

7.3.4. **Definition.** By a *Fisher estimate* of random vector $\mathbf{x}$ based on $\mathbf{y}$, we mean an estimate of the form

$$\hat{\mathbf{x}} = K\mathbf{y},$$

where the following conditions hold:

$$\mathbf{y} = H\mathbf{x} + \mathbf{w}, \tag{7.3-24}$$

$$H \text{ is known,}$$

$$R \triangleq E(\mathbf{w}\mathbf{w}^T) \text{ is known,} \tag{7.3-25}$$

$$KH = (H^T)'', \tag{7.3-26}$$

$$E(\mathbf{x}\mathbf{w}^T) = 0, \text{ and} \tag{7.3-27}$$

$$K \text{ is chosen so that } \|\hat{\mathbf{x}} - \mathbf{x}\| \text{ is minimized}$$

$$\text{subject to the above constraints.}$$

As indicated above, in this section we will solve this problem for the special case when R^{-1} exists.

From Equations (7.3-24) and (7.3-26), we have

$$\begin{aligned}\hat{\mathbf{x}} &= K\mathbf{y} \\ &= K(H\mathbf{x}+\mathbf{w}) \\ &= KH\mathbf{x}+K\mathbf{w}\end{aligned}$$

or

$$\hat{\mathbf{x}} = (H^T)''\mathbf{x} + K\mathbf{w}. \tag{7.3-28}$$

Subtracting $\mathbf{x}$ from both sides of Equation (7.3-28), we obtain

$$\hat{\mathbf{x}} - \mathbf{x} = (H^T)'\mathbf{x} + K\mathbf{w}, \tag{7.3-29}$$

so that

$$\begin{aligned}||\hat{\mathbf{x}} - \mathbf{x}||^2 &= \operatorname{tr}(E((\hat{\mathbf{x}} - \mathbf{x})(\hat{\mathbf{x}} - \mathbf{x})^T)) \\ &= \operatorname{tr}((H^T)E(\mathbf{x}\mathbf{x}^T)(H^T)') + \operatorname{tr}(KRK^T),\end{aligned} \tag{7.3-30}$$

the cross terms vanishing by (7.3-27). Since the first term is constant with respect to K, $||\hat{\mathbf{x}} - \mathbf{x}||$ is minimized exactly when $\operatorname{tr}(KRK^T)$ is minimized. Thus, we introduce the Lagrangian

$$\mathcal{L}(K, \Lambda) = \operatorname{tr}(KRK^T) + \operatorname{tr}[((H^T)'' - KH)\Lambda^T],$$

where the Lagrange multiplier Λ is square. Taking the Gateaux (directional) derivative of $\mathcal{L}$ in the direction of the matrix B, we have

$$\begin{aligned}\mathcal{L}'(K, \Lambda)(B) &= \operatorname{tr}(BRK^T + KRB^T - BH\Lambda^T) \\ &= \operatorname{tr}(2KRB^T - \Lambda H^T B^T) \\ &= \operatorname{tr}((2KR - \Lambda H^T)B^T).\end{aligned}$$

Setting this equal to zero for all B^T, we obtain

$$2KR = \Lambda H^T. \tag{7.3-31}$$

Note that up to this point we have not used the fact that R is assumed nonsingular. Now, assuming R^{-1} exists, we multiply through (7.3-31) by $\frac{1}{2}R^{-1}H$ and obtain

$$KH = \frac{\Lambda}{2}H^T R^{-1} H,$$

which by (7.3-26) is

$$(H^T)'' = \frac{\Lambda}{2}H^T R^{-1} H. \tag{7.3-32}$$

Now, since R^{-1} is positive definite and of full rank, there exists a nonsingular matrix S such that $R^{-1} = SS^T$ (just use the spectral theorem to find an orthogonal matrix U such that $U^T R^{-1} U = D$, D diagonal, and then define $S \triangleq U\sqrt{D}$). It follows that

$$(H^T R^{-1} H)'' = (H^T S S^T H)'' = (H^T S)'' = (H^T)'', \qquad (7.3\text{-}33)$$

the second equality following from part (c) of Theorem 4.2.5 and the last equality following from part (e) of the same theorem and the fact that $S'' = I$. Substituting (7.3-33) into (7.3-32) and multiplying the resulting expression on the right by $(H^T R^{-1} H)^+$, we obtain

$$(H^T R^{-1} H)''(H^T R^{-1} H)^+ = \frac{\Lambda}{2}(H^T R^{-1} H)''$$

or

$$(H^T R^{-1} H)^+ = \frac{\Lambda}{2}(H^T)'',$$

the last expression following from (7.3-32). Multiplying this last expression on the right by $H^T R^{-1}$, we obtain

$$(H^T R^{-1} H)^+ H^T R^{-1} = \frac{\Lambda}{2} H^T R^{-1}. \qquad (7.3\text{-}34)$$

But from (7.3-31), we see that

$$K = \frac{\Lambda}{2} H^T R^{-1},$$

and so (7.3-34) implies that

$$K = (H^T R^{-1} H)^+ H^T R^{-1}. \qquad (7.3\text{-}35)$$

Note that this is the same result as the classical Fisher case except that H is not assumed to have a left inverse and $H^T R^{-1} H$ is not assumed invertible.

Next let us attempt to calculate the error covariance of such an estimate. Recalling our calculation $\|\hat{\mathbf{x}} - \mathbf{x}\|^2$ using (7.3-29), we have

$$P = E((\hat{\mathbf{x}} - \mathbf{x})(\hat{\mathbf{x}} - \mathbf{x})^T) = (H^T)' E(\mathbf{x}\mathbf{x}^T)(H^T)' + KRK^T. \qquad (7.3\text{-}36)$$

Even though (7.3-35) was obtained without any knowledge of $\mathbf{x}$, we are unable to calculate the error covariance of the estimate $\hat{\mathbf{x}}$ because of the presence of the term $(H^T)' E(\mathbf{x}\mathbf{x}^T)(H^T)'$ in (7.3-36). This is not really surprising since, as pointed out earlier, there is no way that $\mathbf{y}$ in (7.3-24) could give us any information about $(H^T)'\mathbf{x}$. In fact, the best we can possibly do is to define the *Fisher covariance* P_f by

$$P_f = E((\hat{\mathbf{x}} - (H^T)''\mathbf{x})(\hat{\mathbf{x}} - (H^T)''\mathbf{x})^T), \qquad (7.3\text{-}37)$$

and note that by (7.3-28)

$$P_f = KRK^T, \qquad (7.3\text{-}38)$$

which combined with (7.3-32) yields

$$P_f = (H^T R^{-1} H)^+ H^T R^{-1} R R^{-1} H (H^T R^{-1} H)^+$$

or

$$P_f = (H^T R^{-1} H)^+. \tag{7.3-39}$$

In summary, therefore, we define a Fisher estimate for $\mathbf{x}$ to be an estimate of the form

$$\hat{\mathbf{x}} = K\mathbf{y},$$

such that the six items in Definition 7.3.4 are satisfied. We further define the Fisher covariance by (7.3-37) and understand that this is the portion of the true error covariance that is attributable to the estimation error in $\mathcal{R}(H^T)$. As solved above, the Fisher estimate for nonsingular $\mathcal{R}$ is given by

$$\hat{\mathbf{x}}_f = (H^T R^{-1} H)^+ H^T R^{-1}, \tag{7.3-40}$$

and the Fisher covariance is given by

$$P_f = (H^T R^{-1} H)^+. \tag{7.3-41}$$

In terms of initializing Kalman filters, the improper prior approach strongly suggests that Fisher estimates constitute a reasonable choice of priors. Nevertheless, the Bayesian and Fisher estimates are fundamentally different, and this is illustrated in Exercise 7.5.6.

7.4 Fisher Estimation with Singular Measurement Noise

In Section 7.3, we defined what was meant by a Fisher estimate (Definition 7.3.4) and derived formulas for such an estimate when R^{-1} existed. We now remove this assumption and allow R to be singular. It will be convenient for us to first suppose that the noisy measurements can be "decoupled" from the noiseless measurements, that is, that the measurement equation (7.3-24) can be written in the form

$$\mathbf{y} = \begin{bmatrix} \mathbf{y}_1 \\ \mathbf{y}_2 \end{bmatrix} = \begin{bmatrix} H_1 \\ H_2 \end{bmatrix} \mathbf{x} + \begin{bmatrix} \mathbf{w}_1 \\ 0 \end{bmatrix}, \tag{7.4-1}$$

where

$$R_1 \triangleq E(\mathbf{w}_1 \mathbf{w}_1^T) \tag{7.4-2}$$

is invertible and

$$E(\mathbf{x}\mathbf{w}_1^T) = 0. \tag{7.4-3}$$

The argument following Definition 7.3.4 in the last section and leading up to Equation (7.3-28) was completely general and did not depend on the nonsingularity of R. Hence, we can apply (7.3-28) in the present situation. Writing

$$K \triangleq [k_1 \,|\, k_2] \tag{7.4-4}$$

and replacing Λ by 2Λ in (7.3-31), we have

$$[K_1 \,|\, K_2] \begin{bmatrix} R_1 & 0 \\ 0 & 0 \end{bmatrix} = \Lambda [H_1^T \,|\, H_2^T]$$

or

$$K_1 R_1 = \Lambda H_1^T \tag{7.4-5}$$

and

$$0 = \Lambda H_2^T. \tag{7.4-6}$$

The condition in (7.3-26) becomes

$$K_1 H_1 + K_2 H_2 = (H^T)''. \tag{7.4-7}$$

Equation (7.4-7) is a particularly difficult equation to utilize as it stands because it involves expressing $(H^T)''$ in terms of H_1 and H_2 (or their transposes). This is essentially the classical problem of expressing the lattice theoretic join of two projections in terms of themselves, an as yet unsolved problem as far as we know.[4] For our purposes, we will engage the trick of multiplying through on the right by H^T to obtain

$$K_1 H_1 [H_1^T \,|\, H_2^T] + K_2 H_2 [H_1^T \,|\, H_2^T] = [H_1^T \,|\, H_2^T].$$

This in turn implies the pair of equations

$$K_1 H_1 H_1^T + K_2 H_2 H_1^T = H_1^T$$

and

$$K_1 H_1 H_2^T + K_2 H_2 H_2^T = H_2^T. \tag{7.4-8}$$

Since R_1 is invertible, Equation (7.4-5) implies that

$$K_1 H_1 = \Lambda H_1^T R_1^{-1} H_1. \tag{7.4-9}$$

Recalling our results in Section 7.3, it seems prudent at this point to define

$$P_1 \triangleq (H_1^T R_1^{-1} H_1)^+ \tag{7.4-10}$$

so that (7.4-9) can be written

$$K_1 H_1 = \Lambda P_1^+. \tag{7.4-11}$$

[4]There are expressions, but they are not computationally useful.

At this point, P_1 is simply a mathematical abstraction with no physical or statistical interpretation; we made the definition in Equation (7.4-10) by analogy to previous results, nothing else. Now, from (7.4-11), we can rewrite the equations in (7.4-8) as

$$\Lambda P_1^+ H_1^T + K_2 H_2 H_1^T = H_1^T \tag{7.4-12}$$

$$\Lambda P_1^+ H_2^T + K_2 H_2 H_2^T = H_2^T. \tag{7.4-13}$$

Multiplying (7.4-12) through by $R_1^{-1} H_1$ on the right and recalling Equation (7.4-10), we can rewrite (7.4-12) as

$$\Lambda (P_1^+)^2 + K_2 H_2 P_1^+ = P_1^+.$$

If we successively multiply this expression on the right by P_1, we obtain

$$\Lambda P_1^+ + K_2 H_2 P_1'' = P_1'' \tag{7.4-14}$$

and

$$\Lambda P_1'' + K_2 H_2 P_1 = P_1. \tag{7.4-15}$$

Solving (7.4-14) for ΛP_1^+ and substituting this into (7.4-13), we obtain

$$[P_1'' - K_2 H_2 P_1''] H_2^T + K_2 H_2 H_2^T = H_2^T$$

or

$$K_2 H_2 [I - P_1''] H_2^T = H_2^T - P_1'' H_2^T$$

and finally

$$K_2 H_2 P_1' H_2^T = P_1' H_2^T. \tag{7.4-16}$$

This equation will be utilized later on.

Since P_1 is symmetric, $\mathcal{R}(P_1) = \mathcal{R}(P_1^+)$, so from Equation (7.4-10) and the fact that R_1^{-1} is nonsingular, we have

$$P_1'' = (H_1^T)'' \tag{7.4-17}$$

(see Equation (7.3-30)). It follows from this and Equation (7.4-15) that

$$\Lambda (H_1^T)'' = P_1 - K_2 H_2 P_1,$$

$$\Lambda H_1^T R_1^{-1} = P_1 H_1^T R_1^{-1} - K_2 H_2 P_1 H_1^T R_1^{-1}.$$

Referring to Equation (7.4-5), this implies that

$$K_1 = P_1 H_1^T R_1^{-1} - K_2 H_2 P_1 H_1^T R_1^{-1}. \tag{7.4-18}$$

Now,

$$\hat{\mathbf{x}} = K\mathbf{y} = [K_1 \,|\, K_2] \begin{bmatrix} \mathbf{y}_1 \\ \mathbf{y}_2 \end{bmatrix} = K_1 \mathbf{y}_1 + K_2 \mathbf{y}_2,$$

so from (7.4-18)

$$\hat{\mathbf{x}} = P_1 H_1^T R_1^{-1} \mathbf{y}_1 - K_2 H_2 P_1 H_1^T R_1^{-1} \mathbf{y}_1 + K_2 \mathbf{y}_2. \tag{7.4-19}$$

Defining

$$\hat{\mathbf{x}}_1 \triangleq P_1 H_1^T R_1^{-1} \mathbf{y}_1, \tag{7.4-20}$$

this last equation can be written as

$$\hat{\mathbf{x}} = \hat{\mathbf{x}}_1 + K_2[\mathbf{y}_2 - H_2 \hat{\mathbf{x}}_1]. \tag{7.4-21}$$

Equation (7.4-20) is exactly a Fisher estimate of $\mathbf{x}$ based on the noisy measurement $\mathbf{y}_1$ and has Fisher error covariance P_1 (this is why we studied the nonsingular noise case first). Equation (7.4-21) has the familiar form of a recursive estimate based on the prior $\hat{\mathbf{x}}_1$. Clearly, our problem now is to calculate the gain matrix K_2. One might conjecture that K_2 is simply a Kalman gain based on the prior covariance P_1 and the noiseless measurement $\mathbf{y}_2 = H_2\mathbf{x}$, that is, has the form $P_1 H_2 (H_2 P_1 H_2^T)^+$. However, such a judgment would be premature since the estimate $\hat{\mathbf{x}}$ we are after is a Fisher estimate and is not a Bayesian estimate. In fact, we will see that this conjecture is correct only if P_1 is invertible.

Since we have reached a hiatus in our analysis, this will be a good place to stop and prove a rather specialized theorem about pseudoinverses that we will need shortly.

7.4.1. **Theorem.** *Let A and B be range-closed operators such that $\mathcal{R}(A^*) \perp \mathcal{R}(B^*)$. Then*

$$(A + B)^+ B = (B^*)''.$$

Proof. The hypothesis is equivalent to the statement

$$\ker(A)^\perp \perp \ker(B)^\perp,$$

and this in turn implies

$$\ker(A)^\perp \subset \ker(B) \tag{7.4-22}$$

and

$$\ker(B)^\perp \subset \ker(A). \tag{7.4-23}$$

Let $\mathbf{x}$ be an arbitrary vector and write

$$\mathbf{x} = \mathbf{x}_1 + \mathbf{x}_2,$$

where $\mathbf{x}_1 \in (\ker B)^\perp$ and $\mathbf{x}_2 \in \ker B$. Then

$$(A + B)^+ B\mathbf{x} = (A + B)^+ B\mathbf{x}_1. \tag{7.4-24}$$

But by (7.4-23), $A\mathbf{x}_1 = \mathbf{0}$, so (7.4-24) can be written

$$(A + B)^+ B\mathbf{x} = (A + B)^+ (A + B)\mathbf{x}_1$$

or

$$(A+B)^{+}B\mathbf{x} = ((A+B)^{*})''\mathbf{x}_1. \tag{7.4-25}$$

But since $\mathcal{R}(A^*) \perp \mathcal{R}(B^*)$, it follows that $((A+B)^*)'' = (A^*)'' + (B^*)''$, and so (7.4-25) becomes

$$(A+B)^{+}B\mathbf{x} = (A^*)''\mathbf{x}_1 + (B^*)''\mathbf{x}_1. \tag{7.4-26}$$

But since $\mathbf{x}_1 \in \mathcal{R}(B^*) = (\ker B)^{\perp}$, it follows that $(B^*)''\mathbf{x}_1 = \mathbf{x}_1$ and from (7.4-23) that $(A^*)''\mathbf{x}_1 = \mathbf{0}$. Hence, (7.4-26) becomes

$$(A+B)^{+}B\mathbf{x} = \mathbf{x}_1$$

or equivalently

$$(A+B)^{+}B\mathbf{x} = (B^*)''\mathbf{x}.$$

This completes the proof. □

We now return to our estimation problem and the determination of K_2. We note that since both $\mathbf{y}_2$ and $H_2\mathbf{x}_1$ are in the range of H_2, it suffices to find K_2H_2'' (see Equation (7.4-21)). Multiplying (7.4-15) through on the right by H_2^T, we obtain

$$\Lambda P_1''H_2^T + K_2H_2P_1H_2^T = P_1H_2^T.$$

Adding this expression to (7.4-16), we obtain

$$\Lambda P_1''H_2^T + K_2H_2(P_1+P_1')H_2^T = (P_1+P_1')H_2^T. \tag{7.4-27}$$

Defining

$$S \triangleq P_1 + P_1', \tag{7.4-28}$$

we see that (7.4-27) can be written as

$$\Lambda P_1''H_2^T + K_2H_2SH_2^T = SH_2^T. \tag{7.4-29}$$

Since $P_1'' = I - P_1'$, it follows from (7.4-6) that (7.4-29) can be rewritten as

$$-\Lambda P_1'H_2^T + K_2H_2SH_2^T = SH_2^T$$

or

$$K_2H_2SH_2^T = SH_2^T + \Lambda P_1'H_2^T. \tag{7.4-30}$$

Since S is invertible ($S^{-1} = P_1^{+} + P_1'$), it follows that $(H_2SH_2^T)'' = H_2''$. Hence, multiplying (7.4-30) on the right by $(H_2SH_2^T)^{+}$, we obtain

$$K_2H_2'' = SH_2^T(H_2SH_2^T)^{+} + \Lambda P_1'H_2^T(H_2SH_2^T)^{+}. \tag{7.4-31}$$

Since P_1 is positive semidefinite, we can find a matrix U such that

$$P_1 = U^2 \tag{7.4-32}$$

and

$$P_1'' = U''.^5 \tag{7.4-33}$$

We then define

$$A = H_2 U \tag{7.4-34}$$

$$B = H_2 P_1' \tag{7.4-35}$$

and note that $\mathcal{R}(A^T) \perp \mathcal{R}(B^T)$. Also note that

$$\begin{aligned}(A+B)(A+B)^T &= AA^T + BA^T + AB^T + BB^T \\ &= AA^T + BB^T \\ &= H_2 U^2 H_2^T + H_2 (P_1')^2 H_2^T \\ &= H_2 P_1 H_2^T + H_2 P_1' H_2^T\end{aligned}$$

$$(A+B)(A+B)^T = H_2 S H_2^T, \tag{7.4-36}$$

the second equality following from the fact that (7.4-22) and (7.4-23) hold. Now, multiplying (7.4-31) on the right by $H_2 P_1' H_2^T$ and noting (7.4-16), we obtain

$$SH_2^T(H_2SH_2^T)^+ H_2 P_1' H_2^T + \Lambda P_1' H_2^T (H_2 S H_2^T)^+ H_2 P_1' H_2^T = P_1' H_2^T.$$

Since S is invertible, we can rewrite this as

$$SH_2^T(H_2SH_2^T)^+ H_2 P_1' H_2^T + \Lambda P_1' S^{-1} S H_2^T (H_2 S H_2^T)^+ H_2 P_1' H_2^T = P_1' H_2^T. \tag{7.4-37}$$

Now from (7.4-34), (7.4-35), and (7.4-36), we note that

$$\begin{aligned}SH_2^T(H_2SH_2^T)^+ &= (U^2 + P_1')H_2^T((A+B)(A+B)^T)^+ \\ &= (U + P_1')(UH_2^T + P_1' H_2^T)((A+B)(A+B)^T)^+ \\ &= (U + P_1')(A+B)^T((A+B)(A+B)^T)^+\end{aligned}$$

$$SH_2^T(H_2SH_2^T)^+ = (U + P_1')(A+B)^+, \tag{7.4-38}$$

the last equality following from Corollary 4.3.6, part (a). Thus, (7.4-37) can be rewritten as

$$(U + P_1')(A+B)^+ BB^T + \Lambda P_1' S^{-1}(U + P_1')(A+B)^+ BB^T = B^T$$

and by Theorem 7.4.1 as

$$(U + P_1')(B^T)'' B^T + \Lambda P_1' S^{-1}(U + P_1')(B^T)'' B^T = B^T$$

or equivalently

$$(U + P_1')B^T + \Lambda P_1' S^{-1}(U + P_1')B^T = B^T.$$

---

[5] Use the spectral theorem to find an orthogonal matrix V such that $VP_1V^T = D$, D diagonal. Then define $U \triangleq V^T\sqrt{D}V$.

Substituting for B^T using (7.4-35), this expression becomes

$$(U + P_1')P_1'H_2^T + \Lambda P_1'(P_1^+ + P_1')(U + P_1')P_1'H_2^T = P_1'H_2^T$$

or

$$P_1'H_2^T + \Lambda P_1'H_2^T = P_1'H_2^T. \tag{7.4-39}$$

But (7.4-39) implies that

$$\Lambda P_1'H_2^T = 0 \tag{7.4-40}$$

and so (7.4-31) becomes

$$K_2H_2'' = SH_2^T(H_2SH_2^T)^+. \tag{7.4-41}$$

Hence, we finally have the result

$$\hat{\mathbf{x}} = \hat{\mathbf{x}}_1 + SH_2^T(H_2SH_2^T)^+[\mathbf{y}_2 - H_2\hat{\mathbf{x}}_1], \tag{7.4-42}$$

where

$$\hat{\mathbf{x}}_1 = P_1H_1^TR_1^{-1}\mathbf{y}_1 \tag{7.4-43}$$

and

$$P_1 = (H_1^TR_1^{-1}H_1)^+. \tag{7.4-44}$$

Note that if P_1 is invertible, then $P_1' = 0$ and $S = P_1$. In this case, the gain matrix used to multiply the residual $\mathbf{y}_2 = H_2\hat{\mathbf{x}}_1$ in (7.4-42) is the Kalman gain for a noiseless measurement. However, if $H_1 = 0$, then $\hat{\mathbf{x}}_1 = 0$ and $S = I$, in which case

$$\hat{\mathbf{x}} = H_2^T(H_2H_2^T)^+\mathbf{y}_2$$

or

$$\hat{\mathbf{x}} = H_2^+\mathbf{y}_2,$$

which is consistent with our motivating requirements. Finally, the Fisher error covariance matrix can be calculated from Equations (7.4-18) and (7.4-40) and the easily established fact that

$$P_f = K_1RK_1^T. \tag{7.4-45}$$

From (7.4-40), (7.4-34), (7.4-35), and (7.4-38), we have

$$\begin{aligned} K_2H_2P_1 &= SH_2^T(H_2SH_2^T)^+H_2P_1 \\ &= (U + P_1')(A + B)^+AU \\ &= (U + P_1')(A^T)''U \\ &= U(A^T)''U \quad (\text{since } \mathcal{R}(A^T) \perp \mathcal{R}(P_1')) \\ &= UA^T(AA^T)^+AU \end{aligned}$$

or

$$K_2H_2P_1 = P_1H_2^T(H_2P_1H_2^T)^+H_2P_1. \tag{7.4-46}$$

From (7.4-18), we have

$$K_1 = P_1 H_1^T R_1^{-1} - P_1 H_2^T (H_2 P_1 H_2^T)^+ H_2 P_1 H_1^T R_1^{-1}.$$

Substituting this result into (7.4-45), we have

$$P_f = P_1 - P_1 H_2^T (H_2 P_1 H_2^T)^+ H_2 P_1. \tag{7.4-47}$$

We now proceed to the general case of singular measurement noise, that is, the case where (7.4-1) is not assumed. In this case, we have

$$\mathbf{y} = H\mathbf{x} + \mathbf{w}, \tag{7.4-48}$$

where

$$E(\mathbf{x}\mathbf{w}^T) = 0 \tag{7.4-49}$$

and

$$R \triangleq E(\mathbf{w}\mathbf{w}^T) \tag{7.4-50}$$

is singular. Since R is symmetric, we can apply the spectral theorem and obtain an orthogonal matrix V such that

$$\Gamma \triangleq \left[\begin{array}{c|c} \Gamma_1 & 0 \\ \hline 0 & 0 \end{array}\right] = VRV^T \tag{7.4-51}$$

and

$$\Gamma_1 = \begin{bmatrix} \gamma_1 & & & 0 \\ & \gamma_2 & & \\ & & \ddots & \\ 0 & & & \gamma_p \end{bmatrix} \tag{7.4-52}$$

is nonsingular. We note that

$$\Gamma'' = \left[\begin{array}{c|c} I_1 & 0 \\ \hline 0 & 0 \end{array}\right], \tag{7.4-53}$$

$$\Gamma' = \left[\begin{array}{c|c} 0 & 0 \\ \hline 0 & I_2 \end{array}\right], \tag{7.4-54}$$

and hence that

$$\Gamma^+ = \left[\begin{array}{c|c} \Gamma_1^{-1} & 0 \\ \hline 0 & 0 \end{array}\right]. \tag{7.4-55}$$

Setting

$$C = V^T \Gamma'' V, \tag{7.4-56}$$

we see at once that $C = C^T$ and $C^2 = C$, that is, C is a projection. Moreover, from (7.4-51), we have that

$$R = V^T \Gamma V, \tag{7.4-57}$$

and so by Theorem 4.2.5, part (e), that

$$\begin{aligned} R'' &= (V^T \Gamma V)'' \\ &= (V^T \Gamma V'')'' \\ &= (V^T \Gamma)'' \\ &= (V^T \Gamma'')'' \\ &= (V^T \Gamma'' V'')'' \\ &= (V^T \Gamma'' V)'' \\ &= C''. \end{aligned}$$

Hence,

$$R'' = C = V^T \Gamma'' V, \tag{7.4-58}$$

the first equality following from the fact that C was a projection to start with. Similarly,

$$R' = V^T \Gamma' V. \tag{7.4-59}$$

From (7.4-57) and Theorem 4.43, we have

$$R^+ = V^T \Gamma^+ V. \tag{7.4-60}$$

Now multiplying (7.4-48) on the left by V, we obtain

$$V\mathbf{y} = VH\mathbf{x} + V\mathbf{w}$$

or

$$\mathbf{z} = B\mathbf{x} + \mathbf{u}, \tag{7.4-61}$$

where

$$\mathbf{z} = V\mathbf{y}, \quad B = VH, \quad \text{and} \quad \mathbf{u} = V\mathbf{w}. \tag{7.4-62}$$

Next, note that

$$\begin{aligned} E(\mathbf{u}\mathbf{u}^T) &= VE(\mathbf{w}\mathbf{w}^T)V^T \\ &= VRV^T \end{aligned}$$

$$E(\mathbf{u}\mathbf{u}^T) = \Gamma. \tag{7.4-63}$$

Writing $\mathbf{u}$ as

$$\mathbf{u} = \begin{bmatrix} \mathbf{u}_1 \\ \mathbf{u}_2 \end{bmatrix}, \quad \mathbf{u}_1 \in \mathrm{R}^P,$$

it follows from (7.4-63) that $E(\mathbf{u}_2\mathbf{u}_2^T) = 0$, whence $\mathbf{u}_2 = \mathbf{0}$. Thus, (7.4-61) can be rewritten in the form

$$\begin{bmatrix} \mathbf{z}_1 \\ \mathbf{z}_2 \end{bmatrix} = \begin{bmatrix} B_1 \\ B_2 \end{bmatrix} \mathbf{x} + \begin{bmatrix} \mathbf{u}_1 \\ \mathbf{0} \end{bmatrix},$$

where

$$E(\mathbf{u}_1\mathbf{u}_1^T) = \Gamma_1.$$

Making a Fisher estimate of $\mathbf{x}$ based on $\mathbf{z}$ using our previous results, we have

$$\hat{\mathbf{x}} = \hat{\mathbf{x}}_1 + K_2[\mathbf{z}_2 - B_2\hat{\mathbf{x}}_1], \tag{7.4-64}$$

where

$$\hat{\mathbf{x}}_1 = P_1 B_1^T \Gamma_1^{-1}\mathbf{z}_1, \tag{7.4-65}$$

$$P_1 = [B_1^T\Gamma_1^{-1}B_1]^+, \tag{7.4-66}$$

$$K_2 = SB_2^T(B_2SB_2^T)^+, \tag{7.4-67}$$

and

$$S = P_1 + P_1'. \tag{7.4-68}$$

Now,

$$\begin{aligned} H^TR^+H &= H^TV^TVR^+V^TVH \\ &= B^T(VR^+V^T)B \\ &= B^T\Gamma^+B \\ &= [B_1^T \mid B_2^T] \left[\begin{array}{c|c} \Gamma_1^{-1} & 0 \\ \hline 0 & 0 \end{array}\right] \begin{bmatrix} B_1 \\ B_2 \end{bmatrix} \\ &= B_1^T\Gamma_1^{-1}B_1. \end{aligned}$$

We have thus shown that

$$P_1 = [H^TR^+H]^+. \tag{7.4-69}$$

From (7.4-65) and the observation that

$$\begin{bmatrix} \Gamma_1^{-1}\mathbf{z}_1 \\ \mathbf{0} \end{bmatrix} = \Gamma^+\mathbf{z},$$

we have

$$\begin{aligned} \hat{\mathbf{x}}_1 &= P_1B_1^T\Gamma_1^{-1}\mathbf{z}, \\ &= P_1B^T\Gamma^+\mathbf{z} \\ &= P_1B^TVV^T\Gamma^+VV^T\mathbf{z} \end{aligned}$$

$$\hat{\mathbf{x}}_1 = P_1H^TR^+\mathbf{y}. \tag{7.4-70}$$

It remains to express K_2 in terms of our original matrices. To this end we first note that

$$\begin{bmatrix} 0 \\ B_2 \end{bmatrix} = \Gamma' \begin{bmatrix} B_1 \\ B_2 \end{bmatrix} = \Gamma' B,$$

so that

$$V^T \begin{bmatrix} 0 \\ B_2 \end{bmatrix} = V^T \Gamma' B = V^T \Gamma' V V^+ B$$

or

$$V^T \begin{bmatrix} 0 \\ B_2 \end{bmatrix} = R' H. \tag{7.4-71}$$

Using the fact that

$$[VAV^T]^+ = VA^+V^T$$

for any matrix A, we have

$$\begin{aligned} H^T[R'HSH^TR']^+V^T &= B^TV[R'HSH^TR']^+V^T \\ &= B^T[VR'HSH^TR'V^T]^+ \\ &= B^T\left[VV^T \begin{bmatrix} 0 \\ B_2 \end{bmatrix} S[0 \mid B_2^T]VV^T\right]^+ \\ &= B^T \left[\begin{array}{c|c} 0 & 0 \\ \hline 0 & B_2SB_2^T \end{array}\right]^+ \end{aligned}$$

or

$$H^T[R'HSH^TR']^+V^T = B^T \left[\begin{array}{c|c} 0 & 0 \\ \hline 0 & (B_2SB_2^T)^+ \end{array}\right]. \tag{7.4-72}$$

Hence,

$$\begin{aligned} SH^T[R'HSH^TR']^+[\mathbf{y} - H\hat{\mathbf{x}}_1] &= SH^T[R'HSH^TR']^+V^T[V\mathbf{y} - VH\hat{\mathbf{x}}_1] \\ &= SB^T \left[\begin{array}{c|c} 0 & 0 \\ \hline 0 & (B_2SB_2^T)^+ \end{array}\right] [\mathbf{z} - B\hat{\mathbf{x}}_1] \\ &= S[B_1^T \mid B_2^T] \left[\begin{array}{c|c} 0 & 0 \\ \hline 0 & (B_2SB_2^T)^+ \end{array}\right] \left[\begin{bmatrix} z_1 \\ z_2 \end{bmatrix} \right. \\ &\quad \left. - \begin{bmatrix} B_1 & \hat{\mathbf{x}}_1 \\ B_2 & \hat{\mathbf{x}}_1 \end{bmatrix}\right] \\ &= S[B_1^T \mid B_2^T] \begin{bmatrix} 0 \\ (B_2SB_2^T)^+(\mathbf{z}_2 - B_2\hat{\mathbf{x}}_1) \end{bmatrix} \\ &= SB_2^T(B_2SB_2^T)^+(\mathbf{z}_2 - B_2\hat{\mathbf{x}}_1) \\ &= K_2(\mathbf{z}_2 - B_2\hat{\mathbf{x}}_1). \end{aligned}$$

Thus, from (7.4-64), we have

$$\hat{\mathbf{x}} = \hat{\mathbf{x}}_1 + SH^T[R'HSH^TR']^+[\mathbf{y} - H\hat{\mathbf{x}}_1]. \tag{7.4-73}$$

We leave it to the reader to verify that

$$P_f - P_1 - P_1H^T[R'HP_1H^TR']^+HP_1. \tag{7.4-74}$$

At long last we reach the main result in this section.

7.4.2. **Theorem.** *If $\hat{\mathbf{x}}$ represents the Fisher estimate as defined in* 7.3.4, *then $\hat{\mathbf{x}}$ is given by*

$$\hat{\mathbf{x}} = \hat{\mathbf{x}}_1 + K_2[\mathbf{y} - H\hat{\mathbf{x}}_1],$$

where

$$\begin{aligned} \hat{\mathbf{x}}_1 &= P_1H^TR^+\mathbf{y}, \\ P_1 &= [H^TR^+H]^+, \\ K_2 &= SH^T[R'HSH^TR']^+, \end{aligned}$$

and

$$S = P_1 + P_1',$$

Moreover, the Fisher error covariance is given by

$$P_f = P_1 - P_1H^T[R'HP_1H^TR']HP_1.$$

Note that $\hat{\mathbf{x}}_1$ is the Fisher estimate of $\mathbf{x}$ based on that portion of $\mathbf{y}$ that is in the range of R and P_1 is the covariance of this estimate. Since one can easily show that $K_2 = K_2R'$, the term $K_2[\mathbf{y} - H\hat{\mathbf{x}}_1]$ uses that portion of $\mathbf{y}$ in the range of R' to improve the estimate $\hat{\mathbf{x}}_1$. The second term in the expression for P_f represents the reduction in the covariance of P_1 because of this improvement. As pointed out earlier, P_f represents the covariance of the error $\hat{\mathbf{x}} - (H^T)''\mathbf{x}$. The true error covariance P is related to P_f via

$$P = P_f + (H^T)'E(\mathbf{x}\mathbf{x}^T)(H^T)',$$

and the second term is impossible to determine without further information. For discussion of problems similar to the one above, see Rao [21].

7.5 Exercises

1. Explain, with appropriate equations, how one could obtain a p-step predictor, that is, at each time step the algorithm predicts the state $\mathbf{x}$ p-steps into the future.

2. Prove that the algorithm used in the Kalman filter recursively provides affine minimum variance estimates (recall Problem 6.3.1).

3. Consider the scalar system

$$\begin{aligned} x(k+1) &= ax(k) + u(k) \\ z(k) &= hx(k) + w(k), \end{aligned}$$

where a and h are fixed real numbers and u and w represent the process and measurement noise (with the usual assumptions).

   (a) If $q = E(u(k))^2$ and $r = E(w(k)^2)$ for all k, derive a recursion relation for the Kalman gain in terms of a, h, q, and r, the last two appearing as a ratio $q/r \triangleq \alpha$. What conclusion can you draw about the effect of noise on the gain?

   (b) Suppose $\alpha = 1$ and $P(0 \mid) = 0$. Express $K(k)$ in terms of the numbers in the Fibonacci sequence.

   (c) If $\hat{\mathbf{x}}(0) = 10$, $z(1) = 10.1$, $z(2) = 9.8$, $z(3) = 10.3$, and $z(4) = 9.9$, find $\hat{\mathbf{x}}(k \mid k)$ for $k = 1, 2, 3, 4$, (assume $\alpha = 1$).

4. Derive a dynamic recursive estimator in terms of the Fisher information vector of Problem 6.3.3. You may assume that $\phi(k)^{-1}$ exists. Discuss the computational advantages of this form when the process noise is zero.

5. This exercise is designed to show how measurements of one random variable can improve another. Let

$$\mathbf{x} = \begin{bmatrix} x_1 \\ x_2 \end{bmatrix}$$

and suppose that we have an initial estimate $\begin{bmatrix} \hat{x}_1 \\ \hat{x}_2 \end{bmatrix}$ of $\mathbf{x}$ whose confidence is expressed by the covariance matrix

$$p_1 = \begin{bmatrix} p_{11} & p_{12} \\ p_{21} & p_{22} \end{bmatrix}.$$

Now suppose a measurement of x_2 is taken with noise w satisfying $r = E(w^2)$ and the customary assumptions. In other words,

$$z = x_2 + w.$$

Find the estimate of x_1 based on z and the new error covariance of this estimate. Compare this new error covariance with p_{11}. What happens if $p_{12} = 0$? What happens when $p_{12} = \sqrt{p_{11}}\sqrt{p_{22}}$ (analogous to having the correlation coefficient $\rho = \pm 1$)?

6. (a) Show that $E(\hat{\mathbf{x}}_f - \mathbf{x})\hat{\mathbf{x}}_f^T) = KRK^T$. How does this differ from a Bayesian estimate?

(b) Using Equation (7.3-9), show that if P is the error covariance of the BLMVE of $\mathbf{x}$ based on $\mathbf{y}$ (as calculated by a Bayesian update for example) and P_f is the Fisher estimate of $\mathbf{x}$ given $\mathbf{y}$, then

$$\operatorname{tr}(P) \leq \operatorname{tr}(P_f).$$

7. Consider the system of 7.5.3 when $a = h = 1$ and $u(k) = 0$ for all k. In this case, x is called a *random constant*. Show that the Kalman gain in this case is given by

$$K(k) = \frac{p_0/r}{1 + p_0/r\, k},$$

where p_0 is the error covariance of the initial estimate $\hat{\mathbf{x}}(0)$.

Though the term random constant simply means that the random variable $\mathbf{x}(k)$ does not change in time, argue, by letting $p_0 = 0$ or k become large, that one can think of a random constant as a constant that is simply initially unknown.

8. This exercise is designed to show how Fisher estimates can be deduced from the maximum likelihood principle. We do a discrete example first to aid in an intuitive understanding.

(a) Suppose we have a discrete sample space with two random variables z and x defined on it. If the distributions of z and x are (possibly) unknown, but $p(z \mid x)$ is known, we can envoke the following principle. If $\hat{z}$ is a measurement of z, that is, a fixed real number, the maximum likelihood principle estimates x as $\hat{x}$, where $\hat{x}$ is chosen so that $p(\hat{z} \mid \hat{x})$ is maximized, that is,

$$p(\hat{z} \mid x) \leq p(\hat{z} \mid \hat{x})$$

for all x.

To illustrate this idea, suppose we consider a certain lake containing an unknown number x of fish. We consider x as a random variable since it is completely unknown. We then dump into the lake a known quantity m of tagged fish. We wish to estimate x by taking samples of size n. If z represents the number of tagged fish that occur in such a sample, then the probability $p(z)$ is also unknown. However, $p(z \mid x)$ can be calculated. If any one fish is chosen, the probability that it will be tagged is m/x. Using the binomial distribution, write an expression for $p(z \mid x)$ and then find the estimate $\hat{x}$ for a given sample $\hat{z}$ of tagged fish using the above idea.

(b) The continuous analog of the above principle is simply to maximize the conditional density function $p(z \mid x)$ for a given $\hat{z}$. That is, for a given $\hat{z}$, we choose $\hat{x}$ so the conditional density function $p(\hat{z} \mid x)$ satisfies

$$p(\hat{z} \mid x) \leq p(\hat{z} \mid \hat{x}).$$

Apply this to the situation where

$$\mathbf{z} = H\mathbf{x} + \mathbf{v}$$

and the following hold.

(1) $R = E(\mathbf{v}\mathbf{v}^T)$ is invertible,
(2) $E(\mathbf{x}\mathbf{v}^T) = 0$,
(3) $E(\mathbf{v}) = 0$, and
(4) $\mathbf{x}$ is completely unknown.

If a measurement $\mathbf{z}$ is secured, we wish to estimate $\mathbf{x}$ using the maximum likelihood principle as follows.

(1) Use the maximum entropy principle to deduce a density for $\mathbf{v}$.
(2) Using step (1) and the measurement equation, deduce the conditional density function $p(\mathbf{z} \mid \mathbf{x})$.
(3) Argue that the maximum likelihood estimate $\mathbf{x}$ based on the measurement $\hat{\mathbf{z}}$ is found by minimizing the expression

$$(\hat{z} - H\mathbf{x})R^{-1}(\hat{z} - H\mathbf{x}).$$

(4) Then carry out the minimization in (3) and show that

$$\hat{\mathbf{x}} = (H^T R^{-1} H)^{-1} H^T R^{-1} \hat{\mathbf{z}}$$

and

$$P = (H^T R^{-1} H)^{-1} \quad \text{(error covariance)}.$$

Note that the maximum likelihood is not a theorem, it is another inference scheme based on belief. The underlying rationale for the belief is to choose $\hat{x}$ so as to maximize the probability that we saw what we saw.

9. Consider the measurement equation

$$\begin{bmatrix} y_1 \\ y_2 \end{bmatrix} = \begin{bmatrix} 0 & 1 & 1 \\ 1 & 0 & 0 \end{bmatrix} \begin{bmatrix} x_1 \\ x_2 \\ x_3 \end{bmatrix} + \begin{bmatrix} w_1 \\ w_2 \end{bmatrix},$$

where $E(w_1^2) = 1$, $E(w_2^2) = 1$, and $E(w_1 w_2) = -1$. Express the Fisher estimate $\hat{\mathbf{x}}_f$ of $\mathbf{x}$ in terms of y_1 and y_2.

Answer.

$$\hat{\mathbf{x}}_f = \frac{1}{7} \begin{bmatrix} -y_1 + 5y_2 \\ 4y_1 + y_2 \\ 4y_1 + y_2 \end{bmatrix}.$$

8

The Linear Quadratic Tracking Problem

In this chapter we will study a very important problem in the field of optimal control theory. In general, control theory is concerned with using the measurements in a dynamical system to "control" the state vector $\mathbf{x}$. The precise meaning of the word control will be made clear as we proceed. The word "quadratic" in the title of this chapter refers to a particular class of control problems that use a quadratic form to measure the performance of a system. The reason we choose this particular performance index is that in the stochastic case it leads to a tractable solution. For reasons that will be clear later, we begin with the deterministic problem.

8.1 Control of Deterministic Systems

We begin our study by considering a discrete system of the form

$$\mathbf{x}(k+1) = \phi(k)\mathbf{x}(k) + C(k)\mathbf{v}(k), \tag{8.1-1}$$

where $\mathbf{v}(k)$ is a deterministic input called the *control vector.* Ideally, the matrix $C(k)$ would be the identity, and we could directly effect the time evolution of each state variable over the next time step by simply adjusting $\mathbf{v}$ at will. In practice, this is frequently impossible. For example, when guiding an aircraft, one can directly affect changes in acceleration (thrust) but not in position or velocity. Certainly one can change position and velocity over some given interval of time, but not arbitrarily, and certainly not independently from one another. Rather, one must settle for changing these quantities by affecting the quantity acceleration alone. Thus, in (8.1-1), we feel obliged to include the matrix $C(k)$.

8.1.1. **Definition.**

$$\Phi(k,k) = I \quad \text{for all} \quad k,$$

$$\Phi(k,s) = \phi(k-1)\phi(k-2)\cdots\phi(s) \quad \text{for} \quad k > s.$$

Here, Φ is called the *state transition matrix.*

8.1.2. **Lemma.**

(a) $\Phi(k+1,s) = \phi(k)\Phi(k,s)$ *for* $k > s$.

(b) $\Phi(k, s-1) = \Phi(k, s)\phi(s-1)$ *for* $k > s > 1$.

8.1.3. **Theorem.** *For any* $n \geq 1$,

$$\mathbf{x}(n) = \Phi(n,0)\mathbf{x}(0) + \sum_{j=0}^{n-1} \Phi(n, j+1)C(j)\mathbf{u}(j).$$

Proof. We use induction on n. The theorem is clearly true for $n = 1$. Assume true for n and check the $n+1$st case:

$$\begin{aligned}
\mathbf{x}(n+1) &= \phi(n)\mathbf{x}(n) + C(n)\mathbf{v}(n) \\
&= \phi(n)\left[\Phi(n,0)\mathbf{x}(0) + \sum_{j=0}^{n-1} \Phi(n, j+1)C(j)\mathbf{v}(j)\right] + C(n)\mathbf{v}(n) \\
&= \Phi(n+1,0)\mathbf{x}(0) + \sum_{j=0}^{n-1} \Phi(n+1, j+1)C(j)\mathbf{v}(j) + C(n)\mathbf{v}(n) \\
&= \Phi(n+1,0)\mathbf{x}(0) + \sum_{j=0}^{n} \Phi(n+1, j+1)C(j)\mathbf{v}(j),
\end{aligned}$$

the last equality making use of $\Phi(n+1, n+1) = I$. □

It will be convenient for us to introduce some new notation.

8.1.4. **Definition.**

(a) $X(p) = \begin{bmatrix} \mathbf{x}(1) \\ \mathbf{x}(2) \\ \vdots \\ \mathbf{x}(p) \end{bmatrix}, \quad V(p) = \begin{bmatrix} \mathbf{v}(0) \\ \mathbf{v}(1) \\ \vdots \\ \mathbf{v}(p-1) \end{bmatrix};$

(b) $\Gamma(p) = \begin{bmatrix} C(0) & 0 & & 0 \\ 0 & C(0) & & \\ & & \ddots & \\ 0 & & & C(p-1) \end{bmatrix};$

(c) $F(p) = \begin{bmatrix} I & 0 & 0 & 0 & 0 \\ \Phi(2,1) & I & 0 & 0 & 0 \\ \Phi(3,1) & \Phi(3,2) & I & 0 & 0 \\ \vdots & & & \vdots & \vdots \\ & & & I & 0 \\ \Phi(p,1) & \Phi(p,2) & \Phi(p,3)\cdots & \Phi(p,p-1) & I \end{bmatrix};$

(d) $E(p) = \begin{bmatrix} \Phi(1,0) \\ \Phi(2,0) \\ \vdots \\ \Phi(p,0) \end{bmatrix}, \quad G(p) = [\Phi(p,1), \Phi(p,2), \ldots, I].$

We will generally suppress the argument p when writing these matrices unless clarity is sacrificed.

8.1.5. **Theorem.**

$$X = E\mathbf{x}(0) + F\Gamma V.$$

Proof. This follows from 8.1.3 and direct calculation of the right-hand side. □

8.1.6. **Theorem.** *F is invertible and*

$$F^{-1} = \begin{bmatrix} I & 0 & 0 & 0 & & 0 & 0 \\ -\phi(1) & I & 0 & 0 & & 0 & 0 \\ 0 & -\phi(2) & I & 0 & & & \\ 0 & 0 & -\phi(3) & I & & & \\ \vdots & & & & & & \\ & & & & & I & 0 \\ 0 & 0 & 0 & 0 & \cdots & -\phi(p-1) & I \end{bmatrix}.$$

Proof. Lemma 8.1.2 and direct calculation. □

8.1.7. **Definition.**

$$S_p \triangleq G\Gamma = [\Phi(p,1)C(0) \,|\, \Phi(p,2)C(1) \,|\, \cdots \,|\, \Phi(p,p)C(p-1)].$$

8.1.8. **Corollary** (to 8.1.3).

$$\mathbf{x}(p) = \Phi(p,0)\mathbf{x}(0) + S_p V.$$

8.1.9. **Definition.** The system (8.1-1) is said to be *stationary* provided $\phi(k) = \phi$ and $C(k) = \mathrm{C}$ for all k, that is, they are constant matrices.

In the case of a stationary system, $\phi(k,s) = \phi^{k-s}$, and the various matrices in 8.1.4 take on the corresponding form. Also, for example, S_p has the form

$$S_p = [\phi^{p-1}C \,|\, \phi^{p-2}\mathrm{C} \,|\, \cdots \,|\, \phi C \,|\, \mathrm{C}]. \tag{8.1-2}$$

8.1.10. **Definition.** The dynamical system (8.1-1) is said to be *controllable from* 0 *in* p *steps* if and only if there exists a $p \in \mathrm{N}$ such that for all choices of $\mathbf{x}_0$ and $\mathbf{x}_f$, there is a V such that $\mathbf{x}(0) = \mathbf{x}_0$ and $\mathbf{x}(p) = \mathbf{x}_f$.

8.1.11. **Theorem.** *The system* (8.1-1) *is controllable if and only if* S_p *has rank* n *for some* p.

Proof. Suppose first that the system is controllable from 0 in p steps and that rank $(S_p) < n$. Then, there exists a vector $\boldsymbol{\alpha} \in \mathrm{R}^n$, $\boldsymbol{\alpha} \neq \mathbf{0}$, such that

$$\boldsymbol{\alpha}^T S_p = \mathbf{0},$$

this following because the rows of S_p are dependent. Let $\mathbf{x}_0 = \mathbf{0}$ and $\mathbf{x}_f = \boldsymbol{\alpha}$. Since the system is controllable, it follows from Corollary 8.1.8 that there exists a V such that

$$\mathbf{x}_f = \boldsymbol{\alpha} = S_p V.$$

Hence,

$$||\boldsymbol{\alpha}||^2 = \boldsymbol{\alpha}^T \boldsymbol{\alpha} = \boldsymbol{\alpha}^T S_p V = 0^T V = 0,$$

and so $\boldsymbol{\alpha} = \mathbf{0}$, a contradiction.

Conversely, suppose that S_p has rank n for some p. Then, since

$$\operatorname{rank}(S_p S_p^T) = \operatorname{rank}(S_p),$$

we have that $S_p S_p^T$ is an $n \times n$ matrix of rank n, hence invertible. Let

$$W \triangleq S_p S_p^T$$

and define

$$V = S_p^T W^{-1}[\mathbf{x}_f - \Phi(p,0)\mathbf{x}_0],$$

then, from Corollary 8.1.8,

$$\begin{aligned} \mathbf{x}(p) &= \Phi(p,0)\mathbf{x}_0 + S_p V \\ &= \Phi(p,0)\mathbf{x}_0 + [\mathbf{x}_f - \Phi(p,0)\mathbf{x}_0] \\ &= \mathbf{x}_f. \quad \square \end{aligned}$$

The number of steps p necessary to insure controllability is not clear in the general case. Also, controllability from 0 may require a different number of steps than controllability from some other point. These observations lead to the introduction of various definitions of controllability used to restrict attention to systems with various desirable features. In the stationary case, things are much simpler. First of all, from Equation (8.1-2), it is clear that $\operatorname{rank}(S_p) \leq \operatorname{rank}(S_{p+1})$ since the change from S_p to S_{p+1} adds columns to S_p thereby (possibly) raising the dimension of the column space. Also, by the Cayley–Hamilton theorem, $\operatorname{rank}(S_p) = \operatorname{rank}(S_{p+1})$ for $p \geq n$ since ϕ^n is a linear combination of $\phi^{n-1}, \phi^{n-2}, \ldots, \phi, I$. Finally, since $\operatorname{rank}(S_p) \leq n$ for all p, we have the result that we never need consider the case $p > n$ since if $\operatorname{rank}(S_n) < n$ then $\operatorname{rank}(S_p) < n$ for all p. However, if $\operatorname{rank}(S_p) = n$ for some $p < n$, then $\operatorname{rank}(S_n) = n$ also. Hence, we have the following.

8.1.12. **Theorem.** *If* (8.1-1) *is time invariant, it is controllable (from any point) if and only if* $\operatorname{rank}(S_n) = n$.

We are now ready to state the general linear quadratic tracking problem of which the linear quadratic regulator problem is a special case. The generality here is simply for the reader's information.

8.1.13. **The General Linear Quadratic Tracking Problem.** Suppose we have a discrete deterministic dynamical system with a control vector as in (8.1-1). To simplify the notation, we will now use subscripts for the time steps so that (8.1-1) becomes

$$\mathbf{x}_{k+1} = \phi_k \mathbf{x}_k + C_k \mathbf{v}_k. \tag{8.1-3}$$

We wish to find a control sequence $\mathbf{v}_0, \mathbf{v}_1, \ldots, \mathbf{v}_p$ such that

$$J_p = \sum_{k=0}^{p-1} (\mathbf{y}_{k+1} - \mathbf{r}_{k+1})^T T_{k+1} (\mathbf{y}_{k+1} - \mathbf{r}_{k+1}) + \mathbf{v}_k^T L_k \mathbf{v}_k, \tag{8.1-4}$$

where T_k and L_k are symmetric, and

$$\mathbf{y}_k = B_k \mathbf{x}_k, \quad k = 1, 2, \ldots, p \tag{8.1-5}$$

is minimized with respect to the constraint (8.1-3).

Before we begin, a few words of explanation about terminology is necessary. For instance, the vectors $\mathbf{r}_1, \mathbf{r}_2, \ldots, \mathbf{r}_p$ are specified before the problem is addressed and these vectors are called the *track*. Then matrix T_k is called the *tracking demand matrix* or simply *tracking matrix*. Clearly, for large T_k (measured by eigenvalues for example), the term $(\mathbf{y}_k - \mathbf{r}_k)^T T_k (\mathbf{y}_k - \mathbf{r}_k)$ will drive the value of J_p high unless $\mathbf{y}_k$ is near r_k. In other words, the larger T_k is, the more ambitious is our control scheme. The matrix L_k is called a *cost* or *liability* matrix and measures the cost (liability, effort, energy) of using the control $\mathbf{v}_k$. Clearly, T_k and L_k are assumed to be positive semidefinite. In addition, since we want to guard against the possibility of infinite costs, we also assume that L_k is strictly positive definite. Finally, since there are some states in $\mathbf{x}_k$ that we may not wish to control, we use B_k to select only those states that we do wish to control. In particular, then B_k will most likely be a matrix containing only 0's and 1's and will have full row rank.

Now, there are two separate problems that one can address in the format of (8.1-4). One is as we have stated it above, the other is to set $T_p = 0$ and impose the additional restriction that

$$\mathbf{y}_p = \mathbf{r}_p.$$

The former problem is called the free final state problem and the latter is called the fixed final state problem. Of the two, the latter is much more difficult to solve. Fortunately, for us, it is the free final state problem that will necessarily be of interest to us in the stochastic control problem and so that is the one we will address. In doing so, however, we will obtain an interesting special case of the fixed final state problem.

We are reading to begin. Rather than simply state and prove the theorem, we wish to obtain the results as we proceed with the solution. First of all, in the spirit of Definition 8.1.4, we introduce some new matrices.

8.1.14. **Definition.**

$$B = \begin{bmatrix} B_1 & 0 & & \\ 0 & B_2 & & 0 \\ & 0 & & \\ & & \ddots & 0 \\ 0 & & 0 & B_p \end{bmatrix}, \quad L = \begin{bmatrix} L_0 & & \\ & L_1 & 0 \\ & & \ddots \\ 0 & & \\ & & L_{p-1} \end{bmatrix},$$

$$T = \begin{bmatrix} T_1 & & & 0 \\ & T_2 & & \\ & & \ddots & \\ 0 & & & T_p \end{bmatrix}, \quad Y = \begin{bmatrix} y_1 \\ y_2 \\ \vdots \\ y_p \end{bmatrix},$$

$$R = \begin{bmatrix} \mathbf{r}_1 \\ \mathbf{r}_2 \\ \vdots \\ \mathbf{r}_p \end{bmatrix}.^1$$

8.1.15. **Problem Formulation.** Using the notation just introduced, we can write

$$J = (Y - R)^T T (Y - R) + V^T L V, \tag{8.1-6}$$

where

$$Y = BX \tag{8.1-7}$$

and

$$X = E\mathbf{x}(0) + F\Gamma V, \tag{8.1-8}$$

this latter relation being Theorem 8.1.5. Rather than use (8.1-7) as a constraint, we simply substitute it into (8.1-6) and obtain

$$J = (BX - R)^T T (BX - R) + V^T L V. \tag{8.1-9}$$

To minimize J subject to the constraint (8.1-8), we form the Lagrangian

$$\mathcal{L}(X, V, \Lambda) = (BX - R)^T T (BX - R) + V^T L V + \Lambda^T [X - E\mathbf{x}(0) - F\Gamma V],$$

where $\Lambda \in \mathrm{R}^n$. Taking the directional derivative in the direction (Z, W) we obtain

$$\begin{aligned} \mathcal{L}'(X, V, \Lambda)(Z, W) &= (BZ)^T T (BX - R) + (BX - R)^T T B Z \\ &\quad + W^T L V + V^T L W + \Lambda^T [Z - F\Gamma W]. \end{aligned}$$

Since this is a scalar quantity, we can take transposes as needed to obtain

$$\mathcal{L}'(X, V, \Lambda)(Z, W) = 2Z^T B^T T (BX - R) + Z^T \Lambda + 2W^T L V - W^T \Gamma^T F^T \Lambda.$$

Since this expression must be zero for all choices of Z and W, we have

$$2B^T T [BX - R] + \Lambda = 0 \tag{8.1-10}$$

and

$$2LV - \Gamma^T F^T \Lambda = 0. \tag{8.1-11}$$

---

[1] By the time this notation conflicts with the measurement noise covariance matrix, we will no longer need it.

Solving (8.1-11) for V, we obtain

$$V = L^{-1}\Gamma^T F^T \frac{\Lambda}{2}.$$

Defining $\Lambda' \triangleq -F^T\frac{\Lambda}{2}$, this expression becomes

$$V = -L^{-1}\Gamma^T\Lambda'. \tag{8.1-12}$$

Since F is invertible (Theorem 8.1.6), we have

$$\Lambda = -2(F^T)^{-1}\Lambda',$$

so (factoring out 2) (8.1-10) becomes

$$(F^T)^{-1}\Lambda' = B^T TBX - B^T TR. \tag{8.1-13}$$

Since Λ will no longer appear in subsequent calculations, we can drop the prime in (8.1-12) and (8.1-13) and simply write Λ.

It is clear from (8.1-12) that if we can calculate Λ, then the control sequence V is completely determined. We now show that this is always possible under the assumptions we have made. The following lemma will be of use to us.

8.1.16. **Lemma.**

(a) *If A is symmetric and positive semidefinite, then for any $\mathbf{x}$ with $A\mathbf{x} \neq \mathbf{0}$, it follows that $\langle A\mathbf{x}, \mathbf{x}\rangle > 0$.*

(b) *If A_1 and A_2 are positive semidefinite matrices of the same size, then $I + A_1A_2$ is always invertible.*

Proof. (a) From the spectral theorem for symmetric matrices, we can assume that A is a diagonal matrix with positive entries, that is, they represent A using an eigen basis $\{\mathbf{f}_1, \mathbf{f}_2, \mathbf{f}_3, \ldots \mathbf{f}_n\}$. Thus, if $\mathbf{x} = \sum_{i=1}^n x_i\mathbf{f}_i$, we have

$$\langle A\mathbf{x}, \mathbf{x}\rangle = \sum_{i=1}^n \lambda_i x_i^2 \geq 0,$$

and this expression is zero exactly when the x_i's corresponding to nonzero λ_i's are zero. But

$$A\mathbf{x} = (\lambda_1 x_1, \lambda_2 x_2, \ldots, \lambda_n x_n),$$

so this condition holds exactly when $A\mathbf{x} = \mathbf{0}$.

(b) If $I + A_1A_2$ is not invertible, then it has a nontrivial kernel, that is, there is a nonzero $\mathbf{x}$ such that

$$(I + A_1A_2)(\mathbf{x}) \neq \mathbf{0}.$$

This is equivalent to saying that

$$A_1 A_2 \mathbf{x} = -\mathbf{x}. \tag{8.1-14}$$

Let $\mathbf{y} \triangleq A_2\mathbf{x}$. If $\mathbf{y} = \mathbf{0}$, then clearly (8.1-14) fails, so we suppose $\mathbf{v} \neq \mathbf{0}$. Then, by part (a),

$$\langle \mathbf{y}, \mathbf{x} \rangle > 0.$$

Now, if $A_1\mathbf{y} = -\mathbf{x} \neq \mathbf{0}$, again by part (a), we have

$$\langle A_1\mathbf{y}, \mathbf{y} \rangle > 0$$

or

$$\langle -\mathbf{x}, \mathbf{y} \rangle > 0,$$

so

$$-\langle \mathbf{x}, \mathbf{y} \rangle > 0,$$

a contradiction. This completes the proof. □

Returning to our control problem, (8.1-8) and (8.1-13) imply

$$\Lambda = F^T B^T T B E \mathbf{x}(0) + F^T B^T T B F \Gamma V - F^T B^T T R,$$

and using (8.1-12),

$$\Lambda = F^T B^T T B E \mathbf{x}(0) - F^T B^T T B F \Gamma L^{-1} \Gamma^T \Lambda - F^T B^T T R$$

or

$$[I + F^T B^T T B F \Gamma L^{-1} \Gamma^T]\Lambda = F^T B^T T B E \mathbf{x}_0 - F^T B^T T R.$$

Letting $A_1 = (BF)^T T(BF)$ and $A_2 = \Gamma L^{-1} \Gamma^T$ in Lemma 8.1.16, we have

$$\Lambda = [I + A_1 A_2]^{-1}[A_1 F^{-1} E \mathbf{x}_0 - (BF)^T T R]. \tag{8.1-15}$$

Equation (8.1-15) makes it clear that by specifying $\mathbf{x}_0$ and the track R, we can calculate Λ and in turn calculate the control sequence V from (8.1-12). Thus, for the deterministic situation, we can determine the control sequence prior to any implementation. This is known as an open loop control. For deterministic systems this solution is always feasible in principle. However, in a noisy operating environment, a situation we will study momentarily, this is not a useful solution. What we really want is a solution that is sensitive to perturbations in $\mathbf{x}_k$, a closed loop solution. The reason for the terminology "closed loop" will be explained later.

We begin by writing Equation (8.1-13) in terms of components. Recalling Theorem 8.1.6,

$$\begin{aligned}\lambda_0 - \phi_1^T \lambda_1 &= B_1^T T_1 B_1 \mathbf{x}_1 - B_1^T T_1 \mathbf{r}_1 \\ \lambda_1 - \phi_2^T \lambda_2 &= B_2^T T_2 B_2 \mathbf{x}_2 - B_2^T T_2 \mathbf{r}_2 \\ &\vdots \\ \lambda_{p-2}\phi_{p-1}^T \lambda_{p-1} &= B_{p-1}^T T_{p-1} B_{p-1} \mathbf{x}_{p-1} - B_{p-1}^T T_{p-1} \mathbf{r}_{p-1} \\ \lambda_{p-1} &= B_p^T T_p B_p \mathbf{x}_p - B_p^T T_p \mathbf{r}_p,\end{aligned} \tag{8.1-16}$$

where

$$\Lambda = \begin{bmatrix} \lambda_0 \\ \lambda_1 \\ \vdots \\ \lambda_{p-1} \end{bmatrix}.$$

For $k < p$, we have

$$\lambda_{k-1} - \phi_k^T \lambda_k = B_k^T T_k B_k \mathbf{x}_k - B_k^T T_k \mathbf{r}_k. \tag{8.1-17}$$

Also note that (8.1-12) can be written

$$\mathbf{v}_k = L_k^{-1} C_k^T \lambda_k. \tag{8.1-18}$$

Now, using Equation (8.1-18), the last of the equations in (8.1-16) can be written in terms of $\mathbf{x}_{p-1}$ and λ_{p-1} by rewriting Equation (8.1-1) in the form

$$\mathbf{x}_{k+1} = \phi_k \mathbf{x}_k - C_k L_k^{-1} C_k^T \lambda_k \tag{8.1-19}$$

and letting $p = k + 1$. The result, assuming the appropriate inverses exist, can be solved for $\boldsymbol{\lambda}_{p-1}$ to obtain an equation of the form

$$\boldsymbol{\lambda}_{p-1} = \Omega_{p-1}\mathbf{x}_{p-1} - \boldsymbol{\nu}_{p-1}.$$

This expression in turn can be substituted into the second last of Equations (8.1-16) to obtain an expression of the form

$$\lambda_{p-2} = W_{p-1}\mathbf{x}_{p-1} - \boldsymbol{\theta}_{p-1} \tag{8.1-20}$$

for suitable W_{p-1} and θ_{p-1}. Note that the last of the equations in (8.1-16) already has the form in (8.1-20). Continuing along these lines, we conjecture that there is a sequence $W_1, W_2, \ldots, W_p$ of matrices and $\boldsymbol{\theta}_1, \ldots, \boldsymbol{\theta}_p$ of vectors such that

$$\boldsymbol{\lambda}_{k-1} = W_k \mathbf{x}_k - \boldsymbol{\theta}_k; \quad 1 \le k \le p. \tag{8.1-21}$$

The proof of this conjecture lies in our ability to produce recursive formulas for W_k and $\boldsymbol{\theta}_k$ in which the appropriate operations can be carried out under our assumptions. The "proof is in the pudding" so to speak.

Substituting (8.1-21) into (8.1-19), we obtain

$$\mathbf{x}_{k+1} = \phi_k \mathbf{x}_k - C_k L_k^{-1} C_k^T [W_{k+1}\mathbf{x}_{k+1} - \boldsymbol{\theta}_{k+1}]$$

or

$$\mathbf{x}_{k+1} + C_k L_k^{-1} C_k^T W_{k+1}\mathbf{x}_{k+1} = \phi_k \mathbf{x}_k + C_k L_k^{-1} C_k^T \boldsymbol{\theta}_{k+1}.$$

If W_{k+1} turns out to be positive semidefinite, this last equation can be expressed as

$$\mathbf{x}_{k+1} = [I + C_k L_k^{-1} C_k^T W_{k+1}]^{-1}[\phi_k \mathbf{x}_k + C_k L_k^{-1} C_k^T \boldsymbol{\theta}_{k+1}], \qquad (8.1\text{-}22)$$

where we have used Lemma 8.1.16. Now, substituting (8.1-21) into (8.1-17), replacing k by $k+1$, we obtain

$$W_k \mathbf{x}_k - \boldsymbol{\theta}_k - \phi_k^T [W_{k+1}\mathbf{x}_{k+1} - \boldsymbol{\theta}_{k+1}] = B_k^T T_k B_k \mathbf{x}_k - B_k^T T_k \mathbf{r}_k.$$

Substituting (8.1-22) into this last expression

$$W_k \mathbf{x}_k - \theta_k - \phi_k^T W_{k+1}[I + C_k L_k^{-1} C_k^T W_{k+1}]^{-1}[\phi_k \mathbf{x}_k + C_k I_k^{-1} C_k^T \boldsymbol{\theta}_{k+1}]$$
$$= -\phi_k^T \boldsymbol{\theta}_{k+1} + B_k^T T_k B_k \mathbf{x}_k - B_k^T T_k \mathbf{r}_k.$$

This can be rearranged to form

$$[W_k - \phi_k^T W_{k+1}[I + C_k L_k^{-1} C_k^T W_{k+1}]^{-1}\phi_k - B_k^T T_k B_k]\mathbf{x}_k$$
$$= \; [\theta_k + \phi_k^T W_{k+1}[I + C_k L_k^{-1} C_k^T W_{k+1}]^{-1} C_k L_k^{-1}\theta_{k+1} - \phi_k^T \theta_{k+1} - B_k^T T_k r_k].$$

The expression on the right is constant and that on the left depends upon all possible sequences $\mathbf{x}_k$ that are initialized by $\mathbf{x}_0$. The only solution for such a situation is that both sides are zero. Hence,

$$W_k = \phi_k^T W_{k+1}[I + C_k L_k^{-1} C_k^T W_{k+1}]^{-1}\phi_k + B_k^T T_k B_k, \qquad (8.1\text{-}23)$$

and

$$\boldsymbol{\theta}_k = [\phi_k^T - \phi_k^T W_{k+1}[I + C_k L_k^{-1} C_k^T W_{k+1}]^{-1} C_k L_k^{-1} C_k^T]\theta_{k+1} + B_k^T T_k \mathbf{r}_k. \qquad (8.1\text{-}24)$$

Note, that if W_{k+1} is symmetric and positive semidefinite, the necessary inverses exist. Applying the matrix inversion lemma[2] to (8.1-23), we obtain

$$W_k = \phi_k^T [W_{k+1}[I - C_k[C_k^T W_{k+1} C_k + L_k]^{-1} C_k^T W_{k+1}]]\phi_k + B_k^T T_k B_k$$

or

$$W_k = \phi_k^T [W_{k+1} - W_{k+1} C_k [C_k^T W_{k+1} C_k + L_k]^{-1} C_k^T W_{k+1}]\phi_k + B_k^T T_k B_k. \qquad (8.1\text{-}25)$$

[2]Group $C_k^T W_{k+1}$ as A_{21}, C_k as A_{12}, and L_k^{-1} as A_{22}.

Equation (8.1-25) is a recursive Riccati-type equation enabling one to compute W_k given W_{k+1} for $k < p$. The relation also makes it clear that if W_p is symmetric then each W_k will also be symmetric. If W_p is positive semidefinite, then W_k will also be positive semidefinite for each k (Exercise 8.4.1). The matrix inversion lemma can also be applied to (8.1-24) to obtain

$$\boldsymbol{\theta}_k = \phi_k^T [I + W_{k+1} C_k I_k^{-1} C_k^T]^{-1} \boldsymbol{\theta}_{k+1} + B_k^T T_k \mathbf{r}_k. \tag{8.1-26}$$

From the above remarks, the existence of the necessary inverses in both (8.1-25) and (8.1-26) depend upon W_p being symmetric and positive semidefinite. But from the last of equations (8.1-16), we see that (8.1-21) in the case $k = p$ implies that

$$W_p = B_p^T T_p B_p \tag{8.1-27}$$

and

$$\boldsymbol{\theta}_p = B_p^T T_p \mathbf{r}_p. \tag{8.1-28}$$

Thus, given the conditions in (8.1-27) and (8.1-28), (8.1-25) and (8.1-26) can be used to calculate the sequences $W_1, W_2, \ldots, W_p$ and $\boldsymbol{\theta}_1, \boldsymbol{\theta}_2, \ldots, \boldsymbol{\theta}_p$. Thus, from Equations (8.1-18) and (8.1-21), we see that our optimal control can be expressed as

$$\mathbf{v}_k = -L_k^{-1} C_k^T [W_{k+1} \mathbf{x}_{k+1} - \boldsymbol{\theta}_{k+1}]$$

or

$$\mathbf{v}_k = -L_k^{-1} C_k^T [W_{k+1} [\phi_k \mathbf{x}_k + C_k \mathbf{v}_k] - \boldsymbol{\theta}_{k+1}].$$

This last equation can be resolved for $\mathbf{v}_k$ to obtain

$$\mathbf{v}_k = [C_k^T W_{k+1} C_k + L_k]^{-1} C_k^T [-W_{k+1} \phi_k \mathbf{x}_k + \boldsymbol{\theta}_k]. \tag{8.1-29}$$

This is the result we seek (note that the indicated inverse does exist). All of the coefficients in (8.1-29) can be precomputed using (8.1-25) and (8.1-27). If the track $\mathbf{r}_1, \ldots, \mathbf{r}_p$ is known prior to operation, then all of the $\boldsymbol{\theta}_k$'s can also be precomputed, otherwise they can be computed "on-line," both cases utilizing (8.1-26) and (8.1-28). The control $\mathbf{v}_k$ can then be calculated using the current state $\mathbf{x}_k$ in Equation (8.1-29). The reader is advised to work Exercise 8.4.2 at this point.

There are some interesting special cases of the above problem. From the matrix inversion lemma, we conclude from Equation (8.1-25) that if W_{k+1} is invertible, then

$$W_k = \phi_k^T [W_{k+1}^{-1} + C_k L_k^{-1} C_k^T]^{-1} \phi_k + B_k^T T_k B_k. \tag{8.1-30}$$

Thus, if ϕ_k is also invertible, W_k is the sum of a positive definite matrix and a positive semidefinite matrix and is hence also invertible. It follows that if each ϕ_k is invertible and B_p is chosen so that $B_p^T T_p B_p$ is invertible, then all of the W_k's are invertible and Equation (8.1-30) holds. This is of no particular advantage unless $T_1, T_2, \ldots, T_{p-1}$ are all zero, in which case (8.1-30) becomes

$$W_k^{-1} = \phi_k W_{k+1}^{-1}(\phi_k^T)^{-1} + \phi_k^{-1} C_k L_k^{-1} C_k^T (\phi_k^T)^{-1}, \tag{8.1-31}$$

which holds for $k < p-1$. This is known as a *Lyapunov equation.* The initialization of (8.1-31) is obtained using (8.1-27) (see Exercise 8.4.3 for an interesting extension of this).

8.2 Stochastic Control with Perfect Observations

We now take up the problem of controlling a linear system of the form

$$\mathbf{x}(k+1) = \phi(k)\mathbf{x}(k) + C(k)\mathbf{v}(k) + \mathbf{u}(k), \tag{8.2-1}$$

where $\mathbf{u}$ is a discrete white process satisfying the usual assumptions from Chapter 7. The control criterion (8.1-4) is now replaced by the criterion

$$J_p = E\left[\sum_{k=0}^{p-1} (\mathbf{y}_{k+1} - \mathbf{r}_{k+1})^T T_{k+1} (\mathbf{y}_{k+1} - \mathbf{r}_{k+1}) + \mathbf{v}_k^T L_k \mathbf{v}_k\right]. \tag{8.2-2}$$

We now wish to minimize J_p subject to (8.2-1) by choosing the random sequence $\mathbf{v}_0, \ldots, \mathbf{v}_{p-1}$ based on perfect observations of $\mathbf{x}_0, \mathbf{x}_1, \ldots, \mathbf{x}_{p-1}$. To this end, we will use the definitions in 8.1.4 and 8.1.14 with the understanding that now X and V are random vectors. We also introduce the random vector

$$U = \begin{bmatrix} \mathbf{u}(0) \\ \mathbf{u}(1) \\ \mathbf{u}(2) \\ \vdots \\ \mathbf{u}(p-1) \end{bmatrix} = \begin{bmatrix} \mathbf{u}_0 \\ \mathbf{u}_1 \\ \mathbf{u}_2 \\ \vdots \\ \mathbf{u}_{p-1} \end{bmatrix}, \tag{8.2-3}$$

and note that Theorem 8.1.5 becomes

$$X = E\mathbf{x}(0) + F\Gamma V + U. \tag{8.2-4}$$

The objective function (8.2-2) becomes

$$J = E[(BX - R)^T T(BX - R) + V^T LV], \tag{8.2-5}$$

which is identical to (8.1-9). In order to minimize (8.2-5) subject to (8.2-4), we form the Lagrangian

$$\mathcal{L}(X, V, \Lambda) = E[(BX-R)^T T(BX-R) + V^T LV] + [\Lambda, X - E\mathbf{x}(0) - F\Gamma V - U],$$

where $\Lambda \in \mathcal{L}_2^n(\Omega, P)$ and $[\cdot,\cdot]$ in the last term is the inner product in $\mathcal{L}_2^n(\Omega, P)$. We then take the directional derivative in the direction (Z, W) and obtain

$$\begin{aligned}\mathcal{L}'(X, V, \Lambda)(Z, W) &= E[2Z^T B^T T(BX - R) + Z^T \Lambda] \\ &\quad + E[2W^T LV - W^T \Gamma^T F^T \Lambda],\end{aligned}$$

where we have used the linearity of the expectation operator in the calculation. Setting this equal to zero, and writing the resulting equation in terms of the inner product in $\mathcal{L}_2^n(\Omega, P)$, we obtain

$$\left[Z, B^T T(BX - R) + \frac{1}{2}\Lambda\right] + \left[W, LV - \Gamma^T F^T \frac{1}{2}\Lambda\right] = 0.$$

Since this must hold for all Z, W, it follows that

$$B^T T(BX - R) + \frac{1}{2}\Lambda = 0 \tag{8.2-6}$$

and

$$LV - \Gamma^T F^T \frac{1}{2}\Lambda = 0. \tag{8.2-7}$$

Note that even though X, V, and Λ are random vectors rather than constant vectors, the form of Equations (8.2-6) and (8.2-7) is identical to that in (8.1-10) and (8.1-11). If we solve these for the open loop control law as in (8.1-15), we must now use (8.2-4) instead of (8.1-8) and the resulting solution is

$$\Lambda = [I + A_1 A_2]^{-1}[A_1 F^{-1}(E\mathbf{x}_0 + U) - (BF)^T TR].$$

Unlike the deterministic situation, this equation cannot be implemented since U is a random sequence which, in general, is unknown. Because of this it is reasonable to try and find a closed loop solution following the calculations in the last section. Unfortunately, in trying to carry out such a program, one must add the term $\mathbf{u}_k$ to Equation (8.1-19), and the resulting analysis produces a control wherein $\mathbf{v}_k$ is a linear combination of $\mathbf{x}_k$, a constant vector, and the vectors $\mathbf{u}_k, \mathbf{u}_{k+1}, \ldots, \mathbf{u}_p$. Again, this cannot be implemented since the $\mathbf{u}_k$'s are unknown, especially those in the future. All is not lost, however, for linearity still seems to be the "rule of the day." Thus, we impose an additional restriction on our problem and require that our control be an affine function of $\mathbf{x}_k$.[3] To this end we introduce some additional notation. Specifically, we define

$$\tilde{X} = \begin{bmatrix} \mathbf{x}_0 \\ \mathbf{x}_1 \\ \vdots \\ \mathbf{x}_{p-1} \end{bmatrix}, \tag{8.2-8}$$

[3]There are alternatives to this restriction that involve assuming the normal distribution throughout [3].

$$\tilde{X}_0 = \begin{bmatrix} \mathbf{x}_0 \\ 0 \\ \vdots \\ 0 \end{bmatrix}, \tag{8.2-9}$$

and

$$A = \begin{bmatrix} 0 & 0 & --- & 0 & 0 \\ I & 0 & --- & 0 & 0 \\ 0 & I & --- & 0 & 0 \\ \vdots & \vdots & & & \\ 0 & 0 & --- & I & 0 \end{bmatrix}. \tag{8.2-10}$$

Then

$$\tilde{X} = AX + \tilde{X}_0. \tag{8.2-11}$$

Requiring our control to be an affine function of $\tilde{X}$, we assume

$$\Gamma V = D\tilde{X} + \Pi, \tag{8.2-12}$$

where D is a diagonal matrix of matrices $D_0, \ldots, D_{p-1}$ and Π is a constant vector. Substituting (8.2-11) into (8.3-12), we obtain

$$\Gamma V = DAX + D\tilde{X}_0 + \Pi. \tag{8.2-13}$$

Thus, our new Lagrangian is of the form

$$\begin{aligned} \mathcal{L}(X, V, \Lambda, M) &= E[(BX - R)^T T(BX - R) + V^T LV] \\ &\quad + [\Lambda, X - E\mathbf{x}(0) - F\Gamma V - U] \\ &\quad + [M, \Gamma V - DAX - D\tilde{X}_0 - \Pi], \end{aligned}$$

where $\Lambda, M \in \mathcal{L}_2^n(\Omega, P)$ and $[\cdot,\cdot]$ represents the inner product in this space. Taking the directional derivative of $\mathcal{L}$ in the direction (Z, W), we obtain

$$\begin{aligned} \mathcal{L}'(X, V, \Lambda, M)(Z, W) &= E(2Z^T B^T T(BX - R) + Z^T\Lambda - Z^T A^T D^T M) \\ &\quad + E(2W^T LV - W^T\Gamma^T F^T\Lambda + W^T\Gamma^T M). \end{aligned}$$

If this expression is to be zero for all choices of Z and W, we infer that

$$2B^T T(BX - R) + \Lambda - A^T D^T M = 0$$

$$2LV - \Gamma^T F^T + \Gamma^T M = 0. \tag{8.2-14}$$

Solving the second equation in (8.2-14) for V, and making the substitutions

$$\Lambda = -2(F^T)^{-1}\Lambda'; \qquad M = 2M',$$

the system (8.3-14) becomes

$$2B^T T(BX - R) - 2(F^T)^{-1}\Lambda' - 2A^T D^T M' = 0$$

$$V = -L^{-1}\Gamma^T(\Lambda' + M').$$

Since Λ and M are no longer needed, we will (for notational convenience) drop the primes in the above system. Dividing the top equation by 2, we obtain

$$\begin{aligned} B^T T(BX - R) - 2(F^T)^{-1}\Lambda - 2A^T D^T M &= 0 \\ V &= -L^{-1}\Gamma^T(\Lambda + M). \end{aligned} \tag{8.2-15}$$

Writing

$$\Lambda = \begin{bmatrix} \boldsymbol{\lambda}_0 \\ \boldsymbol{\lambda}_1 \\ \vdots \\ \boldsymbol{\lambda}_{p-1} \end{bmatrix}; \quad M = \begin{bmatrix} \boldsymbol{\mu}_0 \\ \boldsymbol{\mu}_1 \\ \vdots \\ \boldsymbol{\mu}_{p-1} \end{bmatrix},$$

the first equation in (8.2-15) can be written out in terms of components as

$$\begin{aligned} \boldsymbol{\lambda}_0 - \phi_1^T \boldsymbol{\lambda}_1 + D_1^T \boldsymbol{\mu}_1 &= B_1^T T_1 B_1 \mathbf{x}_1 - B_1^T T_1 \mathbf{r}_1 \\ \vdots \quad &\vdots \\ \boldsymbol{\lambda}_{p-2} - \phi_{p-1}^T \boldsymbol{\lambda}_{p-1} + D_{p-1}^T \boldsymbol{\mu}_{p-1} &= B_{p-1}^T T_{p-1} B_{p-1} \mathbf{x}_{p-1} - B_{p-1}^T T_{p-1} \mathbf{r}_{p-1} \\ \boldsymbol{\lambda}_{p-1} &= B_p^T T_p B_p \mathbf{x}_p - B_p^T T_p \mathbf{r}_p. \end{aligned} \tag{8.2-16}$$

The second of Equations (8.2-15) can be written in terms of components as

$$\mathbf{v}_k = -L_k^{-1} C_k^T (\boldsymbol{\lambda}_k + \boldsymbol{\mu}_k); \qquad k = 0, 1, \ldots, p-1. \tag{8.2-17}$$

For notational convenience, we will introduce the matrices

$$H_k = B_k^T T_k B_k; \quad S_k = C_k L_k^{-1} C_k^T; \quad G_k = B_k^T T_k. \tag{8.2-18}$$

This done, we note that (8.2-1) can be rewritten as

$$\mathbf{x}_{k+1} = \phi_k \mathbf{x}_k - S_k(\boldsymbol{\lambda}_k + \boldsymbol{\mu}_k) + \mathbf{u}_k. \tag{8.2-19}$$

If we let $k = p - 1$ in (8.2-19) and substitute the result into the last of Equations (8.2-16), we obtain

$$\boldsymbol{\lambda}_{p-1} = H_p[\phi_{p-1}\mathbf{x}_{p-1} - S_{p-1}\boldsymbol{\lambda}_{p-1} - S_{p-1}\boldsymbol{\mu}_{p-1} + \mathbf{u}_{p-1}] - G_p \mathbf{r}_p,$$

which can be solved for $\boldsymbol{\lambda}_{p-1}$ as

$$\boldsymbol{\lambda}_{p-1} = (I + H_p S_{p-1})^{-1}[H_p \phi_{p-1} \mathbf{x}_{p-1} - H_p S_{p-1} \boldsymbol{\mu}_{p-1} + H_p \mathbf{u}_{p-1} - G_p \mathbf{r}_p]. \tag{8.2-20}$$

From (8.2-17) and (8.2-18), it follows that

$$C_{p-1}\mathbf{v}_{p-1} = -S_{p-1}(\boldsymbol{\lambda}_{p-1} + \boldsymbol{\mu}_{p-1}),$$

so

$$C_{p-1}\mathbf{v}_{p-1} = -S_{p-1}(I + H_pS_{p-1})^{-1}[H_p\phi_{p-1}\mathbf{x}_{p-1} - H_pS_{p-1}\boldsymbol{\mu}_{p-1} + H_p\mathbf{u}_{p-1} - G_p\mathbf{r}_p] - S_{p-1}\boldsymbol{\mu}_{p-1}. \tag{8.2-21}$$

But by (8.2-13), it follows that

$$C_{p-1}\mathbf{v}_{p-1} = D_{p-1}\mathbf{x}_{p-1} + \boldsymbol{\pi}_{p-1},$$

which contains no noise terms. The only way this is possible is if

$$S_{p-1}(I + H_pS_{p-1})^{-1}[H_pS_{p-1}\boldsymbol{\mu}_{p-1} - H_p\mathbf{u}_{p-1}] - S_{p-1}\boldsymbol{\mu}_{p-1} = 0$$

or

$$[S_{p-1}(I + H_pS_{p-1})^{-1}H_pS_{p-1} - S_{p-1}]\boldsymbol{\mu}_{p-1} = S_{p-1}(I + H_pS_{p-1})^{-1}H_p\mathbf{u}_{p-1}.$$

Using the matrix inversion lemma, this can be rewritten as

$$-(I + S_{p-1}H_p)^{-1}S_{p-1}\boldsymbol{\mu}_{p-1} = (I + S_{p-1}H_p)^{-1}S_{p-1}H_p\mathbf{u}_{p-1}$$

or

$$S_{p-1}\boldsymbol{\mu}_{p-1} = S_{p-1}H_p\mathbf{u}_{p-1}.$$

Clearly, if we set

$$\boldsymbol{\mu}_{p-1} = H_p\mathbf{u}_{p-1}, \tag{8.2-22}$$

the above expression is satisfied. Moreover, if we now substitute (8.2-20) into the next to the last of Equations (8.2-16), and then substitute (8.2-22) into the result, we obtain

$$\begin{aligned}\boldsymbol{\lambda}_{p-2} \quad &- \quad \phi_{p-1}^T(I + H_pS_{p-1})^{-1}(H_p\phi_{p-1}\mathbf{x}_{p-1} - H_pS_{p-1}H_p\mathbf{u}_{p-1} \\ &+ \quad H_p\mathbf{u}_{p-1} - G_p\mathbf{r}_p) + D_{p-1}^TH_p\mathbf{u}_{p-1} = H_{p-1}\mathbf{x}_{p-1} - G_{p-1}\mathbf{r}_{p-1}.\end{aligned}$$

We conclude, therefore that

$$\boldsymbol{\lambda}_{p-2} = W_{p-1}\mathbf{x}_{p-1} + \boldsymbol{\omega}_{p-1} - \boldsymbol{\theta}_{p-1}, \tag{8.2-23}$$

where W_{p-1} is a matrix of constants, $\boldsymbol{\theta}_{p-1}$ is a constant vector, and $\boldsymbol{\omega}_{p-1}$ is a vector function of $\mathbf{u}_{p-1}$ alone. If we now begin with (8.2-23) and repeat the above arguments, we obtain the result that $\boldsymbol{\mu}_{p-2}$ is a function of $\boldsymbol{\omega}_{p-1}$ and $\mathbf{u}_{p-2}$ (this will be clear from arguments to follow), and so we conjecture that

$$\boldsymbol{\lambda}_{k-1} = W_k\mathbf{x}_k + \boldsymbol{\omega}_k - \boldsymbol{\theta}_k, \tag{8.2-24}$$

where W_k is a matrix of constants, $\boldsymbol{\theta}_k$ is a constant vector, and $\boldsymbol{\omega}_k$ is a function of $\mathbf{u}_k, \mathbf{u}_{k+1}, \ldots, \mathbf{u}_{p-1}$. This is analogous to our conjecture (8.1-21) in the last section. Just as there, the proof of this lies in our ability to generate recursive relations to generate W_k, $\boldsymbol{\omega}_k$, and $\boldsymbol{\theta}_k$ subject to the constraints of our problem. The procedure we use is very similar to that used in Section 8.1.

To begin with, we substitute (8.2-24) into (8.2-19) (appropriately adjusting the subscripts) and obtain

$$\mathbf{x}_{k+1} = \phi_k \mathbf{x}_k - S_k[W_{k+1}\mathbf{x}_{k+1} + \boldsymbol{\omega}_{k+1} - \boldsymbol{\theta}_{k+1}] - S_k \boldsymbol{\mu}_k + \mathbf{u}_k.$$

Solving this for $\mathbf{x}_{k+1}$, we obtain

$$\mathbf{x}_{k+1} = [I + S_k W_{k+1}]^{-1}[\phi_k \mathbf{x}_k - S_k \boldsymbol{\omega}_{k+1} + S_k \boldsymbol{\theta}_{k+1} - S_k \boldsymbol{\mu}_k + \mathbf{u}_k]. \quad (8.2\text{-}25)$$

However, if we substitute (8.2-19) into (8.2-24), we obtain

$$\boldsymbol{\lambda}_{k-1} = W_k[\phi_{k-1}\mathbf{x}_{k-1} - S_k \boldsymbol{\lambda}_{k-1} - S_{k-1}\boldsymbol{\mu}_{k-1} + \mathbf{u}_{k-1}] + \boldsymbol{\omega}_{k-1} - \boldsymbol{\theta}_{k-1},$$

which can be solved for $\boldsymbol{\lambda}_{k-1}$ as

$$\begin{aligned}\boldsymbol{\lambda}_{k-1} = [I + W_k S_{k-1}]^{-1}[&W_k \boldsymbol{\theta}_{k-1}\mathbf{x}_{k-1} - W_k S_{k-1}\boldsymbol{\mu}_{k-1} \\ &+ W_k \mathbf{u}_{k-1} + \boldsymbol{\omega}_{k-1} - \boldsymbol{\theta}_{k-1}]. \quad (8.2\text{-}26)\end{aligned}$$

From (8.2-17), it follows that

$$C_{k-1}\mathbf{v}_{k-1} = -S_{k-1}(\boldsymbol{\lambda}_{k-1} + \boldsymbol{\mu}_{k-1}), \quad (8.2\text{-}27)$$

and from (8.2-13), it follows that

$$C_{k-1}\mathbf{v}_{k-1} = D_{k-1}\mathbf{x}_{k-1} + \Pi_{k-1}, \quad (8.2\text{-}28)$$

that is, there are no noise terms. If we substitute (8.2-26) into (8.2-27) and impose the constraint that there are no noise terms,

$$S_{k-1}[I + W_k S_{k-1}]^{-1}[W_k S_{k-1}\boldsymbol{\mu}_{k-1} - W_k \mathbf{u}_{k-1} - \boldsymbol{\omega}_{k-1}] - S_{k-1}\boldsymbol{\mu}_{k-1} = 0$$

or

$$\begin{aligned}&S_{k-1}[I + W_k S_{k-1}]^{-1}W_k - I]S_{k-1}\boldsymbol{\mu}_{k-1} \\ &= S_{k-1}[I + W_k S_{k-1}]^{-1}[W_k \mathbf{u}_{k-1} + \boldsymbol{\omega}_{k-1}].\end{aligned}$$

Using the matrix inversion lemma, this can be rewritten as

$$\begin{aligned}-[I \;+\; &S_{k-1}W_k]^{-1}S_{k-1}\boldsymbol{\mu}_{k-1} \\ =\; &[I + S_{k-1}W_k]^{-1}S_{k-1}W_k \mathbf{u}_{k-1} + S_{k-1}[I + W_k S_{k-1}]^{-1}\boldsymbol{\omega}_{k-1}\end{aligned}$$

or

$$S_{k-1}\boldsymbol{\mu}_{k-1} = -S_{k-1}W_k \mathbf{u}_{k-1} - [I + S_{k-1}W_k]S_{k-1}[I + W_k S_{k-1}]^{-1}\boldsymbol{\omega}_{k-1}$$

$$S_{k-1}\boldsymbol{\mu}_{k-1} = -S_{k-1}W_k \mathbf{u}_{k-1} - S_{k-1}\boldsymbol{\omega}_{k-1}. \quad (8.2\text{-}29)$$

Clearly (8.2-29) can be satisfied if we choose $\boldsymbol{\mu}_{k-1}$ to be

$$\boldsymbol{\mu}_{k-1} = -W_k \mathbf{u}_{k-1} - \boldsymbol{\omega}_{k-1}. \tag{8.2-30}$$

Since our hypothesis is that $\boldsymbol{\omega}_{k-1}$ is a function of $\mathbf{u}_{k-1}, \mathbf{u}_k, \ldots, \mathbf{u}_{p-1}$, it follows that $\boldsymbol{\mu}_{k-1}$ is also. Note that from (8.2-26), (8.2-27), and (8.2-30) it follows that

$$C_{k-1}\mathbf{v}_{k-1} = -S_{k-1}[I + W_k S_{k-1}]^{-1}[W_k \phi_{k-1}\mathbf{x}_{k-1} - \boldsymbol{\theta}_{k-1}]. \tag{8.2-31}$$

Equation (8.2-31) is our control law providing, of course, that we can generate the sequence of W_k's and $\boldsymbol{\theta}_k$'s. We now proceed to do this.

Taking an arbitrary equation in (8.2-16) (with $k < p-1$), we have

$$\boldsymbol{\lambda}_{k-1} - \phi_k^T \boldsymbol{\lambda}_k + D_k^T \boldsymbol{\mu}_k = H_{k-k} - G_k \mathbf{r}_k.$$

If we substitute (8.2-24) (replacing k by $k+1$) into this expression along with (8.2-30), we obtain

$$\boldsymbol{\lambda}_{k-1} - \phi_k^T [W_{k+1}\mathbf{x}_{k+1} + \boldsymbol{\omega}_{k+1} - \boldsymbol{\theta}_{k+1}] - D_k^T [W_{k+1}\mathbf{u}_k + \boldsymbol{\omega}_k] = H_k \mathbf{x}_k - G_k \mathbf{r}_k,$$

and replacing $\mathbf{x}_{k+1}$ using (8.2-25), it follows that

$$\begin{aligned}
\boldsymbol{\lambda}_{k-1} \quad &- \quad \phi_k^T [W_{k+1}[I + S_k W_{k+1}]^{-1}[\phi_k \mathbf{x}_k - S_k \boldsymbol{\omega}_{k+1} + S_k \boldsymbol{\theta}_{k+1} \\
&- \quad S_k(W_{k+1}\mathbf{u}_k + \boldsymbol{\omega}_k) + \mathbf{u}_k] + \boldsymbol{\omega}_{k+1} - \theta_{k+1}] \\
&- \quad D_k^T [W_{k+1}\mathbf{u}_k + \boldsymbol{\omega}_k] = H_k \mathbf{x}_k - G_k \mathbf{r}_k.
\end{aligned} \tag{8.2-32}$$

Since (8.2-24) must hold, we can solve (8.2-32) for $\boldsymbol{\lambda}_{k-1}$ and equate coefficients of $\mathbf{x}_k$, constant terms and terms involving noise. The results are

$$W_k = \phi_k^T W_{k+1}[I + S_k W_{k+1}]^{-1}\phi_k + H_k, \tag{8.2-33}$$

$$\theta_k = \phi_k^T - \phi_k^T W_{k+1}[I + S_k W_{k+1}]^{-1} S_k \theta_{k+1} + G_k \mathbf{r}_k, \tag{8.2-34}$$

and

$$\begin{aligned}
\boldsymbol{\omega}_k \quad = \quad & \phi_k^T W_{k+1}[I + S_k W_{k+1}]^{-1}[-S_k \boldsymbol{\omega}_{k+1} - S_k W_{k+1}\mathbf{u}_k - S_k \boldsymbol{\omega}_k + \mathbf{u}_k] \\
& + D^T W_{k+1}\mathbf{u}_k + D_k^T \boldsymbol{\omega}_k.
\end{aligned} \tag{8.2-35}$$

Equation (8.2-35) can be rearranged as

$$\begin{aligned}
[I \quad + \quad & \phi_k^T W_{k+1}[I + S_k W_{k+1}]^{-1} S_k + D_k^T]\boldsymbol{\omega}_k \\
= \quad & \phi_k^T W_{k+1}[I + S_k W_{k+1}]^{-1}[-S_k \boldsymbol{\omega}_{k+1} - S_k W_{k+1}\mathbf{u}_k + \mathbf{u}_k] \\
& + D_k^T W_{k+1}\mathbf{u}_k.
\end{aligned} \tag{8.2-36}$$

Since S_k is symmetric, and W_{k+1} will be shown to be symmetric, we see by comparing (8.2-31) with (8.2-28) that

$$D_k = -S_k[I + W_{k+1}S_k]^{-1}W_{k+1}\phi_k$$

$$D_k^T = -\phi_k^T W_{k+1}^T [I + S_k W_{k+1}]^{-1} S_k.$$

In other words, (8.2-36) becomes

$$\boldsymbol{\omega}_k = D_k^T [\omega_{k+1} + 2W_{k+1}\mathbf{u}_k] + \phi_k^T W_{k+1} [I + S_k W_{k+1}]^{-1} \mathbf{u}_k, \qquad (8.2\text{-}37)$$

which provides us with the inductive step to prove that $\boldsymbol{\omega}_k$ is a function of $\mathbf{u}_k, \mathbf{u}_{k+1}, \ldots, \mathbf{u}_{p-1}$. Although we could, in principle, implement Equation (8.2-37) by taking $\boldsymbol{\omega}_p = 0$,[4] our control, Equation (8.2-31), requires knowledge of only W_k's and θ_k's; (8.2-37) is only needed to prove our contention about the $\boldsymbol{\omega}_k$'s. At last we come to the really interesting conclusion. Equations (8.2-33) and (8.2-34) are identical to (8.1-23) and (8.1-24) of the last section. Hence, Equations (8.1-25) and (8.1-26) also hold here, namely,

$$W_k = \phi_k^T [W_{k+1} - W_{k+1} C_k [C_k^T W_{k+1} C_k + L_k]^{-1} C_k^T W_{k+1}] \phi_k + B_k^T T_k B_k, \qquad (8.2\text{-}38)$$

and

$$\boldsymbol{\theta}_k = \phi_k^T [I + W_{k+1} C_k L_k^{-1} C_k^T]^{-1} \boldsymbol{\theta}_{k+1} + B_k^T T_k \mathbf{r}_k. \qquad (8.2\text{-}39)$$

Equations (8.2-38) and (8.2-39) can be used to recursively calculate the W_k's and θ_k's using the initial conditions

$$W_p = B_p^T T_p B_p \qquad (8.2\text{-}40)$$

and

$$\boldsymbol{\theta}_p = B_p^T T_p \mathbf{r}_p. \qquad (8.2\text{-}41)$$

These initial conditions are derived by noting the last of Equations (8.2-16). Referring to our control law, Equation (8.1-29), from the last section, we have

$$[C_k^T W_{k+1} C_k + L_k]\mathbf{v}_k = C_k^T [-W_{k+1}\phi_k \mathbf{x}_k + \theta_k].$$

Multiplying this through by $C_k L_k^{-1}$, we obtain

$$[S_k W_{k+1} C_k + C_k]\mathbf{v}_k = S_k [-W_{k+1}\phi_k \mathbf{x}_k + \boldsymbol{\theta}_k]$$

or

$$C_k \mathbf{v}_k = [S_k W_{k+1} + I]^{-1} S_k [-W_{k+1}\phi_k \mathbf{x}_k + \theta_k].$$

This expression can be rearranged using the matrix inversion lemma as

$$C_k \mathbf{v}_k = -S_k [I + W_{k+1} S_k]^{-1} [W_{k+1}\phi_k \mathbf{x}_k - \boldsymbol{\theta}_k],$$

which is identical to the control law (8.2-31) (with $k-1$ replaced by k). At long last we have the result, namely, *in a stochastic dynamical system where* (8.2-1) *holds; if one can perfectly observe the state* $\mathbf{x}_k$ *for each* k, *then one can provide a control sequence to minimize* (8.2-2), *subject to the restriction that the control must be of the form in* (8.2-12), *by simply implementing the closed loop control law* (8.1-29) *one would use if the system were deterministic.*

---

[4] Actual implementation is impossible since the noise vectors $\mathbf{u}_k, \mathbf{u}_{k+1}, \ldots, \mathbf{u}_{p-1}$ are unknown.

8.3 Stochastic Control with Imperfect Measurement

We now address the general problem of controlling a sytem of the form

$$\mathbf{x}_{k+1} = \phi_k \mathbf{x}_k + C_k \mathbf{v}_k + \mathbf{u}_k \tag{8.3-1}$$

so as to minimize

$$J_p = E\left[\sum_{k=0}^{p-1}(B_{k+1}\mathbf{x}_{k+1} - \mathbf{r}_{k+1})^T T_{k+1}(B_{k+1}\mathbf{x}_{k+1} - \mathbf{r}_{k+1}) + \mathbf{v}_k^T L_k \mathbf{v}_k\right] \tag{8.3-2}$$

utilizing only the information

$$\mathbf{z}_k = H_k \mathbf{x}_k + \mathbf{w}_k. \tag{8.3-3}$$

Of course, in case $H_k = I$ and $\mathbf{w}_k = 0$, we solved the problem in the last section. With that result in mind, we will stipulate that the control $\mathbf{v}_k$ is to be a linear combination of $\mathbf{z}_0, \mathbf{z}_1, \ldots, \mathbf{z}_k$, this being the aggregate of available information at time k. Since the z_k's are linear functions of $\mathbf{x}_0, \ldots, \mathbf{x}_k$ and $\mathbf{w}_0, \ldots, \mathbf{w}_k$, this assumption assures us that

$$E(\mathbf{v}(k)\mathbf{u}(k)^T) = 0. \tag{8.3-4}$$

In addition, defining $\mathbf{y}_k$ as in Chapter 7, that is,

$$\mathbf{y}_k = \begin{bmatrix} \mathbf{z}_0 \\ \mathbf{z}_n \\ \vdots \\ \mathbf{z}_k \end{bmatrix}, \tag{8.3-5}$$ [5]

it follows from Lemma 5.3.4 and our assumption about the form of $\mathbf{v}_k$ that

$$\mathbf{v}_k = E(\mathbf{v}_k \mathbf{y}_k^T) E(\mathbf{y}_k \mathbf{y}_k^T)^T \mathbf{y}_k. \tag{8.3-6}$$

As a first step in solving our control problem, we address the problem of recursively calculating $\hat{\mathbf{x}}(k \mid k)$ when $\mathbf{x}$ satisfies (8.3-1) with $\mathbf{v}$ a random sequence such that (8.3-4) and (8.3-6) hold. We thus need to calculate both dynamic and static update relations. From our results in Chapter 6 it is clear that the static update equations are identical to the results there since they only involve Equation (8.3-3). Specifically, given $P(k+1 \mid k)$, $\hat{\mathbf{x}}(k+1 \mid k)$, we have

[5] Not to be confused with the earlier use of $\mathbf{y}_k$ in this chapter.

$$P(k+1\,|\,k+1) = P(k+1\,|\,k) - P(k+1\,|\,k)H_{K+1}^T[R_{k+1} + H_{k+1}P(k+1\,|\,k)H_k^T]^+[H_{k+1}P(k+1\,|\,k)], \tag{8.3-7}$$

and

$$\hat{\mathbf{x}}(k+1\,|\,k+1) = \hat{\mathbf{x}}(k+1\,|\,k) + K(k+1)[\mathbf{z}(k+1) - H_{k+1}\hat{\mathbf{x}}(k+1\,|\,k)], \tag{8.3-8}$$

where $K(K+1)$ is as defined in Chapter 7. For the dynamic updates, we note that

$$\begin{aligned}\hat{\mathbf{x}}(k+1\,|\,k) &= E(\mathbf{x}_{k+1}\mathbf{y}_k^T)E(\mathbf{y}_k\mathbf{y}_k^T)^+\mathbf{y}_k \\ &= E((\phi_k\mathbf{x}_k + C_k\mathbf{v}_k + \mathbf{u}_k)\mathbf{y}_k^T)E(\mathbf{y}_k\mathbf{y}_k^T)^+\mathbf{y}_k \\ &= \phi_k\hat{\mathbf{x}}(k\,|\,k) + C_kE(\mathbf{v}_k\mathbf{y}_k^T)E(\mathbf{y}_k\mathbf{y}_k^T)^+\mathbf{y}_k\end{aligned}$$

$$\hat{\mathbf{x}}(k+1\,|\,k) = \phi_k\hat{\mathbf{x}}(k\,|\,k) + C_k\mathbf{v}_k, \tag{8.3-9}$$

the last equality following from (8.3-6). Noting that by (8.3-1) and (8.3-9)

$$\hat{\mathbf{x}}(k+1\,|\,k) - \mathbf{x}(k+1) = \phi_k[\hat{\mathbf{x}}(k\,|\,k) - \mathbf{x}(k)] - \mathbf{u}(k),$$

it follows that the dynamic update for the covariance is exactly as it was in Chapter 7, that is,

$$P(k+1\,|\,k) = \phi_kP(k\,|\,k)\phi_k^T + Q(k). \tag{8.3-10}$$

In summary, the only equation that changed was (8.3-9) through the addition of the term $C_k\mathbf{v}_k$.

We will have occasion to use the following lemma.

8.3.1. **Lemma.** *Let* $\hat{\mathbf{x}}(k+1\,|\,k)$ *be as given above and define*

$$\omega(k) = \mathbf{z}(k+1) - H(k+1)\hat{\mathbf{x}}(k+1\,|\,k) \qquad k = 0, 1, \ldots, p-1.$$

Then $\omega(k)$ *is a white process.*

Proof. Using the notation from Chapter 7, let M_k^n be the n-fold product of the span of the components of $\mathbf{z}(0), \mathbf{z}(1), \ldots, \mathbf{z}(k)$. Thus, by the projection theorem,

$$[\mathbf{x}(k+1) - \hat{\mathbf{x}}(k+1\,|\,k)] \perp M_k^n. \tag{8.3-11}$$

If $j < k$, then $M_{j+1}^n \subset M_k^n$, so (8.3-11) implies

$$[\mathbf{x}(k+1) - \hat{\mathbf{x}}(k+1\,|\,k)] \perp M_{j+1}^n. \tag{8.3-12}$$

However, $H^T(k+1)[\mathbf{z}(j+1) - H(j+1)\hat{\mathbf{x}}(j+1\,|\,j)] \in M_{j+1}^n$, and so (8.3-12) implies that

$$[\mathbf{x}(k+1) - \hat{\mathbf{x}}(k+1\,|\,k)] \perp [H^T(k+1)(\mathbf{z}(j+1) - H(j+1)\hat{\mathbf{x}}(j+1\,|\,j))].$$

Writing this in terms of inner products, we have

$$[\mathbf{x}(k+1) - \hat{\mathbf{x}}(k+1\,|\,k), H^T(k+1)[\mathbf{z}(j+1) - H(j+1)\hat{\mathbf{x}}(j+1\,|\,j)]] = 0$$

or

$$[H(k+1)(\mathbf{x}(k+1) - \hat{\mathbf{x}}(k+1\,|\,k)), \mathbf{z}(j+1) - H(j+1)\hat{\mathbf{x}}(j+1\,|\,j)] = 0$$

$$[\mathbf{z}(k+1) - H(k+1)\hat{\mathbf{x}}(k+1\,|\,k) - \mathbf{w}(k+1), \mathbf{z}(j+1) - H(j+1)\hat{\mathbf{x}}(j+1\,|\,j)] = 0.$$

Since $\mathbf{w}(k+1) \perp \mathbf{z}(s)$ for $s \leq k$, the above equation reduces to

$$[\omega(k), \omega(j)] = 0, \qquad \text{for} \quad j < k.$$

Since the result is obviously symmetric, the lemma is proved. □

8.3.2. **Corollary.**

$$E(\omega(k)\hat{\mathbf{x}}(j\,|\,j)^T) = 0 \quad \text{for} \quad j \leq k$$

$$E(\omega(k)\mathbf{v}(j)^T) = 0 \quad \text{for} \quad j \leq k.$$

Let us now return to our estimation problem. Defining

$$\delta(k) \triangleq \mathbf{x}(k) - \hat{\mathbf{x}}(k\,|\,k),$$

we can rewrite J_p as

$$J_p = E\left(\sum_{k=0}^{p-1} B_{k+1}(\hat{\mathbf{x}}(k+1\,|\,k+1) + \delta(k+1) - \mathbf{r}_{k+1})^T T_{k+1} \times (B_{k+1}(\hat{\mathbf{x}}(k+1\,|\,k+1) + \delta(k+1)) - r_{k+1})) + \mathbf{v}_k^T L_k \mathbf{v}_k\right). \quad (8.3\text{-}13)$$

However, $E(\delta(k+1)\hat{\mathbf{x}}(k+1\,|\,k+1)^T) = 0$ by the projection theorem so (8.3-13) becomes

$$J_p = E\left(\sum_{k=0}^{p-1}(B_{k+1}\hat{\mathbf{x}}(k+1\,|\,k+1) - \mathbf{r}_{k+1})^T T_{k+1}(B_{k+1}\hat{\mathbf{x}}(k+1\,|\,k+1) - \mathbf{r}_{k+1})\right)$$
$$+E\left(\sum_{k=0}^{p-1}\delta(k+1)^T T_{k+1}\delta(k+1)\right) - 2E\left(\sum_{k=0}^{p-1}\delta(k+1)T_{k+1}\mathbf{r}_{k+1}^T\right)$$
$$+ E\left(\sum_{k=0}^{p-1}\mathbf{v}_k^T L_k \mathbf{v}_k\right). \quad (8.3\text{-}14)$$

At this point, it is necessary to make an additional assumption. We must either suppose that all of the r_k's are zero (the regulator problem) or that

$E(\delta(k)) = 0$ for each k, that is, $\hat{x}(k \mid k)$ is unbiased. The latter would be the case if $\hat{\mathbf{x}}(k \mid k)$ were the BAMVE based on $\mathbf{z}_0, \mathbf{z}_1, \ldots, \mathbf{z}_k$. Let us assume this. In this case, (8.3-14) becomes

$$J_p = E\left(\sum_{k=0}^{p-1}(B_{k+1}\hat{\mathbf{x}}(k+1 \mid k+1) - \mathbf{r}_{k+1})^T T_{k+1}(B_{k+1}\hat{\mathbf{x}}(k+1 \mid k+1) - \mathbf{r}_{k+1})\right)$$

$$+ E\left(\sum_{k=0}^{p-1}\mathbf{v}_k^T T_k \mathbf{v}_k\right) + \operatorname{tr}\left(\sum_{k=0}^{p-1} T_{k+1} P(k+1 \mid k+1)\right). \qquad (8.3\text{-}15)$$

From our work at the beginning of this section, we know that the covariance sequence $P(k \mid k)$ does not depend on $\mathbf{v}(k)$ (in fact can be precomputed), so J_p can be replaced by

$$\begin{aligned} J_p' &= E\left(\sum_{k=0}^{p-1}(B_{k+1}\hat{\mathbf{x}}(k+1 \mid k+1) - \mathbf{r}_{k+1})^T T_{k+1}\right. \\ &\left. \times (B_{k+1}\hat{\mathbf{x}}(k+1 \mid k+1) - \mathbf{r}_{k+1}) + \mathbf{v}_k^T L_k \mathbf{v}_k\right), \end{aligned} \qquad (8.3\text{-}16)$$

since J_p is minimized when J_p' is minimized. The stochastic process $\hat{\mathbf{x}}(k+1 \mid k+1)$ satisfies

$$\begin{aligned} \hat{\mathbf{x}}(k+1 \mid k+1) &= \phi(k)\hat{\mathbf{x}}(k \mid k) + C_k\mathbf{v}(k) + K(k+1) \\ &\quad \times [\mathbf{z}(k+1) - H_{k+1}\hat{\mathbf{x}}(k+1 \mid k)], \end{aligned}$$

this following from (8.3-8) and (8.3-9). We can rewrite this as

$$\hat{\mathbf{x}}(k+1 \mid k+1) = \phi(k)\hat{\mathbf{x}}(k \mid k) + C_k\mathbf{v}(k) + K(k+1)\omega(k). \qquad (8.3\text{-}17)$$

Since $\omega(k)$ is a white process (Lemma 8.3.1), $K(k+1)\omega(k)$ is also white. By Corollary 8.3.2, $K(k+1)\omega(k)$ is uncorrelated with both $\hat{\mathbf{x}}(k \mid k)$ and $\mathbf{v}(k)$.

Equation (8.3-16) represents an objective function for the dynamical system $\hat{\mathbf{x}}(k \mid k)$ satisfying (8.3-17). In other words, we wish to minimize J_p' subject to (8.3-17). In this case, the state vector is $\hat{\mathbf{x}}(k \mid k)$, which is completely observable. In other words, we face the problem of the last section. In summary, we have the following.

8.3.3. **Theorem** (The Separation Principle). *If $\mathbf{x}_k$ is a stochastic vector satisfying*

$$\begin{aligned} \mathbf{x}_{k+1} &= \phi_k\mathbf{x}_k + C_k\mathbf{v}_k + \mathbf{u}_k \\ \mathbf{z}_k &= H_k\mathbf{x}_k + \mathbf{w}_k, \end{aligned}$$

where $\mathbf{u}_k$, $\mathbf{w}_k$ are white sequences with the usual assumptions for Kalman filtering, then the control sequence $\mathbf{v}_k$ necessary to minimize

$$J_p = E\left(\sum_{k=0}^{p-1}(B_{k+1}\mathbf{x}_{k+1} - r_{k+1})^T T_{k+1}(B_{k+1}\mathbf{x}_{k+1} - r_{k+1}) + v_k^T L_k v_k\right)$$

subject to the condition that $C_k \mathbf{v}_k$ *contains no noise terms, and given only the measurements* $\mathbf{z}_0, \ldots, \mathbf{z}_k$ *at time* k, *is given by applying the closed loop deterministic regulator* (8.1-29) *to the Kalman estimate* $\hat{\mathbf{x}}(k \mid k)$. *If* $\mathbf{r}_k \neq 0$, $\hat{\mathbf{x}}(k \mid k)$ *must be affine; if* $\mathbf{r}_k = 0$ *then* $\hat{\mathbf{x}}(k \mid k)$ *may be linear.*

8.4 Exercises

1. Referring to Equation (8.1-25), prove that if W_{k+1} is positive semidefinite, then so is W_k. (Hint: Suppose W_{k+1} was a covariance matrix.)

2. (a) Draw a block diagram showing how you would implement the "closed loop" controller (8.1-29). Your picture should illustrate why the words "closed loop" are used to describe this controller.

   (b) Repeat part (a) for a stochastic controller using Theorem 8.3.3.

3. With reference to Equation (8.1-31), show that the fixed final state problem when $T_1, \ldots, T_{p-1}$ are zero and $T_p^{-1} = 0$ implies controllability.

9

Fixed Interval Smoothing

9.1 Introduction

We now come to the final topic in this text, *fixed interval smoothing.* Just as we have seen with other topics in past chapters, this is an easy problem to state and a very complex problem to solve.

Suppose one has a discrete linear system such as that given in Definition 7.1.1. Further suppose that on the basis of m measurements, for example, $\mathbf{z}(0), \mathbf{z}(1), \ldots, \mathbf{z}(m)$ we wish to estimate $\mathbf{x}(s)$, where $s < m$. In principle, it is clear what we wish to do, namely, project $\mathbf{x}(s)$ onto the n-fold Cartesian product of the linear span of the components of $\mathbf{z}(0), \mathbf{z}(1), \ldots, \mathbf{z}(m)$. Letting

$$\mathbf{y}(m) \triangleq \begin{bmatrix} \mathbf{z}(0) \\ \mathbf{z}(1) \\ \vdots \\ \mathbf{z}(m) \end{bmatrix},$$

the "solution" that we seek is the "batch process" represented by the equation

$$\hat{\mathbf{x}}(s \mid m) = E(\mathbf{x}(s)\mathbf{y}(m)^T)E(\mathbf{y}(m)\mathbf{y}(m)^T)^+\mathbf{y}(m).$$

This is fine[1] as long as we have the measurement vector $\mathbf{y}(m)$. But suppose that all we know is that a Kalman filter has been operating and that $\hat{\mathbf{x}}(r \mid r)$, $P(r \mid r)$, and $P(r+1 \mid r)$ for $r = 0, 1, 2, \ldots, m$, have been secured. The question is, if we know the dynamical system upon which the Kalman filter was based, can we construct $\hat{\mathbf{x}}(s \mid m)$? The answer is yes and is provided by the theorem in the following section.

9.2 The Rauch, Tung, Streibel Smoother

9.2.1. **Theorem** (Rauch, Tung, Streibel—1965 [22]). *Suppose that* $\hat{\mathbf{x}}(r \mid r)$, $P(r \mid r)$, *and* $P(r+1 \mid r)$ *have been secured for* $r = 0, 1, \ldots, m$. *Then for* $r < m$,

(a) $\hat{\mathbf{x}}(r \mid m) = \mathbf{x}(r \mid r) + J[\mathbf{x}(r+1 \mid m) - \phi(r)\mathbf{x}(r \mid r)]$,

---

[1] Unless $E(\mathbf{y}(m)\mathbf{y}(m)^T)$ is too large to handle.

(b) $P(r\,|\,m) = P(r\,|\,r) + J[P(r+1\,|\,m) - P(r+1\,|\,r)]J^T$,

(c) $P(r+1\,|\,r) = \phi(r)P(r\,|\,r)\phi(r)^T + Q(r)$,

where

$$J \triangleq P(r\,|\,r)\phi(r)^T P(r+1\,|\,r)^{-1}.$$

Proof. After all measurements have been completed, this can be formulated as a static updating problem as follows. We assume $\hat{x}(s\,|\,i)$, where $i \leq s$ is known and then update $\hat{\mathbf{x}}$ based on $m - i$ more measurements to obtain $\hat{\mathbf{x}}(s\,|\,m)$; $m \geq s$. This is not as simple as it sounds. The problem stems from the fact that the measurement vector $\mathbf{z}(m)$ for $m > s$ does not bear a simple relation to $\mathbf{x}(s)$ since all of the process noise from s onward has been dynamically propagated to form $\mathbf{x}(m)$ and hence $\mathbf{z}(m)$. The real trick in the following proof is to introduce enough notation to suppress the algebraic nightmare while preserving the essential covariance information.

To begin, we introduce the following notation:

$$\begin{aligned} \Phi(i,i) &\triangleq id \\ \Phi(i+1,i) &\triangleq \phi(i) \\ \Phi(s,i) &\triangleq \phi(s-1)\phi(s-2)\cdots\phi(i); \qquad s > i+1. \end{aligned} \tag{9.2-1}$$

Using this notation we express $\mathbf{z}(s)$ in terms of $\mathbf{x}(i)$, $s > i$, as follows:

$$\begin{aligned} \mathbf{z}(s) &= H(s)\mathbf{x}(s) + \mathbf{w}(s) \\ \mathbf{z}(s) &= H(s)[\phi(s-1)\mathbf{x}(s-1) + \mathbf{u}(s-1)] + \mathbf{w}(s) \\ \mathbf{z}(s) &= H(s)[\phi(s-1)]\phi(s-2)\mathbf{x}(s-2) + \mathbf{u}(s-2)] \\ &\quad + \mathbf{u}(s-1)] + \mathbf{w}(s) \\ \mathbf{z}(s) &= H(s)[\phi(s-1)\phi(s-2)[\phi(s-3)\mathbf{x}(s-3) + \mathbf{u}(s-3)] \\ &\quad + \phi(s-1)\mathbf{u}(s-2) + \mathbf{u}(s-1)] + \mathbf{w}(s), \end{aligned}$$

that is,

$$\begin{aligned} \mathbf{z}(s) &= H(s)[\phi(s,s-3)\mathbf{x}(s-3) + \Phi(s,s-2)\mathbf{u}(s-3) \\ &\quad + \Phi(s,s-1)\mathbf{u}(s-2) + \Phi(s,s)\mathbf{u}(s-1)] + \mathbf{w}(s). \end{aligned}$$

In general, then, for $s > i$

$$\mathbf{z}(s) = H(s)\Phi(s,i)\mathbf{x}(i) + H(s)\sum_{j=i}^{s-1}\Phi(s,j+1)\mathbf{u}(j) + \mathbf{w}(s). \tag{9.2-2}$$

(When $s = i$, we eliminate the sum and have the usual measurement equation.)

Let us write out this equation for $i = r+1$ and $s = r+1, \ldots, m$:

$$s=r+1: \quad \mathbf{z}(r+1) = H(r+1)\Phi(r+1,r+1)\mathbf{x}(r+1) + \mathbf{w}(r+1)$$

$$s=r+2: \quad \mathbf{z}(r+2) = H(r+2)\Phi(r+2,r+1)\mathbf{x}(r+1) + H(r+2)\Phi(r+1,r+1)\mathbf{u}(r+1) + \mathbf{w}(r+2)$$

$$\vdots \qquad \vdots \qquad \vdots$$

$$s=m: \quad \underbrace{\mathbf{z}(m)}_{\mathbf{y}(m,r+1)} = \underbrace{H(m)\Phi(m,r+1)}_{B(m,r+1)}\mathbf{x}(r+1) + \underbrace{H(m)\sum_{j=r+1}^{m-1}\Phi(m,j+1)\mathbf{u}(j) + \mathbf{w}(m)}_{\mathbf{v}(m,r+1)}$$

Defining the concatenated vectors and matrices as indicated above, we have

$$\mathbf{y}(m, r+1) = B(m, r+1)\mathbf{x}(r+1) + \mathbf{v}(m, r+1). \tag{9.2-3}$$

Equation (9.2-3) can be thought of as a measurement equation that measures $\mathbf{z}(r+1), \ldots, \mathbf{z}(m)$ in terms of $\mathbf{x}(r+1)$, corrupted by the noise term $\mathbf{v}(m, r+1)$. Now, all of the indices of $\mathbf{w}$ or $\mathbf{u}$ in $\mathbf{v}(m, r+1)$ are $\geq r+1$, so we have

$$E(\mathbf{x}(r+1)\mathbf{v}(m, r+1)^T) = 0$$

that is, (9.2-3) really acts formally like a measurement equation. Furthermore, if $\mathbf{y}(r, 0)^T = [\mathbf{z}(0)^T, \mathbf{z}(1)^T, \ldots, \mathbf{z}(r)^T]$, then

$$E(\mathbf{y}(r, 0)\mathbf{v}(m, r+1)^T) = 0.$$

This means that the "setup" in Definition 7.1.2 is satisfied and since $\hat{\mathbf{x}}(r+0 \mid r)$ is the BLMVE of $\mathbf{x}(r+1)$ based on $\mathbf{y}(r, 0)$, we can use Theorem 7.2.2 to conclude

$$\hat{\mathbf{x}}(r+1 \mid m) = \hat{\mathbf{x}}(r+1 \mid r) + K[\mathbf{y}(m, r+1) - B(m, r+1)\hat{\mathbf{x}}(r+1 \mid r)]$$

$$P(r+1 \mid m) = P(r+1 \mid r) - KB(m, r+1)P(r+1 \mid r)$$

$$K \triangleq P(r+1 \mid r)B(m, r+1)^T[B(m, r+1)P(r+1 \mid r)B(m, r+1)^T + V(r+1)]^+ \tag{9.2-4}$$

$$V(r+1) \triangleq E(\mathbf{v}(m, r+1)\mathbf{v}(m, r+1)^T].$$

Similarly, letting $i = r$ and $s = r+1, \ldots, m$, we have

$$\mathbf{z}(r+1) = H(r+1)\Phi(r+1, r)\mathbf{x}(r) + H(r)\mathbf{u}(r) + \mathbf{w}(r+1)$$

$$\vdots \qquad \vdots$$

$$\mathbf{z}(m) = H(m)\Phi(m, r)\mathbf{x}(r) + H(m)\textstyle\sum_{j=r}^{m-1}\Phi(m, j+1)\mathbf{u}(u) + \mathbf{w}(m),$$

which can be written in a manner analogous to Equation (9.2-3) as

$$\mathbf{y}(m, r+1) = B(m, r)\mathbf{x}(r) + \mathbf{v}(m, r). \tag{9.2-5}$$

Again,

$$E(\mathbf{x}(r)\mathbf{v}(m,r)^T) = 0,$$

and

$$E(\mathbf{y}(r,0)\mathbf{v}(m,r)^T) = 0,$$

so Theorem 7.2.2 applies once again, and we have

$$\begin{aligned} \hat{\mathbf{x}}(r\,|\,m) &= \hat{\mathbf{x}}(r\,|\,r) + K'[\mathbf{y}(m,r+1) - B(m,r)\hat{\mathbf{x}}(r\,|\,r)] \\ P(r\,|\,m) &= P(r\,|\,r) - K'B(m\,|\,r)P(r\,|\,r) \qquad (9.2\text{-}6) \\ K' &\triangleq P(r\,|\,r)B(m,r)^T[B(m\,|\,r)P(r\,|\,r)B(m\,|\,r)^T + V(r)]^+. \end{aligned}$$

Now, from Equation (9.2-3), we have

$$\begin{aligned} \mathbf{y}(m,r+1) &= B(m,r+1)\mathbf{x}(r+1) + \mathbf{v}(m,r+1) \\ \mathbf{y}(m,r+1) &= B(m,r+1)[\Phi(r+1,r)\mathbf{x}(r) + \mathbf{u}(r)] + \mathbf{v}(m,r+1) \\ \mathbf{y}(m,r+1) &= B(m,r+1)\Phi(r+1,r)\mathbf{x}(r) + B(m,r+1)\mathbf{u}(r) \\ &\quad + \mathbf{v}(m,r+1). \end{aligned}$$

Comparing this to Equation (9.2-5), we deduce

$$B(m,r) = B(m,r+1)\Phi(r+1,r), \quad \text{and}$$

$$\mathbf{v}(m,r) = B(m,r+1)\mathbf{u}(r) + \mathbf{v}(m,r+1). \qquad (9.2\text{-}7)$$

Thus,

$$\begin{aligned} V(r) &= E(\mathbf{v}(m,r)\mathbf{v}(m,r)^T) \\ &= B(m,r+1)E(\mathbf{u}(r)\mathbf{u}(r)^T)B(m,r+1)^T \\ &\quad + B(m,r+1)E(\mathbf{u}(r)\mathbf{v}(m,r+1)^T) \\ &\quad + E(\mathbf{v}(m,r+1)\mathbf{u}(r)^T)B(m,r+1)^T + V(r+1). \end{aligned}$$

The second and third terms are zero since all the noise terms in $\mathbf{v}(m,r+1)$ have index $\geq\ r+1$ and, hence, are uncorrelated with $\mathbf{u}(r)$. Hence, we have

$$V(r) = B(m,r+1)Q(r)B(m,r+1) + V(r+1). \qquad (9.2\text{-}8)$$

We next relate K and K' as follows. By (9.2-6)

$$K' = P(r\,|\,r)B(m,r)^T[B(m\,|\,r)P(r\,|\,r)B(m\,|\,r)^T + V(r)]^+,$$

and so using (9.2-7) and (9.2-8), we have

$$\begin{aligned} K' &= P(r\,|\,r)[B(m,r+1)\Phi(r+1,r)]^T \\ &\quad \times [B(m,r+1)\Phi(r+1,r)P(r\,|\,r)\Phi(r+1,r)^T B(m,r+1)^T \\ &\quad + B(m,r+1)Q(r)B(m,r+1)^T + V(r+1)]^+ \end{aligned}$$

or

$$\begin{aligned} K' &= P(r\,|\,r)\Phi(r+1,r)^T B(m,r+1)^T \\ &\quad \times [B(m,r+1)\{\Phi(r+1,r)P(r\,|\,r)\Phi(r+1,r)^T \\ &\quad + Q(r)\}B(m,r+1)^T + V(r+1)]^+. \end{aligned}$$

But from Equation (7.2-2) (Kalman's theorem), it follows that the term in the braces is simply $P(r+1\,|\,r)$, and so

$$\begin{aligned} K' &= P(r\,|\,r)\Phi(r+1,r)^T B(m,r+1)^T \\ &\quad \times [B(m,r+1)P(r+1\,|\,r)B(m,r+1)^T + V(r+1)]^+. \end{aligned}$$

Now, referring to Equations (9.2-4), multiply the expression for K by $P(r+1\,|\,r)^{-1}$ on the left and obtain

$$P(r+1\,|\,r)^{-1}K = B(m,r+1)^T[B(m,r+1)P(r+1\,|\,r)B(m,r+1)^T + V(r+1)]^+.$$

Then, substituting this in the previous expression, we obtain

$$K' = P(r\,|\,r)\Phi(r+1\,|\,r)^T P(r+1\,|\,r)^{-1}K.$$

Defining

$$J \triangleq P(r\,|\,r)\Phi(r+1\,|\,r)^T P(r+1\,|\,r)^{-1}, \tag{9.2-9}$$

we have

$$K' = JK. \tag{9.2-10}$$

Using Equation (9.2-10) in the first of Equations (9.2-6), we have

$$\hat{\mathbf{x}}(r\,|\,m) = \hat{\mathbf{x}}(r\,|\,r) + JK[\mathbf{y}(m,r+1) - B(m,r)\hat{\mathbf{x}}(r\,|\,r)].$$

However, from the first of Equations (9.2-7), this can be rewritten as

$$\hat{\mathbf{x}}(r\,|\,m) = \hat{\mathbf{x}}(r\,|\,r) + JK[\mathbf{y}(m,r+1) - B(m,r+1)\Phi(r+1,r)\hat{\mathbf{x}}(r\,|\,r)].$$

Now, recalling that in the proof of Kalman's theorem it was shown that

$$\hat{\mathbf{x}}(r+1\,|\,r) = \Phi(r+1,r)\hat{\mathbf{x}}(r\,|\,r) \tag{9.2-11}$$

(this was called Equation (7.2-4) in that theorem), we have

$$\hat{\mathbf{x}}(r\,|\,m) = \hat{\mathbf{x}}(r\,|\,r) + JK[\mathbf{y}(m,r+1) - B(m,r+1)\hat{\mathbf{x}}(r+1\,|\,r)]. \tag{9.2-12}$$

From the first of the equations in (9.2-4), it follows that

$$K[\mathbf{y}(m,r+1) - B(m,r+1)\hat{\mathbf{x}}(r+1\,|\,r)] = \hat{\mathbf{x}}(r+1\,|\,m) - \hat{\mathbf{x}}(r+1\,|\,r),$$

so substituting this into (9.2-12), we have

$$\hat{\mathbf{x}}(r \mid m) = \hat{\mathbf{x}}(r \mid r) + J[\hat{\mathbf{x}}(r+1 \mid m) - \hat{\mathbf{x}}(r+1 \mid r)].$$

Again applying (9.2-11), we obtain Equation (a):

$$\hat{\mathbf{x}}(r \mid m) = \hat{\mathbf{x}}(r \mid r) + J[\hat{\mathbf{x}}(r+1 \mid m) - \phi(r)\hat{\mathbf{x}}(r \mid r)].$$

Equation (b) is obtained as follows. From (9.2-9)

$$J^T = P(r+1 \mid r)^{-1}\Phi(r+1 \mid r)P(r \mid r). \tag{9.2-13}$$

From the second equation in (9.2-4), we have

$$P(r+1 \mid m) - P(r+1 \mid r) = -KB(m, r+1)P(r+1 \mid r).$$

Multiply this on the left by J and on the right by J^T to obtain

$$J[P(r+1 \mid m) - P(r+1 \mid r)]J^T = -JKB(m, r+1)P(r+1 \mid r)J^T.$$

Now, using Equations (9.2-10) and (9.2-13), this becomes

$$J[P(r+1 \mid m) - P(r+1 \mid r)]J^T = -K'B(m, r+1)\Phi(r+1 \mid r)P(r \mid r);$$

but, from the first of Equations (9.2-7), we can write this expression as

$$J[P(r+1 \mid m) - P(r+1 \mid r)]J^T = -K'B(m, r)P(r \mid r).$$

However, comparing this to the second of Equations (9.2-6), we see that

$$J[P(r+1 \mid m) - P(r+1 \mid r)]J^T = P(r \mid m) - P(r \mid r),$$

which is essentially Equation (b).

Equation (c) is exactly Equation (c) from Kalman's theorem and so we are finished. □

9.2.2. **Remarks** (Using the Smoother). We assume $\hat{\mathbf{x}}(r \mid r)$, $P(r \mid r)$, and $P(r+1 \mid r)$ have been calculated and stored for $r = 0, 1, \ldots, m$. Then, with $r = m-1$, we have

$$J = P(m-1 \mid m-1)\phi(m-1)^T P(m \mid m-1)^{-1},$$

whence

$$P(m-1 \mid m) = P(m-1 \mid m-1) + J[p(m \mid m) - P(m \mid m-1)]J^T$$

so

$$\hat{\mathbf{x}}(m-1 \mid m) = \hat{\mathbf{x}}(m-1 \mid m-1) + J[\hat{x}(m \mid m) - \phi(m-1)\hat{\mathbf{x}}(m-1 \mid m-1)].$$

We now know $P(m-1\,|\,m)$ and $\hat{x}(m-1\,|\,m)$. Next let $r = m-2$ and calculate a new J as before. Then

$$P(m-2\,|\,m) = P(m-2\,|\,m-2) + J[P(m-1\,|\,m) - P(m-1\,|\,m-2)]J^T$$

and

$$\hat{\mathbf{x}}(m-2\,|\,m) = \hat{\mathbf{x}}(m-2\,|\,m-2) + J[\hat{\mathbf{x}}(m-1\,|\,m) - \phi(m-2)\hat{\mathbf{x}}(m-2\,|\,m-2)],$$

and so on.

One can also show that smoothing provides an advantage over filtering by showing that

$$P(r\,|\,m) \leq P(r\,|\,r),$$

the (partial) ordering being that induced by positive semidefiniteness.

We show

$$P(r\,|\,m) \leq P(r\,|\,r) \tag{9.2-14}$$

by induction on r. Certainly the above is true for $r = m$. Suppose it is true for r. Then, from Equations (d) and (b) of Kalman's theorem, it follows that

$$P(r\,|\,r-1) - P(r\,|\,r) = P(r\,|\,r-1)H^T(r)[H(r)P(r\,|\,r-1)H^T(r) + R(r)]^+ H(r)P(r\,|\,r-1). \tag{9.2-15}$$

The pseudoinverse of a positive semidefinite matrix is also positive semidefinite (use the spectral theorem). It is also easily proved that if a matrix A is positive semidefinite, then so is BAB^T. These two facts applied to Equation (9.2-15) show that

$$P(r\,|\,r) \leq P(r\,|\,r-1).$$

Combining this with (9.2-14) above, we obtain

$$P(r\,|\,m) \leq P(r\,|\,r-1),$$

and so

$$J[P(r\,|\,m) - P(r\,|\,r-1)]J^T$$

is negative semidefinite; but, referring to Equation (b) of the smoother theorem, this implies (setting r equal to $r-1$)

$$P(r-1\,|\,m) - P(r-1\,|\,r-1) = J[P(r\,|\,m) - P(r\,|\,r-1)]J^T,$$

and so $P(r-1\,|\,m) - P(r-1\,|\,r-1)$ is negative semidefinite, that is,

$$P(r-1\,|\,m) \leq P(r-1\,|\,r-1)$$

(which completes our induction step).

9.3 The Two-Filter Form of the Smoother

In this section we will derive a form of the fixed interval smoother attributable to Fraser and Potter [9]. The Fraser–Potter smoother consists of two Kalman filters, one operating forward in time and the other backward. Our derivation will be from the author's own paper [2]. As usual, we assume the dynamic structure given in Definition 7.1.2, namely,

$$\mathbf{x}(k+1) = \phi(k)\mathbf{x}(k) + \mathbf{u}(k) \tag{9.3-1}$$

and

$$\mathbf{z}(k) = H(k)\mathbf{x}(k) + \mathbf{w}(k), \tag{9.3-2}$$

with the usual assumptions on $\mathbf{u}(k)$ and $\mathbf{w}(k)$. In addition we will *assume* that $\phi(k)^{-1}$ exists for each k. From (9.3-1) and (9.3-2), we can easily derive a set of backward dynamics as follows:

$$\mathbf{x}_B(k+1) = \phi_B(k)\mathbf{x}_B(k) + \mathbf{u}_B(k), \tag{9.3-3}$$

$$\mathbf{z}_B(k) = H_B(k)\mathbf{x}_B(k) + \mathbf{w}_B(k), \tag{9.3-4}$$

where

$$\mathbf{x}_B(k) = \mathbf{x}(m-k), \tag{9.3-5}$$

$$\mathbf{z}_B(k) = \mathbf{z}(m-k), \tag{9.3-6}$$

$$\phi_B(k) = \phi^{-1}(m-k-1), \tag{9.3-7}$$

$$H_B(k) = H(m-k), \tag{9.3-8}$$

$$\mathbf{u}_B(k) = \phi^{-1}(m-k-1)\mathbf{u}(p-k-1), \tag{9.3-9}$$

and

$$\mathbf{w}_B(k) = \mathbf{w}(m-k). \tag{9.3-10}$$

Obviously, the dynamical system, (9.3-3)–(9.3-10), is a convenient fiction—one cannot really reverse time; nor can one simulate these equations and pretend that forward time is backward time because (9.3-9) and (9.3-10) are impossible to generate. If one simply modeled $\mathbf{u}_B(k)$ and $\mathbf{w}_B(k)$ as arbitrary white processes having the correct covariance matricies, it is very, very unlikely that (9.3-5) and (9.3-6) would hold. Thus, we simply regard (9.3-3)–(9.3-5) as a "gedankan" operation; a system we imagine to operate backward in time that happens to take on exactly the same values as we previously obtained when the original system operated forward in time, that is, (9.3-5) and (9.3-6) hold.

Assuming the above imaginary scenario, it follows from (9.3-5) and (9.3-6) that the projection of $\mathbf{x}_B(k)$ and $\mathbf{x}(m-k)$ onto M, the cartesian product of the span of all $p+1$ measurements, will be equal, that is,

$$\hat{\mathbf{x}}_B(k \mid m) = \hat{\mathbf{x}}(m-k \mid m). \tag{9.3-11}$$

This equation implies that the Rauch, Tung, Streibel smoothed estimate is the same whether we use the forward or backward filter equations (providing things are properly initialized of course; see Exercise 9.4.2). The backward filter is, of course, a real operation that can be carried out using forward time as if it were backward time; there is no real-time noise to contend with when operating the backward filter with numerical data.

In what follows, the backward process noise covariance Q_B will be required. This is easily obtained from (9.3-9) and is given by

$$Q_B(k) = \phi^{-1}(m-k-1)Q(m-k-1)\phi^{-T}(m-k-1). \qquad (9.3\text{-}12)$$

Applying Theorem 9.2.1 to the backward filter, one obtains

$$\hat{\mathbf{x}}_B(k \mid m) = \hat{\mathbf{x}}_B(k \mid k) + J_B(k)[\hat{\mathbf{x}}_B(k+1 \mid m) - \phi_B(k)\hat{\mathbf{x}}_B(k \mid k)],$$

where

$$J_B(k) = P_B(k \mid k) + \phi_B^T(k)P_B(k+1 \mid k)^{-1} \qquad (9.3\text{-}13)$$

and P_B represents the backward error covariance. It follows from (9.3-13) and (9.3-11) that

$$\hat{\mathbf{x}}(m-k \mid m) = \hat{\mathbf{x}}_B(k \mid k)+J_B(k)[\hat{\mathbf{x}}(m-k-1 \mid m)-\phi_B(k)\hat{\mathbf{x}}_B(k \mid k)]. \quad (9.3\text{-}14)$$

At this point, we have finished with the notion of indices that increase as running backward in time; this was simply a ploy to obtain (9.3-13). Thus, we will replace $\hat{\mathbf{x}}_B$ in $\hat{\mathbf{y}}$ and P_B by D so as to free $\hat{\mathbf{x}}_B$ and P_B for later use as quantities indexed in forward time. Thus, the second equation in (9.3-13) would contain the symbol D instead of P_B and Eqaution (9.3-14) becomes

$$\hat{\mathbf{x}}(m-k \mid m) = \hat{\mathbf{y}}(k \mid k) + J_B(k)[\hat{\mathbf{x}}(m-k-1 \mid m) - \phi_B(k)\hat{\mathbf{y}}(k \mid k)].$$

Replacing $m-k-1$ by r in this last expression, we obtain

$$\begin{aligned} \hat{x}(r+1 \mid m) \;\; = \;\; & \hat{\mathbf{y}}(m-r-1 \mid m-r-1) + J_B(m-r-1)[\hat{\mathbf{x}}(r \mid m) \\ & - \phi^{-1}(r)\hat{\mathbf{y}}(m-r-1 \mid m-r-1)], \end{aligned}$$

where we have also used (9.3-7). Changing the time index from r to k in the above expression we have

$$\begin{aligned} \hat{x}(k+1 \mid m) = \; & \hat{\mathbf{y}}(m-k-1 \mid m-k-1) + J_B(m-k-1)[\hat{\mathbf{x}}(k \mid m) \\ & - \phi^{-1}(k)\hat{\mathbf{y}}(m-k-1 \mid m-k-1)]. \end{aligned} \qquad (9.3\text{-}15)$$

Note that k is now an index that runs from 0 to m forward in time. Reintroducing the symbols $\hat{\mathbf{x}}_B$ and P_B by

$$\hat{\mathbf{x}}_B(k \mid s) = \mathbf{y}(m-k \mid m-w) \qquad (9.3\text{-}16)$$

$$P_B(k \mid s) = D(m - k \mid m - s), \tag{9.3-17}$$

we now have $\hat{\mathbf{x}}_B$ and P_B representing outputs of the *backward filter* that are indexed in *forward* time. Since Equations (9.3-13) and (9.3-14) will be of no further use to us, this reinterpretation of notation should cause no conflicts. Combining (9.3-16) with (9.3-15), we obtain

$$\hat{\mathbf{x}}(k+1 \mid m) = \hat{\mathbf{x}}_B(k+1 \mid k+1) + J_B(n-k-1)[\hat{\mathbf{x}}(k \mid m) - \phi^{-1}(x)\hat{\mathbf{x}}_B(k+1 \mid k+1)]. \tag{9.3-18}$$

Now, multiplying (9.3-18) through on the left by $J(k)$, substituting the resulting expression for $J(k)\hat{\mathbf{x}}(k+1 \mid m)$ into the appropriate term in part (a) of Theorem 9.2.1, and multiplying the result through by $J_B(m-k-1)^{-1}J(k)^{-1}$, we obtain

$$[J_B(m-k-1)^{-1}J(k)^{-1} - I]\hat{\mathbf{x}}(k \mid m) = [J_B(m-k-1)^{-1}J(k)^{-1} - J_B(m-k-1)^{-1}\phi(k)]\hat{\mathbf{x}}(k \mid k) + [J_B(m-k-1)^{-1} - \phi^{-1}(k)]\hat{\mathbf{x}}_B(k+1 \mid k+1). \tag{9.3-19}$$

Now, from part (c) of Theorem 9.2.1 and the Kalman update equation in Theorem 7.2.2(d), one can easily show that

$$J(k)^{-1} = \phi(k) + Q(k)\phi(k)^{-T}P(k \mid k)^{-1} \tag{9.3-20}$$

so by analogy

$$J_B(m-k-1)^{-1} = \phi_B(m-k-1) + Q_B(m-k-1)\phi_B^{-T}(m-k-1) \times D(m-k-1 \mid m-k-1)^{-1}.$$

Applying (9.3-7), (9.3-12), and (9.3-17) to this last expression, we obtain

$$J_B(m-k-1)^{-1} = \phi(k)^{-1} + \phi(k)^{-1}Q(k)P_B(k+1 \mid k+1)^{-1}. \tag{9.3-21}$$

Substituting (9.3-20) and (9.3-21) into equation (9.3-19) and factoring $\phi^{-1}(k)$ from the left of the resulting expression, we have

$$[\phi(k) + P_B(k+1 \mid k+1)\phi(k)^{-T}P(k \mid k)^{-1} + Q(k)\phi(k)^{-T}P(k \mid k)^{-1}]\hat{\mathbf{x}}(k \mid m) = [P_B(k+1 \mid k+1)\phi(k)^{-T}P(k \mid k)^{-1} + Q(k)\phi(k)^{-T}P(k \mid k)^{-1}]\hat{\mathbf{x}}(k \mid k) + \hat{\mathbf{x}}_B(k+1 \mid k+1). \tag{9.3-22}$$

Note that (9.3-22) was obtained from the sole assumption (9.3-11). We can apply (9.3-22) in the backward direction by replacing $\phi(k)$ by $\phi_B(r)$, $P_B(k+1 \mid k+1)$ by $P(m-r-1 \mid m-r-1)$, $P(k \mid k)$ by $D(r \mid r)$, $\hat{\mathbf{x}}(k \mid m)$ by $\hat{\mathbf{y}}(r \mid m)$, $\hat{\mathbf{x}}_B(k+1 \mid k+1)$ by $\hat{\mathbf{x}}(m-r-1 \mid m-r-1)$, and so on. Reindexing the resulting expression in the forward direction by letting $k = m - r - 1$, we obtain

$$\begin{aligned}
&[\phi^{-1}(k) + P(k\,|\,k)\phi(k)^T P_B(k+1\,|\,k+1)^{-1} \\
&+ \phi(k)^{-1}Q(k)P(k+1\,|\,k+1)^{-1}[\hat{\mathbf{x}}(k+1\,|\,m) \\
&= [P(k\,|\,k)\phi(k)^T P_B(k+1\,|\,k+1)^{-1} \\
&+ \phi(k)^{-1}Q(k)P_B(k+1\,|\,k+1)^{-1}]\hat{\mathbf{x}}_B(k+1\,|\,k+1) + \hat{\mathbf{x}}(k\,|\,k). \qquad (9.3\text{-}23)
\end{aligned}$$

Multiplying Equation (9.3-23) on the left by $\phi(k)$ and using Equations (7.2-4) and (d) of Theorem 7.2.2, it follows that

$$\begin{aligned}
\hat{\mathbf{x}}(k+1\,|\,m) \;=\; & [P(k+1\,|\,k)^{-1} + P_B(k+1\,|\,k+1)]^{-1} \\
& \times [P(k+1\,|\,k)^{-1}\hat{\mathbf{x}}(k+1\,|\,k) \\
& + P_B(k+1\,|\,k+1)^{-1}\hat{\mathbf{x}}_B(k+1\,|\,k+1)]. \qquad (9.3\text{-}24)
\end{aligned}$$

Equation (9.3-24) is the discrete, two-filter form of the smoother. Notice that it has a form that one might describe as a matrix convex combination of the forward and backward filter estimates. A few words about the above derivation are in order.

If one implements a backward Kalman filter as above, then it is necessary to assume that $\mathbf{u}_B(k)$ is uncorrelated with $\mathbf{x}_B(s)$ and $\hat{\mathbf{x}}_B(s)$ for $s \geq k$ (remember time is now indexed in the forward direction). This is inconsistent with the assumptions (9.3-9) and (9.3-10). In fact, it is necessary to drop these expressions once the backward process noise covariance (9.3-12) and the backward measurement noise covariance matrix are calculated (both are necessary to implement the backward filter). The result, (9.3-24), therefore, implicitly assumes that the backward filter comes from a dynamical system described by (9.3-3) and (9.3-8), driven by noises $\mathbf{u}_B$ and $\mathbf{w}_B$ that satisfy the usual correlation properties associated with Kalman filters and have covariances that are calculated using (9.3-9) and (9.3-10); (9.3-9) and (9.3-10), however, are dropped after these calculations are made.

With the above discussion in mind, it would seem reasonable to suppose that the statistics of the forward and backward filter are independent from one another (independence metaprinciple) and, hence, that the forward and backward estimation errors are uncorrelated, specifically, that

$$C(k+1) \triangleq E[(\hat{\mathbf{x}}_B(k+1\,|\,k+1) - \mathbf{x}(k+1))(\hat{\mathbf{x}}(k\,|\,k) - x(k)] \qquad (9.3\text{-}25)$$

is zero. In the usual derivation of the two-filter form of the smoother, this assumption about C is made. In this case, the derivation of (9.3-24) is much, much easier than the above derivation since one can use a theorem about combining two uncorrelated least squares estimates of the same quantity (see [17]). However, if one were to study a dynamical system whose dynamic behavior was completely deterministic ($\mathbf{u}(k) = \mathbf{0}$), then it is not really clear that $C = 0$ is reasonable. This said, perhaps, $C \neq 0$ could occur in other circumstances!

If one makes no assumption about C, then it has already been demonstrated that (9.3-24) holds. In addition, if we define

$$P_s(k) = E((\hat{\mathbf{x}}(k \mid m) - x(k))(\hat{\mathbf{x}}(k \mid m) - \mathbf{x}(k))^T), \tag{9.3-26}$$

then using (9.3-24), it is reasonably easy to show that

$$\begin{aligned} P_s(k+1) = {} & [P(k+1 \mid k)^{-1} + P_B(k+1 \mid k+1)^{-1}]^{-1} \\ & + [P(k+1 \mid k)^{-1} + P_B(k+1 \mid k+1)^{-1}]^{-1} \\ & \times [P(k+1 \mid k)^{-1}\phi(k)C^T P_B(k+1 \mid k+1)^{-1} \\ & + P_B(k+1 \mid k+1)^{-1} C\phi(k)^T P(k+1 \mid k)^{-1}] \\ & \times [P(k+1 \mid k)^{-1} + P_B(k+1 \mid k+1)^{-1}]^{-1}, \end{aligned} \tag{9.3-27}$$

so that the smoothed error covariance expressed in terms of $P(k+1 \mid k)$ and $P_B(k+1 \mid k+1)$ depends upon C. It is reasonable to ask under what circumstances $C = 0$.

The author studied this problem for continuous systems in Reference [2]. The result obtained there was that if Q was nonsingular, then $C = 0$. Moreover, if Q was singular, and the system was separated into noisy and noise-free states (by diagonalizing Q), then the portion of C corresponding to the noisy states was zero. This result is especially gratifying since Fraser has shown that there is no advantage in smoothing noise-free states rather than filtering them [8]. In other words, in the states where it matters, $C = 0$.

It is our contention that the results in Reference [2] should carry over to the discrete case. As of this writing we have no proof of this conjecture, and so this seems to us like an excellent place to stop!

9.4 Exercises

1. Referring to Exercise 7.5.3, part (c), calculate $\hat{\mathbf{x}}(k \mid r)$, for $k = 0, 1, 2, 3,$ 4, using Theorem 9.2.1.

2. Explain how to initialize the backward filter equations in the Fraser–Potter smoother.

3. Repeat Exercise 9.4.1 using the Fraser–Potter form of the smoother. Which form of the smoother seems to be the more reasonable to implement?

Appendix A

Constructing Measures

A1. **The Problem.** Suppose we have a set X and a collection of subsets, denoted $\mathcal{C}_0$, such that for one reason or another each element in $\mathcal{C}_0$ is intuitively measurable and its measure is known. We wish to generate a class of subsets of X, denoted $\mathcal{M}$, such that

(1) $\mathcal{C}_0 \subset \mathcal{M}$;

(2) there is a measure μ on X such that $(X, \mathcal{M}, \mu)$ is a measure space; and

(3) if $A \in \mathcal{C}_0$, then $\mu(A)$ coincides with the (intuitive) measure of A we started with.

Two examples will run throughout the discussion.

A2. **Examples.** (1) Let $X = \mathrm{R}^2$ and let $\mathcal{C}_0$ be the collection of all "half-open" rectangles of the form

$$R = (a_1, b_1] \times (a_2, b_2].$$

For each such rectangle, we define a number

$$\tau(R) = (b_1 - a_1) \times (b_2 - a_2).$$

Here, $\tau(R)$ is clearly the area of R.

(2) Let Ω be some set of outcomes that result from some experiment, operation, random trial, etc. We suppose that for certain subsets of Ω, it is possible to associate a number with each of them that represents the probability of obtaining an outcome in that set. We will call such a set a primitive event and denote the class of all such primitive events by $\mathcal{C}_0$. For $A \in \mathcal{C}_0$, we will denote by $\tau(A)$ the probability that an outcome in A occurs. Clearly, then, the criterion for a set being included in $\mathcal{C}_0$ is that we have some means of measuring its probability. This may be some fancy mathematical construction or simply a belief that $\tau(A)$ makes sense because we can physically describe the event A. More on this idea will be presented in the regular lectures. Note that $\tau(\Omega) = 1$ should be required.

A3. **Definition.** By a *countable covering class* for X we mean a pair $(\mathcal{C}_0, \tau)$ such that

(a) for each $A \subset X$, there is a countable collection $\mathcal{C} \subset \mathcal{C}_0$ such that $\cup\mathcal{C} \supset A$. $\mathcal{C}$ is called a *cover* of A.

(b) for each $C \in \mathcal{C}$, $\tau(C) \in \mathrm{R}$ and $\tau(C) \geq 0$.

(c) $\phi \in \mathcal{C}_0$ and $\tau(\phi) = 0$ (technical consideration).

A4. **Examples.** (1) For R^2, the class $\mathcal{C}_0$ of half-open rectangles clearly satisfies A3.

(2) For Ω, the requirement A3 is essentially that of requiring a stochastic model to be exhaustive, that is, each outcome must be in at least one primitive event.

A5. **Definition.** (1) If $\mathcal{C} = \{C_i, i = 1, 2, \ldots\}$, define

$$\tau(\mathcal{C}) = \sum_{i=1}^{\infty} \tau(C_i).$$

(2) For any set $A \subset X$, define

$$\mu^*(A) = \inf\{\tau(\mathcal{C}) \,|\, \mathcal{C} \quad \text{is a cover of } A\};$$

$\mu^*(A)$ is called the *outer measure* of A.

A6. **Remark.** It is difficult to provide intuitively clear examples of the constructing of outer measures since the above infimum will, even for simple shapes, be an infimum of an uncountable set of numbers. Here is a simple example that might help some.

A7. **Example.** In R^2, let

$$A = \{(x_1, x_2) \,|\, 0 \leq x_1 \leq 1,\ 0 \leq x_2 \leq 1,\ x_1, x_2 - \text{rational}\}.$$

Since the set of pairs of rationals is countable (any real analysis book settles this in Chapter 1), the elements in A can be enumerated, say $p_1, p_2, p_3, \ldots$. For each $\epsilon > 0$, construct a half-open rectangle R_k around p_k whose area is $\epsilon/2^k$. Then, if $\mathcal{C}_\epsilon = \{R_k\}$, we have that $\mathcal{C}_\epsilon$ is a cover of A and

$$\tau(\mathcal{C}_\epsilon) = \sum_{k=1}^{\infty} \epsilon/2^k = \epsilon \frac{1}{1 - 1/2} = 2\epsilon.$$

Then

$$\begin{aligned} \mu^*(A) &= \inf\{\tau(\mathcal{C}_\epsilon) \,|\, \epsilon > 0\} \\ &= \inf\{2\epsilon \,|\, \epsilon > 0\} = 0. \end{aligned}$$

In general, one can think of $\{\tau(\mathcal{C}) \,|\, \mathcal{C}$ covers $A\}$ as the set of all approximations to the "area" of A (if A has an area!). Note that sets always have outer areas.

A8. **Remark.** Two facts about infimums will be used over and over again.

(1) If $A \subset B \subset \mathrm{R}$, then $\inf B \leq \inf A$.

(2) If $x_0 = \inf(A)$, then, for every $\epsilon > 0$, there is a $y \in A$ with $y \leq x_0 + \epsilon$.

$$------|_{x_0}------|_{y}------|_{x_0+\epsilon}$$

A9. **Theorem.** *Here, μ^* as defined above has the following properties.*

(1) $\mu^*(\phi) = 0$.

(2) $A \subset B \Rightarrow \mu^*(A) \leq \mu^*(B)$.

(3) $\mu^*\left(\bigcup_{i=1}^{\infty} A_i\right) \leq \sum_{i=1}^{\infty} \mu^*(A_i)$.

Proof. (1) The first property follows from the fact that $\tau(\phi) = 0$.

(2) If $\mathcal{C}$ is a cover of B and $A \subset B$, then $\mathcal{C}$ is a cover of A. Thus,

$$\{\tau(\mathcal{C}) \,|\, \mathcal{C} \text{ is a cover of } B\} \subset \{\tau(\mathcal{C}) \,|\, \mathcal{C} \text{ is a cover of } A\}.$$

By Remark (1) of A8,

$$\inf\{\tau(\mathcal{C}) \,|\, \mathcal{C} \text{ is a cover of } A\} \leq \inf\{\tau(\mathcal{C}) \,|\, \mathcal{C} \text{ is a cover of } B\},$$

that is,

$$\mu^*(A) \leq \mu^*(B).$$

(3) Let $\epsilon > 0$ be arbitrary. For each positive integer n there is a cover

$$\mathcal{C}_n = \{R_{nk} \,|\, k = 1, 2, \ldots\}$$

such that

$$\bigcup_{k=1}^{\infty} R_{nk} \supset A_n$$

and

$$\tau(\mathcal{C}_n) \leq \mu^*(A_n) + \epsilon/2^n$$

(follows from definition of μ^* and Remark A8 (2)).

$$------|_{\mu^*(A_n)}------|_{\tau(\mathcal{C}_n)}------|_{\epsilon/2^n+\mu^*(A_n)}$$

Thus,

$$\sum_{k=1}^{\infty} \tau(R_{nk}) \leq \mu^*(A_n) + \epsilon/2^n.$$

Now,

$$\bigcup_{n=1}^{\infty} \bigcup_{k=1}^{\infty} R_{nk} \supset \bigcup_{n=1}^{\infty} A_n,$$

so

$$\{R_{nk} \mid k = 1, 2, \ldots; n = 1, 2, \ldots\} \quad \text{is a cover of} \quad \bigcup_{n=1}^{\infty} A_n.$$

Thus

$$\begin{aligned} \mu^*\left(\bigcup_{n=1}^{\infty} A_n\right) &\leq \sum_{n=1}^{\infty}\sum_{k=1}^{\infty} \tau(R_{nk}) \\ &\leq \sum_{n=1}^{\infty}[\mu^*(A_n) + \epsilon/2^n] = \sum_{n=1}^{\infty} \mu^*(A_n) + 2\epsilon. \end{aligned}$$

But ϵ arbitrary $\Rightarrow \mu^*(\cup_{n=1}^{\infty} A_n) \leq \sum_{n=1}^{\infty} \mu^*(A_n)$. □

A10. **Definition.** Any function defined on a collection of subsets of a set satisfying (1), (2), and (3) of A9 is called an *outer measure.* The above construction shows that sequential covering classes always produce outer measures.

A11. **Discussion.** We next wish to define measurability. In Lebesgue's original work [H. Lebesgue, *Lecons sur L'integration,* 2nd ed., Gauthier-Villars, 1928], he defined measurability in a manner much like the following. First, he restricted his attention to subsets of some fixed rectangle R_0. Then, for any subset $E \subset R_0$, he defined the inner measure of E by

$$\mu_*(E) = \mu^*(R_0) - \mu^*(R_0 \setminus E).^1$$

He then said that E is measurable providing

$$\mu^*(E) = \mu_*(E).$$

This is obviously equivalent to saying that

$$\mu^*(E) + \mu^*(R_0 \setminus E) = \mu^*(R_0).$$

Although this definition is adequate for working with areas, volumes, etc., it is not an easy condition with which to work. Moreover, in more abstract settings, for example, probability spaces, it is not strong enough to produce desired results.

The central idea used in general measure theory is attributable to a theorem by Caratheodory that says $E \subset R_0$ is measurable in the above sense iff for every $A \subset R_0$

$$\mu^*(A) = \mu^*(A \cap E) + \mu^*(A \cap E').$$

Note that the statement of the condition makes sense whether E, $A \subset R_0$ or not, and it is this fact that we exploit as follows.

[1] This is not really Lebesgue's definition but is equivalent to it.

A12. **Definition.** Let μ^* be an outer measure on X. [Note that at this point the roles of $\mathcal{C}_0$ and τ are finished—they were used to obtain μ^*]. A subset $E \subset X$ is called *measurable* provided that for every $A \subset X$

$$\mu^*(A) = \mu^*(A \cap E) + \mu^*(A \cap E').$$

In case E is measurable, we define the measure of E, $\mu(E)$, by

$$\mu(E) \triangleq \mu^*(E).$$

Note that by part (3) of A9 one always has

$$\mu^*(A) \leq \mu^*(A \cap E) + \mu^*(A \cap E'),$$

so it suffices when checking measurability to show

$$\mu^*(A) \geq \mu^*(A \cap E) + \mu^*(A \cap E'). \tag{MS}$$

The class of all μ^*-measurable sets will be denoted by $\mathcal{M}$.

A13. **Definition.** A collection $\mathcal{F}$ of subsets of a set X is called a *σ-field* iff

(1) $A \in \mathcal{F} \Rightarrow A' \in \mathcal{F}$.

(2) $\{A_i \mid i = 1, 2, 3, \ldots\} \subset \mathcal{F} \Rightarrow \cup_{i=1}^{\infty} A_i \in \mathcal{F}$.

A14. **Theorem.** *The set $\mathcal{M}$ as defined in* A12 *is a σ-field and, μ satisfies the following two conditions:*

(1) $\mu(\phi) = 0$.

(2) *if* $\{E_i \mid i = 1, 2, \ldots\} \subset \mathcal{M}$, $i \neq j \Rightarrow E_i \cap E_j = \phi$, *then*

$$\mu\left(\bigcup_{i=1}^{\infty} E_i\right) = \bigcup_{i=1}^{\infty} \mu(E_i).$$

Proof. The proof is broken into a series of lemmas.

Lemma A. *If $\mu^*(E) = 0$, then E is measurable. It follows that ϕ is measurable and $\mu(\phi) = 0$.*

Proof. For any set A, $A \cap E \subset E$ and $A \cap E' \subset A$. Hence, $\mu^*(A \cap E) + \mu^*(A \cap E') \leq \mu^*(E) + \mu^*(A) = 0 + \mu^*(A) = \mu^*(A)$ so condition MS is satisfied for E.

Lemma B. *If $E \in \mathcal{M}$, then $E' \in \mathcal{M}$.*

Proof. This follows from $E'' = E$ and Definition A12.

Lemma C. *Any finite union of measurable sets is measurable.*

Proof. We will show this for two sets, the rest follows at once by induction.

Let $E_1, E_2 \in \mathcal{M}$. Since E_1 is measurable, we have that for every A

$$\mu^*(A) = \mu^*(A \cap E_1) + \mu^*(A \cap E_1'). \tag{1}$$

Testing the measurability of E_2 using $A \cap E_1'$, we have

$$\begin{aligned} \mu^*(A \cap E_1') &= \mu^*(A \cap E_1' \cap E_2) + \mu^*(A \cap E_1' \cap E_2') \\ &= \mu^*(A \cap E_1' \cap E_2) + \mu^*(A \cap (E_1 \cup E_2)'). \end{aligned} \tag{2}$$

Now,

$$\begin{aligned} (A \cap E_1' \cap E_2) \cup (A \cap E_1) &= A \cap [(E_1' \cap E_2) \cup E_1] \\ &= A \cap [E_1 \cup E_2]. \end{aligned}$$

So,

$$\mu^*(A \cap E_1' \cap E_2) + \mu^*(A \cap E_1) \geq \mu^*(A \cap (E_1 \cup E_2)). \tag{3}$$

Hence, substituting (2) into (1) and using (3),

$$\begin{aligned} \mu^*(A) &= \mu^*(A \cap E_1) + \mu^*(A \cap E_1' \cap E_2) + \mu^*(A \cap (E_1 \cup E_2') \\ &\geq \mu^*(A \cap (E_1 \cup E_2)) + \mu^*(A \cap (E_1 \cup E_2)'). \end{aligned}$$

This last inequality is condition MS for $E_1 \cup E_2$.

Lemma D. *If $\{E_k\}$ is a sequence of disjoint measurable sets and if for each n*

$$S_n \triangleq \bigcup_{k=1}^{n} E_k,$$

then for each n and any set A, S_n is measurable and

$$\mu^*(A \cap S_n) = \sum_{k=1}^{n} \mu^*(A \cap E_k).$$

Proof. Use induction on n. It is trivial for $n = 1$. Assume true for n and use $A \cap S_{n+1}$ as a test set for S_n:

$$\begin{aligned} \mu^*(A \cap S_{n+1}) &= \mu^*(A \cap S_n \cap S_{n+1}) + \mu^*(A \cap S_{n+1} \cap S_n') \\ &= \mu^*(A \cap S_n) + \mu^*(A \cap E_{n+1}) \\ \text{(induction hypothesis)} &= \sum_{k=1}^{n} \mu^*(A \cap E_k) + \mu^*(A \cap E_{n+1}) \\ &= \sum_{k+1}^{n+1} \mu^*(A \cap E_k). \end{aligned}$$

Thus S_{n+1} measurable follows from Lemma C.

Lemma E. *If $\{E_k\}$ is a sequence of disjoint measurable sets and if $S \triangleq \bigcup_{k=1}^{\infty} E_k$, then for any set A*

$$\mu^*(A \cap S) = \sum_{k=1}^{\infty} \mu^*(A \cap E_k).$$

Proof. Using the notation in Lemma D

$$A \cap S \supset A \cap S_n,$$

so for each n

$$\mu^*(A \cap S) \geq \mu^*(A \cap S_n) = \sum_{k=1}^{n} \mu^*(A \cap E_k).$$

Let $n \to \infty$ to obtain

$$\mu^*(A \cap S) \geq \sum_{k=1}^{\infty} \mu^*(A \cap E_k).$$

The reverse inequality follows from A9, part (3).

Lemma F. *Any countable union of disjoint measurable sets is measurable.*

Proof. We employ the notation of Lemmas D and E. First note $S_n' \subset \mathcal{M}$ and $S_n' \supset S'$. Hence, for any $A \subset X$,

$$\begin{aligned} \mu^*(A) &= \mu^*(A \cap S_n) + \mu^*(A \cap S_n') \\ &= \sum_{k=1}^{n} \mu^*(A \cap E_k) + \mu^*(A \cap S_n') \\ &\geq \sum_{k=1}^{n} \mu^*(A \cap E_k) + \mu^*(A \cap S'). \end{aligned}$$

Let $n \to \infty$ and use Lemma E:

$$\mu^*(A) \geq \sum_{k=1}^{\infty} \mu^*(A \cap E_k) + \mu^*(A \cap S')$$

or

$$\mu^*(A) \geq \mu^*(A \cap S) + \mu^*(A \cap S'),$$

so condition (MS) is satisfied.

Lemma G. *Any countable union of measurable sets is measurable.*

Proof. For each n write

$$F_n \triangleq E_n \setminus \bigcup_{k=1}^{n-1} E_k.$$

Then by Lemmas B and D, each F_n is measurable, the F_n's are disjoint by construction, and

$$\bigcup_{n=1}^{\infty} F_n = \bigcup_{n=1}^{\infty} E_n,$$

so by Lemma F we have proved Lemma G.

Finally, note that if the E_k's are disjoint, set $A = X$ in Lemma E and obtain

$$\mu\left(\bigcup_{k=1}^{\infty} E_k\right) = \mu^*\left(\bigcup_{k=1}^{\infty} E_k\right) = \sum_{k=1}^{\infty} \mu^*(E_k) = \sum_{k=1}^{\infty} \mu(E_k). \qquad \square$$

A15. **Definition.** Let X be a set, $\mathcal{F}$ a σ field of subsets of X, and μ a real valued function defined on $\mathcal{F}$ satisfying

(1) $\mu(\phi) = 0$,

(2) $\mu(F) \geq 0$ for every $F \in \mathcal{F}$, and

(3) $\{F_n\} \subset \mathcal{F}$, $\{F_n\}$ pairwise disjoint

$$\Rightarrow \mu\left(\bigcup_{n=1}^{\infty} F_n\right) = \sum_{n=1}^{\infty} \mu(F_n).$$

Then μ is called a *measure* and $(X, \mathcal{F}, \mu)$ is called a *measure space.* Theorem A14 asserts that if μ^* is any outer measure on X, then one can always construct a measure space using the Caratheodory condition.

A16. **Remark.** The above scenerio can be summarized by the following steps.

(1) In your space X, choose a sequential covering class $\mathcal{C}_0$.

(2) On $\mathcal{C}_0$ define a nonnegative function τ such that $\tau(\phi) = 0$.

(3) Construct μ^* on X using $(\mathcal{C}_0, \tau)$ via Definition A5.

(4) Using μ^*, construct $\mathcal{M}$ via the Caratheodory condition (MS).

(5) Construct μ by restricting μ^* to $\mathcal{M}$. Then $(X, \mathcal{M}, \mu)$ is always a measure space!

There is one missing piece and it is such a gross oversight that it is easily overlooked. In any practical example we would want the elements of $\mathcal{C}_0$ to be measurable. They need not be! The following condition, however, will produce this result.

A17. **Definition.** A sequential covering class $\mathcal{C}_0$ for X is called *regular* iff for each $C \in \mathcal{C}_0$ there exist covers $\mathcal{C}_1$ and $\mathcal{C}_2$ such that

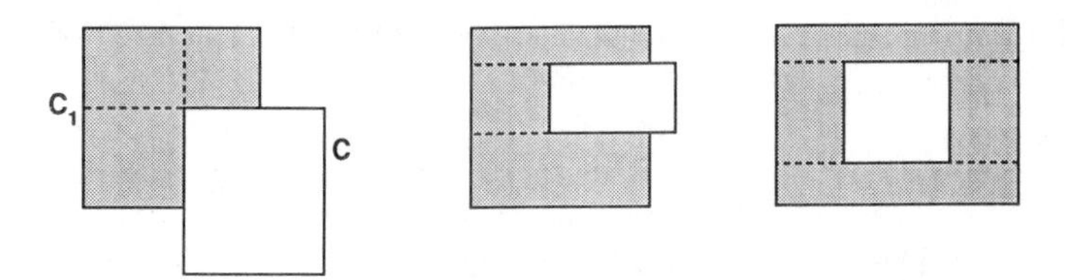

FIGURE A.1. Various configurations that satisfy condition (2) in Definition A17.

(1) $(\cup\mathcal{C}) \cap C = \cup\mathcal{C}_1$.

(2) $(\cup\mathcal{C}) \setminus C = \cup\mathcal{C}_2$.

(3) $\tau(\mathcal{C}) = T(\mathcal{C}_1) + T(\mathcal{C}_2)$.

A18. **Examples.** (1) For the half-open rectangles in R^2, it is easily seen that the intersection of two such rectangles is again one. Hence, for $\mathcal{C}_1$, just define

$$\mathcal{C}_1 \triangleq \{C_i \cap C \,|\, C_i \in \mathcal{C}\}$$

and (1) holds. Condition (2) is slightly more complicated. One need only note that $C_i \setminus C$ be written as a finite union of rectangles, the number depending upon the configuration, for example, Figure A.1. Hence, for each i, we obtain a finite collection of half-open rectangles $\{B_1^i, \ldots, B_{k_i}^i\}$, $0 \leq k_i \leq 4$ (0 meaning the collection is the empty rectangle). If we let

$$\mathcal{C}_2 \triangleq \{B_1^i, \ldots, B_{k_i}^i \,|\, i = 1, 2, \ldots\},$$

then $\mathcal{C}_2$ is countable, $\cup_{j=1}^{k_i} B_j^i = C_i \setminus C$, whence $\cup_{i=1}^{\infty} \cup_{j=1}^{k_i} B_i^j = \cup_{i=1}^{\infty}(C_i \setminus C) = \cup\mathcal{C}\setminus C$ as required by condition (2). Finally $\tau(\mathcal{C}_1)+\tau(\mathcal{C}_2) = \sum_{i=1}^{\infty} \tau(C_i \cap C) + \sum_{i=1}^{\infty} \sum_{j=1}^{k_i} \tau(B_j^i) = \sum_{i=1}^{\infty} \tau(C_i \cap C) + \sum_{i=1}^{\infty}[\tau(C_i) - \tau(C \cap C_i)] = \sum_{i=1}^{\infty} \tau(C_i) = \tau(\mathcal{C})$.

(2) Whereas the arguments in (1) are technical and depend upon the shape of the C_i's for Ω, we simply argue that if A and B are primitive events, then $A \cap B$ (A and B) and $A \setminus B$ (A but not B) are events. If this seems a bit glib, we will have more to say in the regular lectures.

A19. **Theorem.** *If $\mathcal{C}_0$ is a regular covering class, then each $C \in \mathcal{C}_0$ is measurable.*

Proof. Let $C \in \mathcal{C}_0$ and let A be any set X; $\epsilon > 0$. Let $\mathcal{C}$ be a cover of A such that

$$\tau(\mathcal{C}) \leq \mu^*(A) + \epsilon.$$

Then by A17 (use that notation), $\mathcal{C}_1$ is a cover of $A \cap C$ (since $A \cap C \subset (\cup\mathcal{C}) \cap C$) and $\mathcal{C}_2$ is a cover of $A \cap C'$. Thus, $\mathcal{C}_1 \cup \mathcal{C}_2$ is a cover of A and

$$\tau(\mathcal{C}_1 \cup \mathcal{C}_2) = \tau(\mathcal{C}_1) + \tau(\mathcal{C}_2) = \tau(\mathcal{C})$$

by (3) of A17. Hence,

$$\tau(C_1) + \tau(C_2) \leq \mu^*(A) + \epsilon,$$

so

$$\mu^*(A \cap C) + \mu^*(A \cap C') \leq \mu^*(A) + \epsilon.$$

Since ϵ is arbitrary, the result follows. □

A20. **Remark.** One can easily construct other regular measures on R^2. For example, if $f \geq 0$ is Riemann integrable (even locally), one could define

$$\tau((a_1, b_1] \times (a_2, b_2]) = \int_{a_1}^{b_1} \int_{a_2}^{b_2} f(x_1, x_2)\, dx_1\, dx_2$$

and obtain a regular cover.

In the two examples in this appendix, the first represents the plane R^2, $\mathcal{M}$ all of those sets in R that have area, and for $A \in \mathcal{M}$, $\mu(A)$ represents the area of A. This type of area is called *Lebesgue area* (or *Lebesgue measure*) after its inventor. If one weakens covering classes from countable to finite and goes through the above constructions, one obtains a finitely additive area function called *Jordan content* or *Jordan area.*

In the second example, Ω represents a set of outcomes and $\mathcal{E}$, the measurable sets, are all those subsets of Ω that have a probability associated with them. Also, $\mu(A)$ (usually written $p(A)$) represents the probability of obtaining an outcome in A. Here, $\mathcal{E}$ is called the set of events.

Appendix B

Two Examples from Measure Theory

In this appendix we will construct two examples that are very, very useful for the construction of counterexamples. The first is the so-called Cantor Set.

B1. **The Cantor Set.** We will construct this example inductively on the closed interval $[0,1]$. We first construct the set M_0, which we will refer to as the middle third set.

The set M_0 will be a union of a countable collection of open sets M_i; $i = 1, 2, \ldots$; each M_i being a finite union of open intervals. To begin with

$$M_1 = \left(\frac{1}{3}, \frac{2}{3}\right).$$

If we look at the complement of M_1 in $[0,1]$, we have two sets, $[0,1/3]$ and $[2/3,1]$. The set M_2 will be formed by taking the "middle third" of each of these sets. Thus,

$$M_2 = (1/9, 2/9) \cup (7/9, 8/9).$$

The complement of $M_1 \cup M_2$ in $[0,1]$ consists of four intervals: $[0,1/9]$, $[2/9,3/9]$, $[6/9,7/9]$, and $[8/9,9/9]$. Therefore, M_3 will consist of the middle third of each of these intervals, that is,

$$M_3 = (1/27, 2/27) \cup (7/27, 8/27) \cup (19/27, 20/27) \cup (25/27, 26/27).$$

The induction should now be clear. Given M_n, the set $[0,1] \setminus \cup_{k=1}^{n} M_k$ consists of 2^k disjoint closed intervals, each of length $1/3^k$. M_{n+1} is formed by taking the open middle third of each of these closed intervals and forming the union. We can make three observations:

(1) the M_k's are pairwise disjoint,

(2) $\mu(M_k) = \dfrac{2^{k-1}}{3^k} = \dfrac{1}{2}\left(\dfrac{2}{3}\right)^k$, and

(3) if $A_n \triangleq \cup_{k=1}^{n} M_k$, then the distance between adjacent intervals in A_n is $1/3^n$.

We define

$$M_0 = \bigcup_{k=1}^{\infty} M_k$$

and

$$C_0 = [0,1] \setminus M_0.$$

Since M_0 is open, C_0 is closed. By (1) and (2) above,

$$\begin{aligned}
\mu(M_0) &= \mu\left(\bigcup_{k=1}^{\infty} M_k\right) \\
&= \sum_{k=1}^{\infty} \mu(M_k) \\
&= \frac{1}{2}\sum_{k=1}^{\infty}\left(\frac{2}{3}\right)^k \\
&= \frac{1}{2}\cdot\frac{2/3}{1-(2/3)} \\
&= 1.
\end{aligned}$$

Hence, we have

$$\mu(C_0) = 1 - \mu(M_0) = 0,$$

and so C_0 cannot contain any intervals. If $x \in C_0$ and $\epsilon > 0$, then by (3) above the open interval $(x-\epsilon, x+\epsilon)$ contains points in at least two subintervals of M_0. Finally, since C_0 cannot contain intervals, $(x-\epsilon, x+\epsilon)$ must contain other points in C_0, so that every point in C_0 is a limit point of C_0. Thus, C_0 is a so-called perfect set (a set that is equal to its set of limit points).

Finally, we can give a very interesting description of C_0 by using ternary fractions. Each real number in $[0,1]$ can be expressed in the form

$$0 \,.\, a_1 a_2 a_3 \cdots,$$

where $a_i = 0, 1$, or 2, and in a manner that is analogous to decimal fractions

$$0 \,.\, a_1 a_2 a_3 \cdots = \sum_{k=1}^{\infty} a_k \cdot \frac{1}{3^k}.$$

With a bit of thought, the reader should be able to see that if we let

$$x = 0 \,.\, a_1 a_2 a_3 \cdots,$$

then $x \in M_1$ implies that $a_1 = 1$. Similarly if $x \in M_2$, then $a_2 = 1$, and in general that if $x \in M_k$, then $a_k = 1$. Hence, it follows that C_0 consists of all those numbers in $[0,1]$ whose ternary fraction expansion contains no 1's.

This is clearly an uncountable set since such numbers can obviously be put into one-to-one correspondence with all binary decimals on $[0,1]$ (simply replace 2's with 1's and vice versa).

For future reference, we summarize the above observations in the next theorem.

B2. Theorem.

(1) C_0 *is closed;*

(2) C_0 *contains no intervals;*

(3) *each point in* C_0 *is a limit point of* C_0;

(4) C_0 *is measurable and* $\mu(C_0) = 0$;

(5) C_0 *is uncountable;*

(6) *if* $x \in C_0$ *and* $\epsilon > 0$, *then* $(x - \epsilon, x + \epsilon)$ *contains points in at least two subintervals of* M_0.

The next example shows that when one constructs measures on a set Σ, then consideration of the class $\mathcal{M}$ of measurable sets is unavoidable. Of course, in very simple situations, such as discrete measures on finite or countable sets, $\mathcal{M}$ turns out to be just the class of all subsets of Σ. In more general situations, and specifically when one constructs Lebesgue measure, there do exist nonmeasurable sets. We begin with a lemma.

B3. Lemma. *If* $A, B \subset \mathrm{R}$ *and* $B = A + a$ *for some* $a \in \mathrm{R}$, *then* $\mu^*(B) = \mu^*(A)$.

Proof. The sequential covering class for Lebesgue measure can be taken to be the collection of half-open intervals in R. Now, if

$$\mathcal{C}_1 = \{I_\alpha \mid \alpha \in \mathcal{A}\}$$

is a countable cover for A, then clearly

$$\mathcal{C}_2 = \{I_\alpha + a \mid \alpha \in \mathcal{A}\}$$

is a countable cover for B. In the notation of Appendix A we have

$$\tau(\mathcal{C}_1) = \tau(\mathcal{C}_2)$$

since the translation of a half-open interval does not change its length. Likewise, for each cover $\mathcal{C}_3$ of B, there is a cover $\mathcal{C}_4$ for A such that

$$\tau(\mathcal{C}_3) = \tau(\mathcal{C}_4).$$

Hence,

$$\inf\{\tau(\mathcal{C}) \mid \mathcal{C} \text{ is a countable cover for } A\}$$

$$= \inf\{\tau(\mathcal{C}) \,|\, \mathcal{C} \text{ is a countable cover for } B\}$$

or in other words

$$\mu^*(A) = \mu^*(B). \qquad \square$$

B4. **Theorem.** *If* $E \subset \mathrm{R}$, $\mu(E) > 0$, *then there is a set* A_0 *such that*

(1) $A_0 \subset E$, *and*

(2) A_0 *is nonmeasurable.*

Proof. We can always find some closed interval F such that $\mu(E \cap F) > 0$, so without loss of generality, we will suppose that $E \subset [-1/2, 1/2]$.

We begin by defining a relation R on E as follows. If $x, y \in E$, then

$$xRy \Longleftrightarrow x - y \quad \text{is rational.}$$

It is easily seen that R is an equivalence relation, and hence, the equivalence classes of R form a partition of E, that is,

(1) $E = \cup R[x]$;

(2) $R[x] \cap R[y] \neq \emptyset \Longrightarrow R[x] = R[y]$.

Clearly, each $R[x]$ is a countable set, so the union in (1) must be a noncountable union. For, if not, then E would be a countable set and hence we would have $\mu(E) = 0$, contradicting the hypothesis. Choose one point from each equivalence class, and call this set A_0.[1] Next, enumerate all of the rationals in $[-1, 1]$, for example,

$$0 = r_0, r_1, r_2, r_3, \ldots,$$

and for each r_k define

$$A_k = A_0 + r_k.$$

By Lemma B3

$$\mu^*(A_k) = \mu^*(A_0) \triangleq \alpha$$

for each k. We also claim that

$$E \subset \bigcup_{k=0}^{\infty} A_k \subset [-3/2, 3/2].$$

The second inclusion is clear. To see the first, let $x \in E$ be arbitrary. Then $x \in R[x]$ and by the construction of the set A_0, $R[x] = R[y]$ for some $y \in A_0$. But $x \in R[y]$ implies that $x - y = r_k$ for some rational number

[1] This step requires the axiom of choice since there are an uncountable number of distinct equivalence classes.

(necessarily in $[-1,1]$ since two points in $[-1/2,1/2]$ can differ by at most 1). Thus, $x = y + r_k$ and so $x \in A_0 + r_k = A_k$ as required.

With the above inclusion established, it follows that

$$\mu^*(E) \le \mu^*\left(\bigcup_{k=0}^{\infty} A_k\right) \le \mu^*[-3/2,3/2]$$

or

$$0 < \mu(E) \le \mu^*\left(\bigcup_{k=0}^{\infty} A_k\right) \le 3.$$

The A_k's are pairwise disjoint (why?), so if A_0 is measurable then each A_k is measurable, and we would have

$$0 < \sum_{k=0}^{\infty} \mu(A_k) \le 3,$$

that is,

$$0 < \alpha + \alpha + \alpha + \cdots \le 3.$$

If $\alpha = 0$ or $\alpha > 0$, we contradict this inequality. Thus, A_0 is nonmeasurable. Since $A_0 \subset E$, we are finished. □

Appendix C

Measurable Functions

C1. **Definition.** Let Ω be a set, $\mathcal{F}$ a σ field of subsets of Ω. A function $f : \Omega \to \mathrm{R}$ (or R^n) is said to be *measurable with respect to* $\mathcal{F}$, or simply $\mathcal{F}$ *measurable,* providing the following holds.

$$\text{If } U \subset \mathrm{R}, \quad U \text{ open}, \quad \text{then } f^{-1}(U) \in \mathcal{F}. \tag{M}$$

C2. **Lemma.** *If f and g are measurable functions (meaning $\mathcal{F}$ measurable) and α is a real number, then*

(a) *αf is a measurable function,*

(b) *$\alpha \pm f$ is a measurable function, and*

(c) $\{\omega \mid f(\omega) > g(\omega)\} \in \mathcal{F}$.

Proof. Parts (a) and (b) are quite easy and are left for the reader. For (c), first note that

$$\{\omega \mid f(\omega) > g(\omega)\} = \bigcup_{m=-\infty}^{\infty} \bigcup_{n=-\infty}^{\infty} \left[\left\{\omega \mid f(\omega) > \frac{m}{n}\right\} \cap \left\{\omega \mid g(\omega) < \frac{m}{n}\right\}\right].$$

This identity essentially holds because between every two real numbers there is a rational number. Given this fact, the interested reader may wish to formally prove the above identity. It now follows that

$$\{\omega \mid f(\omega) > g(\omega)\} = \bigcup_{m=-\infty}^{\infty} \bigcup_{n=-\infty}^{\infty} \left[f^{-1}\left(\frac{m}{n}, \infty\right) \cap g^{-1}\left(-\infty, \frac{m}{n}\right)\right],$$

and each set in the double union is in $\mathcal{F}$. Thus, the set on the left is in $\mathcal{F}$. □

C3. **Theorem.** *If f and g are measurable, then so is $f \pm g$.*

Proof. $\{\omega \mid f(\omega) + g(\omega) < b\} = \{\omega \mid f(\omega) < b - g(\omega)\}$.
Now, $-g$ is measurable by part (a) of C2, and so by part (b) of the same lemma, $b - g$ is measurable. Thus, from part (c), the set

$$(f+g)^{-1}(-\infty, b) = \{\omega \mid f(\omega) + g(\omega) < b\}$$

is in $\mathcal{F}$. Simiarly, $(f+g)^{-1}(a, \infty)$ is in $\mathcal{F}$, and so

$$(f+g)^{-1}(a, b) = (f+g)^{-1}(a, \infty) \quad (f+g)^{-1}(-\infty, b)$$

is in $\mathcal{F}$. From the structure of open sets, it follows that $(f+g)^{-1}(U) \in \mathcal{F}$ for any open set U.

The result for $f - g$ follows at once from the fact that $-g$ is measurable and

$$f - g = f + (-g). \quad \square$$

C4. **Theorem.** *f is measurable $\Longleftrightarrow$ $f^{-1}(F) \in \mathcal{F}$ for each closed set F.*

Proof. This follows easily from the identity

$$f^{-1}(\mathrm{R}) \setminus F) = \Omega \setminus f^{-1}(F). \quad \square$$

C5. **Theorem.** If f is measurable, so is $|f|$.

Proof. This follows from the identity

$$|f|^{-1}(U) = [f^{-1}(U) \cap f^{-1}([0,\infty))] \cup [f^{-1}(-U) \cap f^{-1}((-\infty,0))]. \quad \square$$

C6. **Theorem.** *If f and g are measurable, then so is $f \cdot g$.*

Proof. This follows from the identity

$$f \cdot g = \frac{1}{4}[|f+g|^2 - |f-g|^2]$$

and the above theorem. $\square$

C7. **Definition.** Let $p(\omega)$ be a predicate on $E \subset \Omega$, that is, $p(\omega)$ is a proposition that is either true or false depending upon the choice of $\omega \in E$. We say that p is *true almost everywhere* on E provided that there exists a measurable set $E_1 \subset E$ such that

(a) $\mu(E_1) = 0$.

(b) $p(\omega)$ is true for all $\omega \in E \setminus E_1$.

In this case we write p [a.e.]. For example, if $\lim_{n\to\infty} f_n(\omega) = f_0(\omega)$ for all $\omega \in E \setminus E_1$, we write

$$\lim_{n\to\infty} f_n(\omega) = f_0(\omega) \quad \text{[a.e.]}.$$

C8. **Theorem.** *If f_n is a sequence of measurable functions defined on some common domain, and if* $\lim_{n\to\infty} f_n = f_0$ [a.e.], *then f_0 is measurable.*

Outline of Proof. If one can show that $f_0^{-1}(-\infty, b)$ and $f_0^{-1}(a,\infty)$ are in $\mathcal{F}$, then f_0 is measurable (see proof of C3). To show that $f_0^{-1}(a,\infty) \in \mathcal{F}$, for example, proceed as follows. Define

$$A_m^k \triangleq \left\{\omega \mid f_k(\omega) > a + \frac{1}{m}\right\}$$

$$B_m^n \triangleq \bigcap_{k=n}^{\infty} A_m^k.$$

Then show

$$f_0^{-1}(a,\infty) = \{\omega \mid f_0(\omega) > a\} = \bigcup_{n=1}^{\infty} \bigcup_{m=1}^{\infty} B_m^n. \qquad \square$$

C9. **Definition.** (a) Let E be any subset of Ω. By the *characteristic function of* E, we mean the function

$$C_E : \Omega \to \mathrm{R}$$

defined by

$$C_E(\omega) = \begin{cases} 1 & \text{if} \quad \omega \in E \\ 0 & \text{if} \quad \omega \neq E. \end{cases}$$

(b) Let $E_1, E_2, \ldots, E_n$ be any finite collection of (disjoint) sets in $\mathcal{F}$ and let $\alpha_1, \alpha_2, \ldots, \alpha_n$ be any set of real numbers. Then, a function of the form

$$f(\omega) = \sum_{k=1}^{n} \alpha_k C_{E_k}(\omega)$$

is called a *simple function*. Note that simple functions are $\mathcal{F}$ measurable.

C10. **Theorem.** *If f_0 is a nonnegative measurable function defined on a set E, then there is a nondecreasing sequence f_n of nonnegative simple functions such that for every $\omega \in E$*

$$\lim_{n\to\infty} f_n(\omega) = f_0(\omega).$$

Proof. For each integer n and each $\omega \in E$, let

$$f_n(\omega) = \begin{cases} \frac{k-1}{2^n} & \text{for} \quad \frac{k-1}{2^n} \leq f_0(\omega) < \frac{k}{2^n}; \quad k = 1, 2, \ldots, n2^n \\ n & \text{for} \quad f_0(\omega) \geq n. \end{cases}$$

Then f_n is simple, nonnegative, and nondecreasing. Moreover, for any ω, there exists an m such that $f_0(\omega) < m$. Thus, for $n \geq m$, we have

$$0 \leq f_0(\omega) - f_n(\omega) < \frac{1}{2}n. \qquad \square$$

Although the word "measurable" has been used frequently in this appendix, we have not worked with a single measure. For the next theorem, we must suppose that in addition to $\mathcal{F}$ being a σ field, that there is a measure μ such that $(\Omega, \mathcal{F}, \mu)$ is a measure space.

C11. **Theorem.** *Suppose that f_0 satisfies the hypothesis of* C10, *and that*

$$\{\omega \mid f_0(\omega) \neq 0\} = \bigcup_{k=1}^{\infty} E_k,$$

where $\mu(E_k) < \infty$ for each k (for example, any measurable function defined on R^n has the property since R^n itself can be so represented). Then the sequence f_n in C10 can be constructed so as to have the additional property that for each n, $\{\omega \mid f_n(\omega) \neq 0\}$ has finite measure.

Proof. Suppose the E_k's are as above. Define

$$A_n \triangleq \bigcup_{k=1}^{n} E_k.$$

Then, $\{A_n\}$ is an expanding sequence of sets of finite measure, whence C_{A_n} forms a nondecreasing sequence of measurable functions. If f_n is as in C10, define

$$g_n \triangleq f_n \cdot C_{A_n}$$

and note that g_n has the desired features. □

We now look at an example that is used in the main body of the text, specifically in Section 1.1.

C12. **The Cantor Function.** Recall from Appendix B that on $[0,1]$ we can construct two sets C_0 and M_0, where M_0 is open and $C_0 = [0,1] \setminus M_0$. The characteristics of C_0 are summarized in theorem B2, and the characteristics of M_0 are in Section B1.

Using the notation in B1, note that each set M_k is a finite union of 2^{k-1} open intervals, each of length $1/3^k$. Let us denote these open intervals by E_{sk}, $s = 1, 2, \ldots, 2^{k-1}$. For example, if $k = 3$,

$$\begin{aligned} E_{13} &= (1/27, 2/27), \\ E_{23} &= (7/27, 8/27), \\ E_{33} &= (19/27, 20/27), \end{aligned}$$

and

$$E_{43} = (25/27, 26/27).$$

Now define a function g on M_0 by defining g on each E_{sk} as

$$g(x) = \frac{2s-1}{2^k} \quad \text{for} \quad x \in E_{sk}$$

so that g is constant on each E_{sk}. With a bit of work (a picture is a great deal of help), one can see that g is monotone increasing on M_0 (if necessary, a rather messy induction can be carried out). Since the numbers of the form $(2s-1)/2^k$ are dense in $[0,1]$, that is, each point in $[0,1]$ can be written as a limit of such numbers, we can define

$$c(x) \triangleq \lim_{h \to x} g(h).$$

Since g is monotone, the only way the left- and right-hand limits could differ is if there were a jump at x, so the above limit is well defined since no jump can occur (denseness). Thus, c is continuous, $c(0) = 0$, $c(1) = 1$, and c is constant on each of the sets E_{sk}. Next, define

$$f(x) \triangleq c(x) + x.$$

Clearly $f : [0,1] \to [0,2]$ and is strictly increasing, hence, one to one. Moreover, for each E_{sk}, $f(E_{sk}) = a + E_{sk}$, where a is the left end point of E_{sk}. Thus, $\mu(f(E_{sk})) = \mu(E_{sk})$, whence

$$\begin{aligned} \mu(f(M_0)) &= \mu\left(f\left(\bigcup_{k=1}^{\infty}\bigcup_{s=1}^{2^k} E_{sk}\right)\right) \\ &= \mu\left(\bigcup_{k=1}^{\infty}\bigcup_{s=1}^{2^k} f(E_{sk})\right) \\ &= \sum_{k=1}^{\infty}\sum_{s=1}^{2^k} \mu(f(E_{sk})) \\ &= \sum_{k=1}^{\infty}\sum_{s=1}^{2^k} \mu(E_{sk}) \\ &= \mu(M_0) \\ &= 1. \end{aligned}$$

Thus, $[0,2] \setminus f(M_0)$ has measure 1 and so contains a nonmeasurable subset N. Letting

$$\begin{aligned} \Omega &\triangleq [0,2] \\ X &\triangleq f^{-1} \\ M &\triangleq f^{-1}(N), \end{aligned}$$

we see that $M \subset C_0$ so $\mu^*(M) \leq \mu^*(C_0) = 0$, that is, M is measurable, X is continuous and, hence, is measurable, and $X^{-1}(M) = N$, nonmeasurable. This is the example mentioned in Section 1 of Chapter 1.

Appendix D

Integration

D1. **Definition.** Let $(\Omega, \mathcal{M}, \mu)$ be a measure space and let

$$f = \sum_{i=1}^{n} \alpha_i C_{A_i}$$

be a simple function (see Definition C9).

(a) We say that f is *integrable* on $E \in \mathcal{M}$, provided that

$$\mu(E \cap \{\omega \mid f(\omega) \neq 0\}) < \infty.$$

This is equivalent to saying that

$$\mu\left(\bigcup_{k=1}^{n} (E \cap A_k)\right) < \infty.$$

(b) If f is an integrable, simple function, we define

$$\int_E f \, d\mu = \sum_{k=1}^{n} \alpha_k \mu(E \cap A_k).$$

D2. **Theorem.** *Let $E_1, \ldots E_n$ be disjoint sets in $\mathcal{M}$ with $E_0 = \cup_{k=1}^{n} E_k$. Then, if f is a simple function that is integrable on each E_k, f is integrable on E_0 and*

$$\int_{E_0} f \, d\mu = \sum_{k=1}^{n} \int_{E_k} f \, d\mu.$$

Proof. Let $A = \{\omega \mid f(\omega) \neq 0\}$. Then

$$\mu(E_0 \cap A) = \sum_{k=1}^{n} \mu(E_k \cap A) < \infty,$$

so that f is integrable on E_0. Let

$$f = \sum_{i=1}^{m} \alpha_i C_{A_i}.$$

Then

$$\begin{aligned}
\sum_{k=1}^{n} \int_{E_k} f\, d\mu &= \sum_{k=1}^{n} \sum_{i=1}^{m} \alpha_i \mu(E_k \cap A_i) \\
&= \sum_{i=1}^{m} \sum_{k=1}^{n} \alpha_i \mu(E_k \cap A_i) \\
&= \sum_{i=1}^{m} \alpha_i \mu\left(\left(\bigcup_{k=1}^{n} E_k\right) \cap A_i\right) \\
&= \sum_{i=1}^{m} \alpha_i \mu(E_0 \cap A_i) \\
&= \int_{E_0} f\, d\mu. \qquad \square
\end{aligned}$$

D3. **Theorem.** *If f and g are integrable simple functions, $\alpha, \beta \in \mathrm{R}$, then $\alpha f + \beta g$ is an integrable simple function, and*

$$\int_E (\alpha f + \beta g)\, d\mu = \alpha \int_E f\, d\mu + \beta \int_E g\, d\mu.$$

The proof of this is an easy exercise and hence is omitted.

D4. **Theorem.** *If f and g are simple functions, f integrable on E, and $f(\omega) \geq |g(\omega)|$ for $\omega \in E$, then g is integrable on E and*

$$\int_E f\, d\mu \geq \int_E g\, d\mu.$$

This proof will also be left as an exercise. One hint, however, is to note that

$$\{\omega \mid g(\omega) \neq 0\} \subset \{\omega \mid f(\omega) \neq 0\}.$$

D5. **Theorem.** *Let f_n be a nondecreasing sequence of nonnegative, simple functions, each integrable on $E \in \mathcal{M}$, and let f_0 be a nonnegative, simple function such that for each $\omega \in E$, $\lim_{n\to\infty} f_n(\omega)$ exists and*

$$\lim_{n\to\infty} f_n(\omega) \geq f_0(\omega).$$

If f_0 is integrable on E, then

$$\lim_{n\to\infty} \int_E f_n\, d\mu \geq \int_E f_0 d\mu.$$

(*Note: the left side may diverge.*)

Proof. Since f_n is a nondecreasing sequence, it follows from D4 that $\int_E f_n\, d\mu$ is a nondecreasing sequence of nonnegative numbers. If

$\lim_{n\to\infty} \int_E f_n \, d\mu = \infty$, the conclusion is obviously true. Hence, we suppose that $\lim_{n\to\infty} \int_E f_n \, d\mu$ is finite.

If f_0 is integrable on E, and if

$$E_0 \triangleq \{\omega \,|\, f_0(\omega) > 0\},$$

then

$$\mu(E \cap E_0) < \infty.$$

Furthermore, since f_0 is simple, there is a number M such that $f_0(\omega) \leq M$ for all ω, hence, for all $\omega \in E$. If $\mu(E \cap E_0) = 0$, then the theorem is trivial since $\int_E f_0 \, d\mu = 0$. Hence, we will suppose that $\mu(E \cap E_0) > 0$. This done we can proceed as follows.

For any $\epsilon > 0$, define

$$B_n = \{\omega \in E \,|\, f_n(\omega) \geq f_0(\omega) - \epsilon/2\mu(E \cap E_0)\}.$$

Then B_n is an expanding sequence of sets, and one can verify that

$$\bigcup_{n=1}^{\infty} B_n = E \supset E \cap E_0.$$

Thus, $(E \cap E_0) \setminus B_n$ is a contracting sequence of sets, so by Exercise 1.10.3,

$$\begin{aligned}
\lim_{n\to\infty} \mu(E \cap E_0 \setminus B_n) &= \mu\left(\bigcap_{n=1}^{\infty} (E \cap E_0 \setminus B_n)\right) \\
&= \mu\left(\bigcap_{n=1}^{\infty} (E_0 \cap E \cap B_n')\right) \\
&= \mu\left(E_0 \cap E \cap \bigcap_{n=1}^{\infty} B_n'\right) \\
&= \mu\left(E_0 \cap E \cap \left(\bigcup_{n=1}^{\infty} B_n\right)'\right) \\
&= \mu(E_0 \cap E \cap E') \\
&= \mu(\phi) \\
&= 0.
\end{aligned}$$

Hence, there is an integer n_0 such that $n > n_0$ implies

$$\mu(E \cap E_0 \setminus B_n) < \epsilon/2M.$$

Thus, for all $n > n_0$

$$\int_E f_n \, d\mu \geq \int_{E \cap E_0 \cap B_n} f_n \, d\mu$$

$$\geq \int_{E\cap E_0\cap B_n} [f_0 - \epsilon/2\mu(E\cap E_0)]\,d\mu$$
$$= \int_{E\cap E_0\cap B_n} f_0 d\mu - \frac{\epsilon}{2}\frac{\mu[E\cap E_0\cap B_n]}{\mu(E\cap E_0)}$$
$$\geq \int_{E\cap E_0\cap B_n} f_0\,d\mu - \frac{\epsilon}{2}.$$

However, by Theorem D2, we have

$$\int_{E\cap E_0} f_0\,d\mu = \int_{E\cap E_0 B_n} f_0\,d\mu + \int_{E\cap E_0\cap B_n'} f_0\,d\mu,$$

so that the above inequality becomes

$$\begin{aligned}\int_E f_n\,d\mu &\geq \int_{E\cap E_0} f_0\,d\mu - \int_{E\cap E_0\cap B_n'} f_0\,d\mu - \frac{\epsilon}{2}\\ &\geq \int_{E\cap E_0} f_0\,d\mu - M\frac{\epsilon}{2M} - \frac{\epsilon}{2}\\ &= \int_E f_0\,d\mu - \epsilon.\end{aligned}$$

Thus, we have shown that for every $\epsilon > 0$, there is an integer n_0 such that $n > n_0$ implies that

$$\int_E f_n\,d\mu \geq \int_E f_0\,d\mu - \epsilon$$

and so

$$\lim_{n\to\infty}\int_E f_n\,d\mu \geq \int_E f_0\,d\mu. \qquad \square$$

D6. **Corollary.** *Let f_n be a nondecreasing sequence of nonnegative simple functions, each integrable on E and let f_0 be a simple function such that for each $\omega \in E$*

$$\lim_{n\to\infty} f_n(\omega) = f_0(\omega).$$

Then, f_0 is integrable on E if and only if

$$\lim_{n\to\infty}\int_E f_n\,d\mu < \infty,$$

and in this case

$$\lim_{n\to\infty}\int_E f_n\,d\mu = \int_E f_0\,d\mu.$$

Proof. If f_0 is simple and nonintegrable on E, then letting

$$f_0 = \sum_{k=1}^{n} \alpha_k C_{A_k},$$

the condition of nonintegrability,

$$\mu(E \cap \{\omega \mid f_0(\omega) \neq 0\}) = \infty,$$

implies that for some A_{k_0}

$$\mu(E \cap A_{k_0}) = \infty.$$

Let

$$B_m \triangleq \left\{\omega \mid f_m(\omega) > \frac{\alpha_{k_0}}{2}\right\} \cap E.$$

Since f_m is nondecreasing, B_m is an expanding sequence of sets. Moreover,

$$\bigcup_{m=1}^{\infty} B_m \supset A_{k_0} \cap E,$$

this last inclusion following from the hypothesis $\lim_{m\to\infty} f_m(\omega) = f_0(\omega)$. Thus,

$$\begin{aligned}
\lim_{n\to\infty} \int_E f_n \, d\mu &\geq \lim_{n\to\infty} \int_{B_n} f_n \, d\mu \\
&\geq \lim_{n\to\infty} \int_{B_n} \frac{\alpha_{k_0}}{2} d\mu \\
&= \frac{\alpha_{k_0}}{2} \lim_{n\to\infty} \mu(B_n) \\
&= \frac{\alpha_{k_0}}{2} \left(\bigcup_{n=1}^{\infty} B_n\right) \\
&\geq \frac{\alpha_{k_0}}{2} \mu(E \cap A_{k_0}) \\
&= \infty.
\end{aligned}$$

Thus, writing the contrapositive, we have shown that

$$\lim_{n\to\infty} \int_E f_n \, d\mu < \infty \quad \text{implies } f_0 \text{ is integrable.}$$

Conversely, suppose f_0 is integrable. Then, from D4 and the monotonicity of f_n (which implies $f_n \leq f_0$), it follows that

$$\int_E f_n \, d\mu \leq \int_E f_0 \, d\mu$$

for every n, and so

$$\lim_{n\to\infty} \int_E f_n \, d\mu \leq \int_E f_0 \, d\mu < \infty.$$

The reverse inequality follows from D7. □

D7. **Corollary.** *If f_n and g_n are two nondecreasing sequences of nonnegative, integrable, simple functions such that for each $\omega \in E$*

$$\lim_{n\to\infty} f_n(\omega) = \lim_{n\to\infty} g_n(\omega),$$

then

$$\lim_{n\to\infty} \int_E f_n \, d\mu = \lim_{n\to\infty} \int_E g_n \, d\mu.$$

Proof. For any fixed positive integer k

$$\lim_{n\to\infty} f_n(\omega) \geq g_k(\omega), \qquad \text{for all} \quad \omega \in E.$$

Hence, by D5,

$$\lim_{n\to\infty} \int_E f_n \, d\mu \geq \int_E g_k \, d\mu.$$

But by D4, $\int_E g_k \, d\mu$ is a nondecreasing sequence of nonnegative real numbers bounded above, and so

$$\lim_{n\to\infty} \int_E f_n \, d\mu \geq \lim_{k\to\infty} \int_E g_k \, d\mu.$$

To get the reverse inequality, simply reverse the roles of f_n and g_k. □

We are now ready to construct the integral for measurable functions that do not simply take on a finite number of values. We do this in a manner that is analogous to the construction of Riemann integrals (the kind of integral one constructs in freshman calculus). If f is Riemann integrable, then one approximates f by a sequence of step functions f_n and defines $\int f$ to be the limit of $\int f_n$, the approximating areas. Here, we do the same thing except that we use simple functions in place of step functions. Specifically, suppose that $f_0 \geq 0$ is measurable. Then by C10 we know that there is a nondecreasing sequence of nonnegative, simple functions, for instance f_n, such that $\lim_{n\to\infty} f_n = f_0$. We then simply define

$$\int_E f_0 \, d\mu \triangleq \lim_{n\to\infty} \int_E f_n \, d\mu.$$

There are, however, some potential problems we could encounter with this definition. For example, suppose we have two different sequences f_n and g_n both converging to f_0. Could it happen that

$$\lim_{n\to\infty} \int_E f_n \, d\mu \neq \lim_{n\to\infty} \int_E g_n \, d\mu?$$

No, Corollary D7 assures us that this will not happen.

There is one other possible source of difficulty. If f_0 is a simple function then it is certainly measurable, so one can envision a sequence of simple functions f_n, distinct from f_0, and converging monotonically upward to f_0. Then

$$\int_E f_0 \, d\mu = \lim_{n\to\infty} \int_E f_n \, d\mu.$$

But wait! Since f_0 was simple, we already has a meaning for the left-hand side from Definition D1. Are both numbers the same? Yes, Corollary D6 assures us that f_0 is integrable exactly when such a sequence of integrals converge and that they do converge to $\int_E f_0 \, d\mu$.

Everything is consistent, so we now make the definition official.

D8. **Definition.** (a) Let f_0 be any nonnegative measurable function and let E be any measurable set. We say that f_0 is *integrable* on E if and only if there exists a nondecreasing sequence of nonnegative, simple functions, each integrable on E, such that for each $\omega \in E$

$$\lim_{n\to\infty} f_n(\omega) = f_0(\omega)$$

and

$$\lim_{n\to\infty} \int_E f_n \, d\mu < \infty.$$

(b) If f_0 is as in (a) and is integrable on E, we define the *integral* of f on E by

$$\int_E f_0 \, d\mu \triangleq \lim_{n\to\infty} \int_E f_n \, d\mu.$$

(c) If f_0 is any measurable function, define

$$f_0^+(\omega) = \begin{cases} f_0(\omega) & \text{if } \; f_0(\omega) > 0 \\ 0 & \text{if } \; f_0(\omega) \le 0 \end{cases}$$

$$f_0^-(\omega) = \begin{cases} 0 & \text{if } \; f_0(\omega) \ge 0 \\ -f_0(\omega) & \text{if } \; f_0(\omega) < 0. \end{cases}$$

Then, of course, $f_0 = f_0^+ - f_0^-$ and $f_0^+ \ge 0$, $f_0^- \ge 0$.

(d) If f_0 is any measurable function, we say that f_0 is *integrable* on the measurable set E providing that both f_0^+ and f_0^- are integrable on E, and in this case we define

$$\int_E f_0 \, d\mu = \int_E f_0^+ \, d\mu - \int_E f_0^- \, d\mu.$$

The integral we have just defined is called the Lebesgue integral. Its usefulness lies not in its computational aspects, but rather in its theoretical properties and its conceptual features. In fact, it is a theorem (which we will not prove) that whenever f_0 is Riemann integrable on a (suitable) bounded subset E of R^n, then f_0 is Lebesgue integrable there and the two integrals are equal. It is quite unlikely that in practice anyone would encounter a function that is Lebesgue integrable but not Riemann integrable. Why use

Lebesgue integrals at all? Much of Chapter 2 in the main text is devoted to explaining this in the context of estimation theory. Also, in the remainder of this appendix and the next, we will point out certain features of the Lebesgue integral that are conceptually useful for our purposes.

D9. **Theorem.** *If f is any measurable function and $\mu(E) = 0$, then*

$$\int_E f \, d\mu = 0.$$

Proof. The result is obvious for simple functions, so the result follows from D8. □

D10. **Theorem.** *Let $E_1, \ldots, E_n$ be disjoint measurable sets with*

$$E_0 = \bigcup_{k=1}^{n} E_k$$

and let f be integrable on each set E_k. Then f is integrable on E_0 and

$$\int_{E_0} f \, d\mu = \sum_{k=1}^{n} \int_{E_k} f \, d\mu.$$

Proof. By D2 this holds for simple functions. By passing to the limit as in D8(b), it holds for nonnegative integrable functions, and by D8(d) it holds in general. □

D11. **Theorem.** *If f and g are integrable on E, then*

$$\int_E (\alpha f + \beta g) \, d\mu = \alpha \int_E f \, d\mu + \beta \int_E g \, d\mu.$$

The proof of this follows from D3 and a rather lengthy case analysis involving the signs of α, β, f, and g. We omit the proof.

D12. **Theorem.** *If f is integrable on E and $f \geq 0$ [a.e.], then $\int_E f \, d\mu$ is a nonnegative, nondecreasing, set function on the measurable subsets B of E.*

Proof. By Theorem D10, $\int_B f \, d\mu$ is finitely additive, so monotonicity and nonnegativity of the integral are equivalent. We show the latter. Let

$$A = \{\omega \mid f(\omega) < 0\}.$$

Then, by hypothesis,

$$\mu(A) = 0.$$

On $E \cap A'$, $f(\omega) \geq 0$ so $\int_{B \cap A'} f \, d\mu \geq 0$ for all measurable $B \subset E$. Thus, for all measurable $B \subset E$,

$$\int_B f \, d\mu = \int_{B \cap A'} f \, d\mu + \int_{B \cap A} f \, d\mu = \int_{B \cap A'} f \, d\mu \geq 0,$$

the second integral being zero by D9. □

D13. **Corollary.** *If f and g are each integrable on E and if $f \leq g$ [a.e.] on E, then*

$$\int_E f \, d\mu \leq \int_E g \, d\mu.$$

Proof. By D12,

$$\int_E (g - f) \, d\mu \geq 0$$

so the result follows from D11. □

D14. **Theorem.** *If $0 \leq f_0 \leq g$ [a.e.] on E, is measurable, and g is integrable, then f_0 is integrable on E.*

Proof. Define

$$\begin{aligned} B &\triangleq \{\omega \,|\, g(\omega) > 0\} \\ A &\triangleq \{\omega \,|\, f_0(\omega) > 0\}. \end{aligned}$$

Then, clearly $A \subset B$. Now, since g is integrable, there is a sequence g_n of nonnegative, nondecreasing functions, each simple and integrable, such that

$$\lim_{n \to \infty} g_n(\omega) = g(\omega).$$

If $g_n(\omega) = 0$ for each n, then $g(\omega) = 0$, so we have (from the contrapositive) that

$$B \subset \bigcup_{n=1}^{\infty} \{\omega \,|\, g_n(\omega) > 0\}.$$

Since each g_n is integrable, $\mu(\{\omega \,|\, g_n(\omega) > 0\}) < \infty$ and so defining

$$B_n \triangleq B \cap \{\omega \,|\, g_n(\omega) > 0\},$$

we have

$$B = \bigcup_{n=1}^{\infty} B_n$$

and $\mu(B_n) < \infty$ for each n. Let

$$A_n \triangleq A \cap B_n.$$

Then

$$\bigcup_{n=1}^{\infty} A_n = A \cap \bigcup_{n=1}^{\infty} B_n = A \cap B = A$$

and $\mu(A_n) < \infty$ for each n. From this and Theorem C11, it follows that there exists a sequence f_n of nonnegative, nondecreasing, integrable (from C11), simple functions such that

$$\lim_{n\to\infty} f_n(\omega) = f_0(\omega);$$

but, then

$$\lim_{n\to\infty} \int_E f_n \, d\mu \leq \int E g \, d\mu,$$

(use D13) and so f_0 is integrable. □

D15. **Corollary.** *A measurable function f is integrable on E if and only if $|f|$ is integrable on E.*

Proof. Suppose f is integrable on E. Then, by definition, both f^+ and f^- are integrable on E. Since $|f| = f^+ + f^-$, it follows from D11 that $|f|$ is integrable on E.

Conversely, suppose $|f|$ is integrable on E. Then, since f^+ is measurable and since $0 \leq f^+ \leq |f|$, it follows from D14 that f^+ is integrable. Similarly, f^- is integrable, and so by D11, $f = f^+ - f^-$ is integrable. □

D15 has an interesting consequence. If we define $f : [0,1] \to \mathrm{R}$ by

$$f(\omega) = \begin{cases} 1 & \text{if } \omega \text{ is rational} \\ -1 & \text{if } \omega \text{ is irrational,} \end{cases}$$

then $|f(\omega)| = 1$, so $|f|$ is Riemann integrable. However, f is not Riemann integrable (why?) so D15 does not hold for Riemann integrals. Of course, by D15, f is Lebesgue integrable, so combining this example with the remarks following Definition D8, we see that the class of Lebesgue integrable functions is larger than the class of Riemann integrable functions. Although this is worth noting, it is not of practical consequence to us.

D16. **Corollary.** *If f is measurable and g is integrable on E with $|f| \leq g$* [a.e.], *then f is integrable on E.*

Proof. By D14, $|f|$ is integrable on E, so by D14, f is integrable on E. □

D17. **Corollary.** *If $f = g$* [a.e.] *on E and g is integrable on E, then f is integrable on E and*

$$\int_E f \, d\mu = \int_E g \, d\mu.$$

Proof. We have

$$0 \leq f^+ \leq g^+$$

and

$$0 \leq f^- \leq g^-,$$

so by D14, f^+ and f^- are integrable, hence, so is f. Equality of the integrals follows from D13. □

D18. **Theorem.** *If f is integrable on E, then*

$$\left|\int_E f\,d\mu\right| \leq \int_E |f|\,d\mu.$$

Proof. By D12, both $\int_E f^+\,d\mu$ and $\int_E f^-\,d\mu$ are nonnegative. Thus,

$$\begin{aligned}\left|\int_E f\,d\mu\right| &= \left|\int_E f^+ d\mu - \int_E f^-\,d\mu\right| \\ &\leq \max\left\{\int_E f^+\,d\mu, \int_E f^-\,d\mu\right\} \\ &\leq \int_E f^+\,d\mu + \int_E f^-\,d\mu \\ &= \int_E |f|\,d\mu. \qquad \square\end{aligned}$$

D19. **Theorem.** *If $f \geq 0$ [a.e.] in E and if $\int_E f\,d\mu = 0$, then $f = 0$ [a.e.] in E.*

Proof. Let

$$A_0 \triangleq \{\omega \mid \omega \in E, f(\omega) > 0\}$$

and

$$A_n \triangleq \left\{\omega \mid \omega \in E, f(\omega) > \frac{1}{n}\right\}.$$

Then A_n is an expanding sequence of measurable sets and

$$A_0 = \bigcup_{n=1}^{\infty} A_n.$$

Hence, $\mu(A_0) = \lim_{n\to\infty}(A_n)$. It follows that if $\mu(A_0) > 0$, then for some n_0,

$$\mu(A_{n_0}) > 0.$$

Thus,

$$\int_E f\,d\mu \geq \int_{A_{n_0}} f\,d\mu \geq \mu(A_{n_0}) \cdot \frac{1}{n_0} > 0,$$

which is a contradiction. Thus, $\mu(A_0) = 0$. □

D20. **Corollary.** *If f and g are both integrable on E and if*

$$\int_A f\,d\mu = \int_A g\,d\mu$$

for every measurable $A \subset E$, then $f = g$ [a.e.] on E.

Proof. See Theorem 1.6.8 in main text. □

D21. **Theorem** (Lebesgue Monotone Convergence Theorem). *Let f_n be a nondecreasing sequence of nonnegative functions, each integrable on E, and let f_0 be a function such that*

$$\lim_{n\to\infty} f_n = f_0 \quad \text{[a.e.]}.$$

Then, f_0 is integrable on E if and only if

$$\lim_{n\to\infty} \int_E f_n < \infty$$

and in this case

$$\lim_{n\to\infty} \int_E f_n \, d\mu = \int_E f_0 \, d\mu.$$

Proof. For each n, let g_{nm} be a nondecreasing sequence of nonnegative, integral, simple functions converging pointwise to f_n. Then, for each n and m, define h_{nm} by

$$h_{nm}(\omega) = \max_{i\le n} g_{im}(\omega).$$

The double sequence h_{nm} is obviously nondecreasing as a function of n (by construction), and g_{nm} is nondecreasing as a function of m (given), so h_{nm} is also nondecreasing as a function of m. Furthermore, each h_{nm} is a nonnegative, simple, integrable function. Now, since f_m is a nondecreasing sequence, we have for each n and m with $n \le m$ that

$$g_{nm}(\omega) \le h_{nm}(\omega) \le h_{mm}(\omega) \le \max_{i\le m} f_i(\omega) = f_m(\omega). \qquad (*)$$

Letting $m \to \infty$, we then obtain

$$f_n(\omega) \le \lim_{m\to\infty} h_{mm}(\omega) \le f_0(\omega) \quad \text{[a.e.]},$$

so letting $n \to \infty$, we get the result

$$\lim_{m\to\infty} h_{mm}(\omega) = f_0(\omega) \quad \text{[a.e.]}. \qquad (**)$$

Since h_{mm} are simple integrable, we have from Equation (**) and the definition of integral, that if

$$\lim_{m\to\infty} \int_E h_{mm} d\mu < \infty,$$

then f_0 is integrable (Theorem D9 takes care of the exceptional set of measure zeros). Moreover,

$$\int_E f_0 \, d\mu = \lim_{m\to\infty} \int_E h_{mm} \, d\mu$$

in this case. By inequality (*), however,

$$\int_E h_{mm}\, d\mu \leq \int_E f_m\, d\mu$$

for each m. Hence, if $\lim_{m\to\infty} \int_E f_m\, d\mu < \infty$, it follows that f_0 is integrable and

$$\int_E f_0\, d\mu \leq \lim_{m\to\infty} \int_E f_m\, d\mu.$$

The reverse inequality is easy. Since f_n is monotone increasing, $f_n \leq f_0$ [a.e.]. Hence,

$$\int_E f_n\, d\mu \leq \int_E f_0\, d\mu,$$

and so

$$\lim_{n\to\infty} \int_E f_n\, d\mu \leq \int_E f_0\, d\mu,$$

which completes the proof. Note that this latter argument proves the converse of the integrability condition. □

D22. **Theorem** (Countable Additivity). *Let* $\{E_n \mid n = 1, 2, \ldots\}$ *be a sequence of disjoint measurable sets, and let*

$$E_0 = \bigcup_{n=1}^{\infty} E_n.$$

If f_0 *is a nonnegative measurable function, integrable on each* E_n*, then* f_0 *is integrable on* E_0 *if and only if*

$$\sum_{n=1}^{\infty} \int_{E_n} f_0\, d\mu < \infty,$$

and in this case

$$\int_{E_0} f_0\, d\mu = \sum_{n=1}^{\infty} \int_{E_n} f_0\, d\mu.$$

Proof. For each integer k, let

$$A_k = \bigcup_{n=1}^{k} E_n$$

and let the sequence f_k be defined by

$$f_k(\omega) = \begin{cases} f_0(\omega) & \text{for } \omega \in A_k \\ 0 & \text{otherwise.} \end{cases}$$

Then, by D10,

$$\int_{E_0} f_k \, d\mu = \int_{A_k} f_0 \, d\mu = \sum_{n=1}^{k} \int_{E_n} f_0 \, d\mu.$$

The theorem now follows from D21. □

D23. **Corollary.** *If f is integrable on Ω, then $\int_E f \, d\mu$ is an everywhere finite, countably additive set function on the class $\mathcal{M}$ of all measurable subsets of Ω. If $f \geq 0$, then $\int_E d\mu$ is a measure.*

Here is the first result that show the technical advantage of the Lebesgue integral. If we are going to construct probability measures using probability density functions, we want to be assured that such a construction, that is, defining

$$P(E) = \int_E f \, d\mu$$

really does produce a (countably additive) measure. Corollary D23 assures us that this is indeed the case.

D24. **Definition.** We say that a sequence f_n of measurable functions *converges in measure* to f_0 and write

$$\lim_{n \to \infty} f_n = f_0 \quad \text{(meas.)}$$

provided to every pair of positive numbers (ϵ, θ), there corresponds a natural number n_0 such that if $n > n_0$, then

$$\mu(\{\omega \mid |f_n(\omega) - f_0(\omega)| \geq \epsilon\}) < \theta.$$

This is equivalent to saying that

$$\lim_{n \to \infty} \mu(\{\omega \mid |f_n(\omega) - f_0(\omega)| \geq \epsilon\}) = 0.$$

If the particular measure space under scrutiny happens to be the probability space $(\Omega, \mathcal{E}, P)$ from Chapter 1, then the above notion is called *convergence in probability.*

Note that the condition for convergence in measure does not necessarily make sense unless f_0 is measurable. Thus, implicit in Definition D24 is the fact that the limit function is measurable.

D25. **Theorem.** *If* $\lim_{n \to \infty} f_n = f_0$ (meas.) *and* $\lim_{n \to \infty} g_n = g_0$ (meas.), *then*

(a) $\lim_{n \to \infty}(f_n + g_n) = f_0 + g_0$ (meas.), *and*

(b) *if* $\mu(\Omega) < \infty$, *and* $\lim_{n \to \infty} f_n g_n = f_0 g_0$ (meas.).

Proof. (a) This part is easy and we omit the proof.

(b) We prove two lemmas. □

Lemma A. *If* $f_0 = g_0 = 0$, *then* $\lim_{n\to\infty} f_n g_n = 0$ (meas.).

Proof. This follows at once from

$$\{\omega \mid |f_n(\omega)g_n(\omega)| \geq \epsilon\} \subset \{\omega \mid |f_n(\omega)| \geq \sqrt{\epsilon}\} \cup \{\omega \mid |g_n(\omega)| \geq \sqrt{\epsilon}\}. \quad \square$$

Lemma B. *If* $f_0 = 0$ *and* $\mu(\Omega) < \infty$, *then* $\lim_{n\to\infty} f_n g_0 = 0$ (meas.).

Proof. Let

$$A_k = \{\omega \mid |g_0(\omega)| \geq k\}.$$

Since the range of g_0 is R, it follows that

$$\phi = \bigcap_{k=0}^{\infty} A_k$$

or else the very definition of a function is contradicted. It follows from $\mu(\Omega) < \infty$ that A_1 has finite measure, and so by Exercise 1.10.3, part (b),

$$\lim_{k\to\infty} \mu(A_k) = 0.$$

Hence, given $\theta > 0$, there is a k_0 such that

$$\mu(\{\omega \mid |g_0(\omega)| \geq k_0\}) < \theta/2.$$

The lemma then follows by noting that given $\epsilon > 0$,

$$\{\omega \mid |f_n(\omega) \cdot g_0(\omega)| \geq \epsilon\} \subset \{\omega \mid |f_n(\omega)| \geq \epsilon/k_0\} \cup \{\omega \mid |g_0(\omega)| \geq k_0\}.$$

To complete the proof simply apply Lemmas A and B and part (a) to the identity

$$f_n g_n - f_0 g_0 = (f_n - f_0)(g_n - g_0) + f_0(g_n - g_0) + g_0(f_n - f_0). \quad \square$$

D26. **Definition.** Let f_n be a sequence of measurable functions.

(a) f_n is said to be *uniformly Cauchy* on E providing for every $\epsilon > 0$ there is a corresponding n_0 such that $n, m > n_0$ implies that

$$|f_n(\omega) - f_m(\omega)| < \epsilon$$

for all $\omega \in E$.

(b) f_n is said to be *almost uniformly Cauchy* if and only if for each $\eta > 0$, there is a subset $E \subset \Omega$ such that $\mu(\Omega \setminus E) < \eta$ and f_n is uniformly Cauchy on E.

(c) f_n is said to be a *Cauchy sequence in measure* if and only if for every pair of positive numbers (ϵ, η), there is a corresponding n_0 such that $n, m > n_0$ implies that

$$\mu(\{\omega \mid |f_n(\omega) - f_m(\omega)| \geq \epsilon\}) < \eta.$$

D27. **Theorem.** *If f_n is a Cauchy sequence in measure, then there is a subsequence f_{n_k} that is almost uniformly Cauchy.*

Proof. For each positive integer k, there is another positive integer m_k such that

$$\mu(\{\omega \mid |f_m(\omega) - f_n(\omega)| \geq 1/2^k\}) < 1/2^k \qquad (*)$$

whenever $n, m > m_k$. If we set

$$\begin{aligned} n_1 &= m_1 \\ n_2 &= \max\{n_1 + 1, m_2\} \\ n_3 &= \max\{n_2 + 1, m_3\} \end{aligned}$$

and so on, then the indices $n_1, n_2, \ldots$ determine an infinite subsequence f_{n_k} of f_n and relation (*) holds provided $n \geq n_k$ and $m \geq n_k$. For each k, define

$$E_k \triangleq \{\omega \mid |f_{n_{k+1}}(\omega) - f_{n_k}(\omega)| \geq 1/2^k\}.$$

Then defining

$$B_i = \bigcup_{k=i}^{\infty} E_k, \qquad i \text{ any positive integer},$$

we claim that f_{n_k} is uniformly Cauchy on B_i'. For, if i is any positive integer and $\epsilon > 0$ is arbitrary, then choosing r and s so that $r \geq s \geq i$ and $1/2^{s-1} < \epsilon$, we have

$$\begin{aligned} |f_{n_r}(\omega) - f_{n_s}(\omega)| &\leq \sum_{k=s}^{r-1} |f_{n_{k+1}}(\omega) - f_{n_k}(\omega)| \\ &\leq \sum_{k=s}^{r-1} \frac{1}{2^k} \\ &\leq \frac{1}{2^{s-1}} \\ &< \epsilon \end{aligned}$$

for $\omega \in \cap_{k=i}^{\infty}(\Omega \backslash E_k) = \Omega \backslash \cup_{k=i}^{\infty} E_k$. Thus, for any i, the indices n_k determine a subsequence f_{n_k} that is uniformly Cauchy on $\Omega \setminus B_i$. If for any $\eta > 0$ we now choose i so that

$$\frac{1}{2^{i-1}} < \eta,$$

it follows from relation (*) that

$$\mu(B_i) \leq \sum_{k=i}^{\infty} \mu(E_k) \leq \sum_{k=i}^{\infty} \frac{1}{2^k} = \frac{1}{2^{i-1}} < \eta$$

as required. $\square$

D28. **Corollary.** *If f_n is a Cauchy sequence in measure, then there is a subsequence f_{n_k} that is Cauchy almost everywhere (but not uniformly so).*

Proof. If f_n is Cauchy in measure, then by D27 there is a subsequence f_{n_k} that is almost uniformly Cauchy. Hence, for any integer $m > 0$, there is a set B_m with $\mu(B_m) < 1/m$, and f_{n_k} is uniformly Cauchy on $X \setminus B_m$. Let

$$B = \cap_{m=1}^{\infty} B_m.$$

Then, f_{n_k} is Cauchy on

$$\bigcup_{m=1}^{\infty} (\Omega \setminus B_m) = \Omega \setminus B,$$

and by Exercise 1.10.3, part (b),

$$0 = \lim_{m\to\infty} \mu(B_m) = \mu\left(\bigcap_{m=1}^{\infty} B_m\right),$$

that is,

$$0 = \mu(B). \qquad \square$$

D29. **Definition.** Let $\{a_k\}$ be a sequence of real numbers. For each k, define

$$b_k = \inf\{a_n \mid n \geq k\}$$

$$\underline{\lim}_n a_n \triangleq \lim_{k\to\infty} b_k.$$

Since the b_k's are nondecreasing, this last limit exists (although it may be infinite).

D30. **Theorem** (Fatou's Lemma). *Let f_n be a sequence of nonnegative, integrable functions. Define f_0 by*

$$f_0(\omega) = \underline{\lim}_n f_n(\omega) \quad \text{[a.e.]}.$$

If $\underline{\lim}_n \int_\Omega f_n \, d\mu < \infty$, then f_0 is integrable and

$$\int_\Omega f_0 \, d\mu \leq \underline{\lim}_n \int_\Omega f_n \, d\mu.$$

Proof. For each $\omega \in \Omega$, let

$$g_k(\omega) = \inf_{n>k} f_n(\omega).$$

Then $0 \leq g_k(\omega) \leq f_k(\omega)$. Using the identity

$$\{\omega \mid g_k(\omega) \geq a\} = \bigcap_{n=k}^{\infty} \{\omega \mid f_n(\omega) \geq a\}, \qquad \text{for any } a,$$

it follows that g_k is measurable, hence, by D14 each g_k is integrable. Moreover,

$$\lim_k \int_\Omega g_k \, d\mu \leq \underline{\lim}_k \int_\Omega f_k \, d\mu.^1$$

But $\lim_k g_k(\omega) = \underline{\lim}_k f_k(\omega) = f_0(\omega)$, so by D21

$$\int_\Omega f_0 \, d\mu \leq \underline{\lim}_k \int_\Omega f_k \, d\mu. \qquad \square$$

This appendix by no means exhausts the important theorems in Lebesgue integration theory (nor is that our purpose). However, at this point, we have developed enough to describe an important class of Hilbert spaces, and that will be done in Appendix E.

[1] $\int_\Omega g_k \, d\mu$ is nondecreasing, so $\underline{\lim}_k \int_\Omega g_k \, d\mu = \lim_k \int_\Omega g_k \, d\mu$.

Appendix E

Introduction to Hilbert Space

E1. **Definition.** Let $\mathcal{V}$ be a vector space. By an *inner product* on $\mathcal{V}$, we mean a function

$$\langle \cdot, \cdot \rangle : \mathcal{V} \times \mathcal{V} \to \mathrm{F} \qquad (\mathrm{F} = \mathrm{R} \text{ or } \mathrm{C})$$

such that the following hold.

(a) $\langle \mathbf{x}, \mathbf{y} \rangle = \langle \mathbf{y}, \mathbf{x} \rangle^*$ hold for all $\mathbf{x}, \mathbf{y} \in \mathcal{V}$ (* denotes complex conjugation).

(b) $\langle \mathbf{x} + \mathbf{y}, \mathbf{z} \rangle = \langle \mathbf{x}, \mathbf{z} \rangle + \langle \mathbf{y}, \mathbf{z} \rangle$ for all $\mathbf{x}, \mathbf{y}, \mathbf{z} \in \mathcal{V}$.

(c) $\langle \lambda \mathbf{x}, \mathbf{y} \rangle = \langle \langle \mathbf{x}, \mathbf{y} \rangle$ for all $\lambda \in \mathrm{F}$, all $\mathbf{x}, \mathbf{y} \in \mathcal{V}$.

(d) $\langle \mathbf{x}, \mathbf{x} \rangle \geq 0$ (note, then, $\langle \mathbf{x}, \mathbf{x} \rangle$ is real).

(e) $\langle \mathbf{x}, \mathbf{x} \rangle = 0$ if and only if $\mathbf{x} = \mathbf{0}$.

An inner product space is any vector space $\mathcal{V}$ equipped with an inner product.

E2. **Lemma.**

(a) $\langle \mathbf{x}, \mathbf{y} + \mathbf{z} \rangle = \langle \mathbf{x}, \mathbf{y} \rangle + \langle \mathbf{x}, \mathbf{z} \rangle$

(b) $\langle \mathbf{x}, \lambda \mathbf{y} \rangle = \lambda^* \langle \mathbf{x}, \mathbf{y} \rangle$.

E3. **Examples.** (a) On C^3, define $\langle \mathbf{x}, \mathbf{y} \rangle$ as follows. If $\mathbf{x} = (x_1, x_2, x_3)$ and $\mathbf{y} = (y_1, y_2, y_3)$, then

$$\langle \mathbf{x}, \mathbf{y} \rangle \triangleq \sum_{i=1}^{3} x_i y_i^*.$$

We leave it to the reader to show that E1 holds.

(b) Analogous to the definition in part (a), on R^3 we define

$$\langle \mathbf{x}, \mathbf{y} \rangle \triangleq \sum_{i=1}^{3} x_i y_i.$$

(c) Let $(\Omega, \mathcal{M}, \mu)$ be a measure space. We define $\mathcal{L}_2'(\Omega, \mu)$ as

$$\mathcal{L}_2'(\Omega, \mu) \triangleq \left\{ f : \Omega \to \mathrm{R} \,|\, f \text{ measurable, } \int_\Omega f^2 d\mu < \infty \right\}.$$

To study this important example, we will prove a series of lemmas.

Lemma E3-1. *If $f, g \in \mathcal{L}_2'(\Omega, \mu)$, then $f \cdot g$ is integrable.*

Proof. Since $0 \leq (f \pm g)^2 = f \pm 2fg + g^2$, it follows that $\mp 2fg \leq f^2 + g^2$, that is, $|fg| \leq (f^2/2) + (g^2/2)$. The result follows from Corollary D16. □

Lemma E3-2. *If $f, g \in \mathcal{L}_2'(\Omega, \mu)$, then so is $f + g$.*

Proof. Since $(f + g)^2 = f^2 + 2fg + g^2$ and the three terms on the right are integrable, it follows that $(f + g)^2$ is integrable. □

Lemma E3-3. *If $f \in \mathcal{L}_2'(\Omega, \mu)$, then $\alpha f \in \mathcal{L}_2'(\Omega, \mu)$ for any $\alpha \in$ R.*

Proof. Clear. □

So far we have shown that $\mathcal{L}_2'(\Omega, \mu)$ is closed under addition and scalar multiplication.

Lemma E3-4. *If $\mu(\Omega) < \infty$, then $f \in \mathcal{L}_2'(\Omega, \mu) \Rightarrow f$ is integrable.*

Proof. For the proof given in E3-1, let $g = 1$. □

Note that if $\Omega = [1, \infty)$, then for $f(x) = 1/x$, $f^2 \in \mathcal{L}_2(\Omega, \mu)$ but $f \notin \mathcal{L}_2(\Omega, \mu)$ (μ being Lebesgue measure).

Lemma E3-5. *Let $f, g \in \mathcal{L}_2'(\Omega, \mu)$. Then $\int_\Omega (f - g)^2 d\mu = 0$ if and only if $f = g$ [a.e.].*

Proof. If $f = g$ [a.e.], then let

$$E \triangleq \{\omega \mid f(\omega) \neq g(\omega)\}$$
$$E' \triangleq \{\omega \mid f(\omega) = g(\omega)\}.$$

Then $\mu(E) = 0$ and so

$$\int_\Omega (f - g)^2 \, d\mu = \int_E (f - g)^2 \, d\mu + \int_{E'} (f - g) \, d\mu = 0 + 0 = 0.$$

Conversely, if $\int_\Omega (f - g)^2 \, dP = 0$, then from D19, we have that $(f - g)^2 = 0$ [a.e.], whence $f = g$ [a.e.]. □

Lemma and Definition E3-6. *If $f, g \in \mathcal{L}_2'(\Omega, \mu)$, define $f \sim g$ if and only if $\int_\Omega (f - g)^2 \, d\mu = 0$. It follows that $\sim$ is an equivalence relation.*

Definition E3-7. Let $[f]$ denote the equivalence class generated by f using $\sim$. We then define

$$\mathcal{L}_2(\Omega, \mu) \triangleq \{[f] \mid f \in \mathcal{L}_2'(\Omega, \mu)\}$$

equipped with the operations

$$[f] + [g] = [f + g]$$

$$\alpha[f] = [\alpha f]$$

and the inner product

$$\langle [f], [g] \rangle = \int_\Omega f \cdot g \, d\mu.$$

Theorem E3-8. *All of the operations defined in* E3-7 *are well defined, and* $\langle \cdot, \cdot \rangle$ *is indeed an inner product.*

The above construction essentially defines two functions in $\mathcal{L}_2'(\Omega, \mu)$ to be equal iff they are equal almost everywhere. It is customary to simply write f instead of $[f]$ with the understanding that equality really means $\sim$.

The next few definitions and theorems are based on the observation that in R^3 (Example E3(b)), the length of a vector $\mathbf{x}$, denoted by $||\mathbf{x}||$, is given by $||\mathbf{x}|| = \sqrt{\langle \mathbf{x}, \mathbf{x} \rangle}$. This observation generalizes beautifully, even to complex spaces.

E4. **Definition.** For any $\mathbf{x} \in \mathcal{V}$, we define the *norm* of $\mathbf{x}$ to be the real number

$$||\mathbf{x}|| = \sqrt{\langle \mathbf{x}, \mathbf{x} \rangle}.$$

E5. **Theorem** (Cauchy–Schwarz).

$$|\langle \mathbf{x}, \mathbf{y} \rangle| \leq ||\mathbf{x}|| \, ||\mathbf{y}|| \quad \textit{for all} \quad \mathbf{x}, \mathbf{y} \in \mathcal{V}$$

and equality obtains if and only if $\mathbf{x} = \lambda \mathbf{y}$ *for some* λ *or* $\mathbf{y} = \mathbf{0}$, *that is,* $\{\mathbf{x}, \mathbf{y}\}$ *is dependent.*

Proof. First suppose that $||\mathbf{y}|| = 1$. Then

$$\begin{aligned} 0 &\leq ||\mathbf{x} - \langle \mathbf{x}, \mathbf{y} \rangle \mathbf{y}||^2 \\ &= \langle \mathbf{x} - \langle \mathbf{x}, \mathbf{y} \rangle \mathbf{x} - \langle \mathbf{x}, \mathbf{y} \rangle \mathbf{y} \rangle \\ &= \langle \mathbf{x}, \mathbf{x} \rangle - \langle \mathbf{x}, \mathbf{y} \rangle \langle \mathbf{y}, \mathbf{x} \rangle - \langle \mathbf{x}, \mathbf{y} \rangle^* \langle \mathbf{y}, \mathbf{y} \rangle \\ &\quad + \langle \mathbf{x}, \mathbf{y} \rangle \langle \mathbf{x}, \mathbf{y} \rangle^* \langle \mathbf{y}, \mathbf{y} \rangle \\ &= ||\mathbf{x}||^2 = \langle \mathbf{x}, \mathbf{y} \rangle \langle \mathbf{x}, \mathbf{y} \rangle^* \\ &= ||\mathbf{x}|| - |\langle \mathbf{x}, \mathbf{y} \rangle|^2. \end{aligned}$$

Thus, we have shown that if $||\mathbf{y}|| = 1$, then

$$|\langle \mathbf{x}, \mathbf{y} \rangle| \leq ||\mathbf{x}||.$$

Now if $\mathbf{y} \neq \mathbf{0}$, define $\mathbf{z} \triangleq (1/||\mathbf{y}||)\mathbf{y}$ so that $||\mathbf{z}|| = 1$. Substituting $\mathbf{z}$ for $\mathbf{y}$ in the above inequality, we obtain

$$\left| \langle \mathbf{x}, \frac{1}{||\mathbf{y}||} \mathbf{y} \rangle \right| \leq ||\mathbf{x}||.$$

$$\frac{1}{||\mathbf{y}||} |\langle \mathbf{x}, \mathbf{y} \rangle| \leq ||\mathbf{x}||,$$

or

$$|\langle \mathbf{x}, \mathbf{y} \rangle| \leq ||\mathbf{x}|| \, ||\mathbf{y}||.$$

Hence, we have shown that if $\mathbf{y} \neq \mathbf{0}$, then

$$|\langle \mathbf{x}, \mathbf{y} \rangle| \leq ||\mathbf{x}|| \, ||\mathbf{y}||.$$

Clearly, if $\mathbf{y} = \mathbf{0}$, both sides are zero and so equality holds. Suppose therefore that $\mathbf{y} \neq \mathbf{0}$ and $|\langle \mathbf{x}, \mathbf{y} \rangle| = ||\mathbf{x}|| \, ||\mathbf{y}||$. As in the above calculation, this is equivalent to saying that $|\langle \mathbf{x}, \mathbf{z} \rangle| = ||\mathbf{x}||$, where $\mathbf{z} = (1/||\mathbf{y}||)\mathbf{y}$. However, from the first part of the proof, this equality holds exactly when $\mathbf{x} = \langle \mathbf{x}, \mathbf{z} \rangle \mathbf{z}$, or what is the same thing,

$$\mathbf{x} = \frac{1}{||\mathbf{y}||^2} \langle \mathbf{x}, \mathbf{y} \rangle \cdot \mathbf{y}. \qquad \square$$

E6. **Theorem.** $||\cdot||$ *satisfies the following:*

(a) $||\mathbf{x}|| \geq 0$ *for all* $\mathbf{x}$; $||\mathbf{x}|| = 0$ *if and only if* $\mathbf{x} = \mathbf{0}$.

(b) $||\lambda \mathbf{x}|| = |\lambda| \, ||\mathbf{x}||$ *for all scalars* λ, *and all* $\mathbf{x} \in \mathcal{V}$.

(c) $||\mathbf{x} + \mathbf{y}|| \leq ||\mathbf{x}|| + ||\mathbf{y}||$.

Proof. Parts (a) and (b) are easy; we will do (c). By Theorem E5,

$$|\langle \mathbf{x}, \mathbf{y} \rangle| \leq ||\mathbf{x}|| \, ||\mathbf{y}||,$$

so

$$2|\langle \mathbf{x}, \mathbf{y} \rangle| \leq 2||\mathbf{x}|| \, ||\mathbf{y}||,$$

and, thus,

$$2 \operatorname{Re}\langle \mathbf{x}, \mathbf{y} \rangle \leq 2||\mathbf{x}|| \, ||\mathbf{y}||,$$

where

$$\operatorname{Re}\langle \mathbf{x}, \mathbf{y} \rangle = \text{ real part of } \langle \mathbf{x}, \mathbf{y} \rangle = \frac{\langle \mathbf{x}, \mathbf{y} \rangle + \langle \mathbf{x}, \mathbf{y} \rangle^*}{2}.$$

It follows that

$$\langle \mathbf{x}, \mathbf{y} \rangle + \langle \mathbf{x}, \mathbf{y} \rangle^* \leq 2||\mathbf{x}|| \, ||\mathbf{y}||$$

or

$$\langle \mathbf{x}, \mathbf{y} \rangle + \langle \mathbf{y}, \mathbf{x} \rangle \leq 2||\mathbf{x}|| \, ||\mathbf{y}||.$$

Adding $||\mathbf{x}||^2 + ||\mathbf{y}||^2$ to both sides of this, we obtain

$$\langle \mathbf{x}, \mathbf{y} \rangle + \langle \mathbf{x}, \mathbf{y} \rangle + \langle \mathbf{y}, \mathbf{x} \rangle + \langle \mathbf{y}, \mathbf{y} \rangle \leq ||\mathbf{x}||^2 + 2||\mathbf{x}|| \, ||\mathbf{y}|| + ||\mathbf{y}||^2$$

or

$$\langle \mathbf{x} + \mathbf{y}, \mathbf{x} + \mathbf{y} \rangle \leq (||\mathbf{x}|| + ||\mathbf{y}||)^2;$$

but, this is equivalent to

$$||\mathbf{x} + \mathbf{y}||^2 \leq (||\mathbf{x}|| + ||\mathbf{y}||)^2,$$

which implies

$$||\mathbf{x}+\mathbf{y}|| \leq ||\mathbf{x}|| + ||\mathbf{y}||$$

by taking positive roots. □

E7. **Definition and Theorem.** *On $\mathcal{V}$ define*

$$d(\mathbf{x},\mathbf{y}) \triangleq ||\mathbf{x}-\mathbf{y}||.$$

Then $(\mathcal{V}, d)$ is a metric space.

Proof. We must check that d satisfies the three properties of a distance function

(1)

$$\begin{aligned} d(\mathbf{x},\mathbf{y}) &= ||\mathbf{x}-\mathbf{y}|| \\ &= ||(-1)(\mathbf{y}-\mathbf{x})|| \\ &= |-1|\,||\mathbf{y}-\mathbf{x}|| \\ &= ||\mathbf{y}-\mathbf{x}|| \\ &= d(\mathbf{y},\mathbf{x}). \end{aligned}$$

Thus, d is symmetric.

(2) $d(\mathbf{x},\mathbf{y}) = 0 \Longleftrightarrow ||\mathbf{x}-\mathbf{y}|| = 0 \Longleftrightarrow \mathbf{x}-\mathbf{y} = 0 \Longleftrightarrow \mathbf{x} = \mathbf{y}$.

(3)

$$\begin{aligned} d(\mathbf{x},\mathbf{z}) &= ||\mathbf{x}-\mathbf{z}|| \\ &= ||\mathbf{x}-\mathbf{y}+\mathbf{y}-\mathbf{z}|| \\ &= ||(\mathbf{x}-\mathbf{y})+\mathbf{y}-\mathbf{z})|| \\ &\leq ||\mathbf{x}-\mathbf{y}|| + ||\mathbf{y}-\mathbf{z}|| \\ &= d(x,y) + d(y,z). \end{aligned}$$

Thus, the triangle inequality holds. □

E8. **Theorem** (Continuity of the Inner Product). *If $x_n \to x$ and $y_n \to y$ (this makes sense since we now have a metric), then*

$$\lim_{(n,m)\to(\infty,\infty)} \langle \mathbf{x}_n, \mathbf{y}_m \rangle = \langle \mathbf{x}, \mathbf{y} \rangle.$$

E9. **Theorem** (Parallelogram Law).

$$||\mathbf{x}+\mathbf{y}||^2 + ||\mathbf{x}-\mathbf{y}||^2 = 2||\mathbf{x}||^2 + 2||\mathbf{y}||^2.$$

Proof. This is easily proved by simply expanding the left-hand side. □

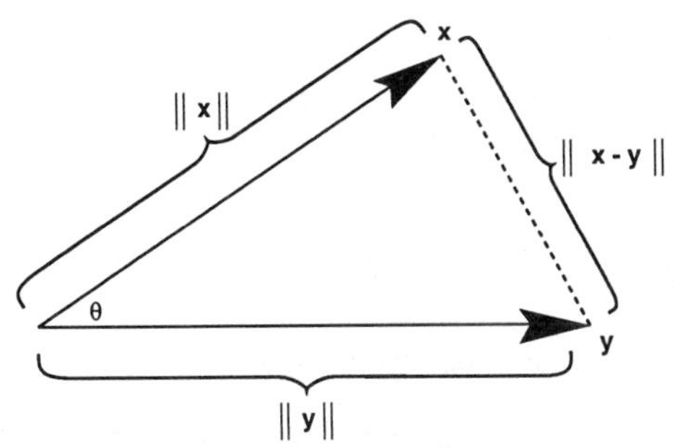

FIGURE E.1. The figure used to interpret $\langle \mathbf{x}, \mathbf{y} \rangle$ geometrically in R^3.

So far, we have defined the notion of an inner product space and have shown that $||\mathbf{x}||$ as defined in E4 produces the usual properties of a norm (Theorem E6) and, hence, provides us with a distance function (Definition and Theorem E7). We have also given three examples, the second of which is the familiar R^3 that we used to motivate the definition of norm and distance. The third example, $\mathcal{L}_2(\Omega, \mu)$, is essentially the space of all square integrable functions, except that we must regard two such functions as being the same if they are equal almost everywhere (this was effected using an equivalence relation). $\mathcal{L}_2(\Omega, \mu)$ is a much different space than R^3, in general it is not even finite dimensional. Nevertheless, from what we have done above, we can sensibly speak of length and distance. What is even more remarkable, is that we can successfully "lift" the concept of orthogonality from R^3 and speak of it in $\mathcal{L}_2(\Omega, \mu)$, in fact, speak of it in any inner product space. To see how this is done, refer to Figure E.1.

From the law of cosines applied in the plane spanned by $\mathbf{x}$ and $\mathbf{y}$, we have

$$\begin{aligned} ||\mathbf{x}-\mathbf{y}||^2 &= ||\mathbf{x}||^2 + ||\mathbf{y}||^2 - 2||\mathbf{x}||\,||\mathbf{y}|| \cos\theta, \\ \langle \mathbf{x}-\mathbf{y}, \mathbf{x}-\mathbf{y} \rangle &= ||\mathbf{x}||^2 + ||\mathbf{y}||^2 - 2||\mathbf{x}||\,||\mathbf{y}|| \cos\theta, \\ ||\mathbf{x}||^2 - 2\langle \mathbf{x}, \mathbf{y} \rangle + ||\mathbf{y}||^2 &= ||\mathbf{x}||^2 + ||\mathbf{y}||^2 - 2||\mathbf{x}||\,||\mathbf{y}|| \cos\theta, \\ -2\langle \mathbf{x}, \mathbf{y} \rangle &= -2||\mathbf{x}||\,||\mathbf{y}|| \cos\theta, \\ \langle \mathbf{x}, \mathbf{y} \rangle &= ||\mathbf{x}||\,||\mathbf{y}|| \cos\theta. \end{aligned}$$

Thus, we have expressed the inner product in R^3 in terms of the norms of $\mathbf{x}$ and $\mathbf{y}$, and the angle θ between them. From this it is easily seen that $\mathbf{x}$ and $\mathbf{y}$ are orthogonal exactly when $\langle \mathbf{x}, \mathbf{y} \rangle = 0$. This, as we see, motivates our definition for orthogonality in general inner product spaces.

E10. **Definition.** (a) In an inner space, two vectors $\mathbf{x}$ and $\mathbf{y}$ are said to be *orthogonal* if and only if $\langle \mathbf{x}, \mathbf{y} \rangle = 0$. We denote this by $\mathbf{x} \perp \mathbf{y}$, read "$\mathbf{x}$ perp $\mathbf{y}$".

(b) A vector $\mathbf{x}$ is said to be *orthogonal* to a set S in case $\mathbf{x} \perp \mathbf{s}$ for every $\mathbf{s} \in S$. We write this as $\mathbf{x} \perp S$.

E11. **Lemma.** $\mathbf{x} \perp \mathbf{y}$ *for all* $\mathbf{y} \Longleftrightarrow \mathbf{x} = \mathbf{0}$.

Proof. If $\mathbf{x} = \mathbf{0}$, then $\langle \mathbf{x}, \mathbf{y} \rangle = 0$ for all $\mathbf{y}$. Conversely, if $\mathbf{x} \perp \mathbf{y}$ for all $\mathbf{y}$, then in particular $\mathbf{x} \perp \mathbf{x}$ and so $||\mathbf{x}||^2 = 0$. Thus, $\mathbf{x} = \mathbf{0}$. □

E12. **Theorem** (Pythagoras).

(a) *If* $\mathbf{x} \perp \mathbf{y}$ *then* $||\mathbf{x}+\mathbf{y}||^2 = ||\mathbf{x}||^2 + ||\mathbf{y}||^2$.

(b) *If* $\mathbf{x}_1, \mathbf{x}_2, \ldots, \mathbf{x}_n$ *are pairwise orthogonal, then*

$$||\mathbf{x}_1 + \ldots + \mathbf{x}_n||^2 = ||\mathbf{x}_1||^2 + ||\mathbf{x}_2||^2 + \ldots + ||\mathbf{x}_n||^2.$$

Proof. (a)

$$\begin{aligned} ||\mathbf{x}+\mathbf{y}||^2 &= \langle \mathbf{x}+\mathbf{y}, \mathbf{x}+\mathbf{y} \rangle \\ &= \langle \mathbf{x}, \mathbf{x} \rangle + \langle \mathbf{y}, \mathbf{x} \rangle + \langle \mathbf{x}, \mathbf{y} \rangle + \langle \mathbf{y}, \mathbf{y} \rangle \\ &= ||\mathbf{x}||^2 + 0 + 0 + ||\mathbf{y}||^2 \\ &= ||\mathbf{x}||^2 + ||\mathbf{y}||^2. \end{aligned}$$

(b) Use part (a) and mathematical induction. □

E13. **Theorem.** *Let* $\mathcal{V}$ *be an inner product space,* $\mathcal{M}$ *a vector subspace of* $\mathcal{V}$, *and* $\mathbf{x}$ *an arbitrary vector in* $\mathcal{V}$. *If there is a vector* $\mathbf{m}_0 \in \mathcal{M}$ *such that* $||\mathbf{x} - \mathbf{m}_0|| \leq ||\mathbf{x} - \mathbf{m}||$ *for all* $\mathbf{m} \in \mathcal{M}$, *then* $\mathbf{m}_0$ *is unique. A necessary and sufficient condition that* $\mathbf{m}_0$ *be such a vector, is that* $(\mathbf{x} - \mathbf{m}_0) \perp \mathcal{M}$.

Proof. We will show that last condition first. Hence, suppose m_0 is a vector such that

$$||\mathbf{x} - \mathbf{m}_0|| \leq ||\mathbf{x} - \mathbf{m}||$$

for all $\mathbf{m} \in \mathcal{M}$. We will show that $(\mathbf{x} - \mathbf{m}_0) \perp \mathcal{M}$. Suppose not. Then there is some $\mathbf{m} \in \mathcal{M}$ that is not orthogonal to $\mathbf{x} - \mathbf{m}_0$. Since $\mathbf{0} \perp \mathcal{V}$, $\mathbf{m} \neq \mathbf{0}$. We can suppose, without loss of generality, that $||\mathbf{m}|| = 1$. Define

$$\delta \triangleq \langle \mathbf{x} - \mathbf{m}_0, \mathbf{m} \rangle.$$

By assumption, $\delta \neq 0$. Let $\mathbf{m}_1 \triangleq \mathbf{m}_0 + \delta\mathbf{m} \in \mathcal{M}$. Then

$$\begin{aligned} ||\mathbf{x} - \mathbf{m}_1||^2 &= ||\mathbf{x} - \mathbf{m}_0 - \delta\mathbf{m}||^2 \\ &= \langle \mathbf{x} - \mathbf{m}_0 - \delta\mathbf{m}, \mathbf{x} - \mathbf{m}_0 - \delta\mathbf{m} \rangle \\ &= \langle \mathbf{x} - \mathbf{m}_0, \mathbf{x} - \mathbf{m}_0 \rangle - \langle \mathbf{x} - \mathbf{m}_0, \delta\mathbf{m} \rangle \\ &\quad - \langle \delta\mathbf{m}, \mathbf{x} - \mathbf{m}_0 \rangle + \langle \delta\mathbf{m}, \delta\mathbf{m} \rangle \\ &= ||\mathbf{x} - \mathbf{m}_0||^2 - \delta^* \langle \mathbf{x} - \mathbf{m}_0, \mathbf{m} \rangle \\ &\quad - \delta \langle \mathbf{m}, \mathbf{x} - \mathbf{m}_0 \rangle + \delta\delta^* ||\mathbf{m}||^2 \\ &= ||\mathbf{x} - \mathbf{m}_0||^2 - |\delta|^2 - |\delta|^2 + |\delta|^2 \\ &= ||\mathbf{x} - \mathbf{m}_0||^2 - |\delta|^2. \end{aligned}$$

Since $\delta \neq 0$, we have

$$\|\mathbf{x} - \mathbf{m}_1\| < \|\mathbf{x} - \mathbf{m}_0\|,$$

which is a contradiction.

Next suppose $(\mathbf{x} - \mathbf{m}_0) \perp \mathcal{M}$. Then for any $\mathbf{m} \in \mathcal{M}$, $\mathbf{m}_0 - \mathbf{m} \in \mathcal{M}$ and by the Pythagorean theorem

$$\|\mathbf{x} - \mathbf{m}\|^2 = \|\mathbf{x} - \mathbf{m}_0 + \mathbf{m}_0 - \mathbf{m}\|^2 = \|\mathbf{x} - \mathbf{m}_0\|^2 + \|\mathbf{m}_0 - \mathbf{m}\|^2.$$

It follows that

$$\|\mathbf{x} - \mathbf{m}_0\| \leq \|\mathbf{x} - \mathbf{m}\|, \quad \text{all } \mathbf{m} \in \mathcal{M}.$$

The above identity says more. If $\mathbf{m} \neq \mathbf{m}_0$, then $\|\mathbf{m}_0 - \mathbf{m}\|^2 \neq 0$ so $\|\mathbf{x} - \mathbf{m}\| > \|\mathbf{x} - \mathbf{m}_0\|$; hence, $\mathbf{m}_0$ is unique.

Note that E13 establishes the uniqueness of $\mathbf{m}_0$ as well as a condition to test it. Existence is missing. It turns out that we do not, as yet, have enough structure to establish existence. We now correct this flaw. □

E14. **Definition.** An inner product space $\mathcal{H}$ is called a *Hilbert space* if and only if $\mathcal{H}$ is complete with respect to the norm, that is, Cauchy sequences converge.

E15. **Examples.** Parts (a) and (b) of E3 are Hilbert spaces. We leave this to the reader. (c) $\mathcal{L}_2(\Omega, \mu)$ is a Hilbert space. We do this in two steps.

Theorem 1. *If f_n is Cauchy in $\mathcal{L}_2(\Omega, \mu)$ then f_n is a Cauchy sequence in measure.*

Proof. Suppose not. Then there exists a pair of numbers (ϵ, η) such that for every n, m

$$\mu\{\omega \mid |f_n(\omega) - f_m(\omega)| \geq \epsilon\} \geq \eta.$$

Thus,

$$\int_\Omega |f_n - f_m|^2 \, d\mu \geq \int_{\{\omega \mid |f_n(\omega) - f_m(\omega)| > \epsilon\}} |f_n - f_m|^2 \, d\mu \geq \epsilon\eta$$

for all n, m, a contradiction. □

Theorem 2. *$\mathcal{L}_2(\Omega, \mu)$ is complete.*

Proof. Let f_n be a Cauchy sequence in $\mathcal{L}_2(\Omega, \mu)$. By Theorem E15-1 above and Corollary D28, there is a subsequence f_{n_k} that is Cauchy almost everywhere (which means Cauchy in $\mathcal{L}_1(\Omega, \mu)$). Since the reals are complete, $f_{n_k}(\omega)$ converges for almost all ω, so there is a function f_0 defined by

$$f_0 = \lim_{k \to \infty} f_{n_k} \quad \text{[a.e.]}.$$

The remainder of the proof is devoted to showing that $f_0 \in \mathcal{L}_2(\Omega, \mu)$ and $f_n \to f_0$ in $\mathcal{L}_2(\Omega, \mu)$.

If we fix an index i, we then have

$$\lim_{k\to\infty} |f_{n_i} - f_{n_k}|^2 = |f_{n_i} - f_0|^2 \quad \text{[a.e.]},$$

so by Fatou's lemma (Theorem D30),

$$\underline{\lim}_k \int_\Omega |f_{n_i} - f_{n_k}|^2\, d\mu \geq \int_\Omega |f_{n_i} - f_0|^2\, d\mu.$$

Since $\{f_n\}$ is Cauchy in $\mathcal{L}_2(\Omega, \mu)$, it follows that there is an integer k_0 such that if $i > k_0$ and $k > k_0$, then

$$\int_\Omega |f_{n_i} - f_{n_k}|^2\, d\mu < \frac{\epsilon^2}{4}.$$

From the previous inequality, we then have that for $i > k_0$

$$\int_\Omega |f_{n_i} - f_0|^2\, d\mu \leq \frac{\epsilon^2}{4}, \tag{$*$}$$

and so $f_{n_i} - f_0 \in \mathcal{L}_2(\Omega, \mu)$. However, $f_{n_i} \in \mathcal{L}_2(\Omega, \mu)$ (for each i) by hypothesis, so it follows that $f_0 \in \mathcal{L}_2(\Omega, \mu)$. Since $\{f_n\}$ is Cauchy in $\mathcal{L}_2(\Omega, \mu)$, there is an n_0 such that $n > n_0$ and $m > m_0$ imply $\|f_n - f_m\| < \frac{\epsilon}{2}$. Thus, if we take $i > k_0$ and choose $n, n_i > n_0$, we have

$$\|f_n - f_{n_i}\| < \frac{\epsilon}{2} \tag{$**$}$$

and by relation $(*)$ that

$$\|f_{n_i} - f_0\| < \frac{\epsilon}{2}. \tag{$***$}$$

Hence, for $n > N_0$, we have by relation $(**)$ and $(***)$ that

$$\|f_n - f_0\| \leq \|f_n - f_{n_i}\| + \|f_{n_i} - f_0\| < \frac{\epsilon}{2} + \frac{\epsilon}{2} = \epsilon,$$

and we are done. □

Note that in Example E3(c) we could just as well have used Riemann integrals to obtain an inner product space, everything works. The problem is, however, in this case we would not have completeness (see Reference [11] for an example). As we see from the next theorem, completeness is essential for our purposes (see Chapter 2, also).

E16. **Theorem** (The Projection Theorem). *Let $\mathcal{H}$ be a Hilbert space and $\mathcal{M}$ a closed subspace of $\mathcal{H}$. Corresponding to any vector $\mathbf{x} \in \mathcal{H}$, there is a unique vector $\hat{\mathbf{x}}$ in $\mathcal{M}$ such that*

$$\|\mathbf{x} - \hat{\mathbf{x}}\| \leq \|\mathbf{x} - \mathbf{m}\|$$

for all $\mathbf{m} \in \mathcal{M}$. *Furthermore, a necessary and sufficient condition that* $\hat{\mathbf{x}} \in \mathcal{M}$ *be the (necessarily unique) minimizing vector is that* $(x - \hat{\mathbf{x}}) \perp \mathcal{M}$.

Proof. Uniqueness and the orthogonality condition were already established in E14. We now establish existence.

If $\mathbf{x} \in \mathcal{M}$, then $\hat{\mathbf{x}} = \mathbf{x}$ will work. Thus, assume that $\mathbf{x} \notin \mathcal{M}$. Define

$$\delta = \inf_{m \in \mathcal{M}} ||\mathbf{x} - \mathbf{m}||.$$

We will produce $\hat{\mathbf{x}} \in \mathcal{M}$ with $||\mathbf{x} - \hat{\mathbf{x}}|| = \delta$. Let $\mathbf{m}_i$ be a sequence of vectors in $\mathcal{M}$ such that $||\mathbf{x} - \mathbf{m}_i|| \to \delta$ (property of infimum). By the parallelogram law (E9),

$$||\mathbf{m}_j - \mathbf{m}_i||^2 + ||\mathbf{m}_j - \mathbf{x} - (\mathbf{x} + \mathbf{m}_i)||^2 = 2||\mathbf{m}_j - \mathbf{x}||^2 + 2||\mathbf{x} - \mathbf{m}_i||^2,$$

and so

$$||\mathbf{m}_j - \mathbf{m}_i||^2 = 2||\mathbf{m}_j - \mathbf{x}||^2 + 2||\mathbf{x} - \mathbf{m}_i||^2 - 4\left\|\mathbf{x} - \frac{\mathbf{m}_i + \mathbf{m}_j}{2}\right\|^2.$$

For all i, j, the vector $\frac{1}{2}(\mathbf{m}_i + \mathbf{m}_j) \in \mathcal{M}$, and so by the definition of δ,

$$\left\|\mathbf{x} - \frac{1}{2}(\mathbf{m}_i + \mathbf{m}_j)\right\| \geq \delta.$$

Hence, it follows that

$$||\mathbf{m}_i - \mathbf{m}_j||^2 \leq 2||\mathbf{m}_j - \mathbf{x}||^2 + ||\mathbf{m}_i - \mathbf{x}||^2 - 4\delta^2.$$

As $i, j \to \infty$, the right-hand side of this expression approaches zero, hence we conclude that $\{\mathbf{m}_i\}$ is a Cauchy sequence. Thus, there is an $\hat{\mathbf{x}} \in \mathcal{M}$ such that $\mathbf{m}_i \to \hat{\mathbf{x}}$ (here is where we used completeness as well as the fact that $\mathcal{M}$ is closed). By continuity of the norm, which follows from E8, $||\mathbf{x} - \hat{\mathbf{x}}|| = \delta$. □

In three dimensions, the above theorem is very clear, and the picture of it, Figure E.2, is a great way to remember what it says.

Note that to apply the projection theorem, one must make sure that the space upon which one is projecting, namely, $\mathcal{M}$, is closed. There is one case in which this situation is automatically satisfied, and fortunately (for us), it is the situation one encounters when proving the Kalman theorem. Here it is.

E17. **Theorem.** *A finite dimensional subspace of a Hilbert space is always closed.*

This is a special case of a more general (and more difficult) theorem. The proof of the above theorem is easy and left to the reader.

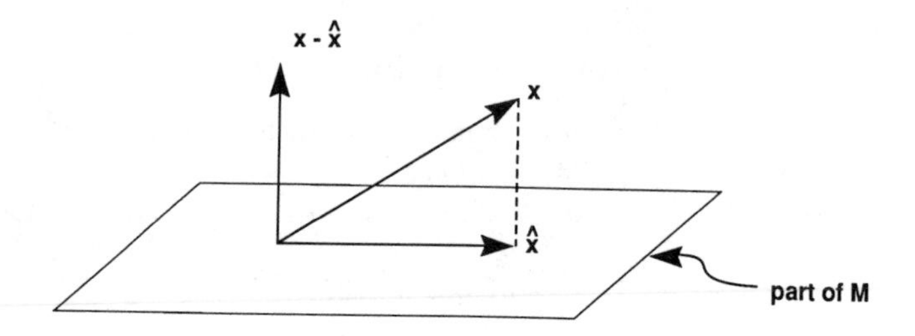

FIGURE E.2. The projection theorem in R^3.

E18. **Definition.** Let $\mathcal{S}$ be any subset of $\mathcal{H}$ ($\mathcal{H}$ will always denote a Hilbert space). We define

$$\mathcal{S}^{\perp} = \{\mathbf{x} \,|\, \mathbf{x} \perp \mathbf{s} \text{ for every } \mathbf{s} \in \mathcal{S}\}.$$

Note that $\mathcal{H}^{\perp} = \{0\}$ and $\{0\}^{\perp} = \mathcal{H}$.

E19. **Theorem.** *Let $\mathcal{M}$ and $\mathcal{N}$ be subsets of $\mathcal{H}$. Then*

(a) $\mathcal{M}^{\perp}$ *is a closed subspace;*

(b) $\mathcal{M} \subset \mathcal{M}^{\perp\perp}$;

(c) *if* $\mathcal{M} \subset \mathcal{N}$, *then* $\mathcal{N}^{\perp} \subset \mathcal{M}^{\perp}$;

(d) $\mathcal{M} = \mathcal{M}^{\perp\perp\perp}$; *and*

(e) $\mathcal{M}^{\perp\perp} = \overline{\mathcal{M}}$ *(the closure of* $\mathcal{M}$*).*

Proof. (a) That $\mathcal{M}^{\perp}$ is a subspace is left to the reader; we show it to be closed.

Let $\{\mathbf{x}_n\}$ be a sequence in $\mathcal{M}^{\perp}$ such that $\mathbf{x}_n \to \mathbf{x}$. For any $\mathbf{z} \in \mathcal{M}$, we have by continuity of the inner product,

$$\langle \mathbf{x}_n, \mathbf{z} \rangle \to \langle \mathbf{x}, \mathbf{z} \rangle;$$

but, $\langle$bf $\mathrm{x}_n, \mathbf{z}\rangle = 0$ for all n, so $\langle \mathbf{x}, \mathrm{z}\rangle = 0$, that is, $\mathbf{x} \in \mathcal{M}^{\perp}$.

(b) If $\mathbf{x} \in \mathcal{M}$, then for all $\mathbf{y} \in \mathcal{M}^{\perp}$, we have $\mathbf{x} \perp \mathbf{y}$. Hence,

$$\mathbf{x} \in \{\mathbf{z} \,|\, \mathbf{z} \perp \mathbf{y} \text{ for every } \mathbf{y} \in \mathcal{M}^{\perp}\},$$

that is,

$$\mathbf{x} \in \mathcal{M}^{\perp\perp}.$$

(c) Suppose $\mathcal{M} \subset \mathcal{N}$ and let $\mathbf{x} \in \mathcal{N}^{\perp}$. Then for all $\mathbf{y} \in \mathcal{N}$, $\mathbf{x} \perp \mathbf{y}$. But since $\mathcal{M} \subset \mathcal{N}$, it is also true that $\mathbf{x} \perp \mathbf{y}$ for all $\mathbf{y} \in \mathcal{M}$. Thus, $\mathbf{x} \in \mathcal{M}^{\perp}$.

(d) Using (b) with $\mathcal{M}^\perp$ replacing $\mathcal{M}$, we obtain

$$\mathcal{M}^\perp \subset \mathcal{M}^{\perp\perp\perp}.$$

Also, applying (c) to (b) we obtain

$$\mathcal{M}^{\perp\perp\perp} \subset \mathcal{N}^\perp$$

and the result follows.

(e) Since $\mathcal{M} \subset \mathcal{M}^{\perp\perp}$ and $\mathcal{M}^{\perp\perp}$ is closed (by (a)) we have that $\overline{\mathcal{M}} \subset \mathcal{M}^{\perp\perp}$ (property of closure being the smallest closed set containing $\mathcal{M}$). Suppose $\mathcal{M}^{\perp\perp} \setminus \overline{\mathcal{M}} \neq \phi$. Then, of course, there exists $\mathbf{x} \in \mathcal{M}^{\perp\perp} \setminus \overline{\mathcal{M}}$. By the projection theorem, there exists a unique $\hat{\mathbf{x}} \in \overline{\mathcal{M}}$ such that $(\mathbf{x}-\hat{\mathbf{x}}) \perp \overline{\mathcal{M}}$, that is, $(\mathbf{x}-\hat{\mathbf{x}}) \in \overline{\mathcal{M}}^\perp$. However, $\mathcal{M} \subset \overline{\mathcal{M}}$ so by (c), $\overline{\mathcal{M}}^\perp \subset \mathcal{M}^\perp$. Hence, $(\mathbf{x}-\hat{\mathbf{x}}) \in \mathcal{M}^\perp$. Since $\mathbf{x} \in \mathcal{M}^{\perp\perp}$, it follows that $(\mathbf{x}-\hat{\mathbf{x}}) \perp \mathbf{x}$. We also have $(\mathbf{x}-\hat{\mathbf{x}}) \perp \hat{\mathbf{x}}$ (since $\hat{\mathbf{x}} \in \mathcal{M}$), and so

$$\begin{aligned} 0 &= \langle \mathbf{x}-\hat{\mathbf{x}}, \mathbf{x}\rangle - \langle \mathbf{x}-\hat{\mathbf{x}}, \hat{\mathbf{x}}\rangle \\ 0 &= \langle \mathbf{x}-\hat{\mathbf{x}}, \mathbf{x}-\hat{\mathbf{x}}\rangle \\ 0 &= \|\mathbf{x}-\hat{\mathbf{x}}\|^2, \end{aligned}$$

which implies $\mathbf{x} = \hat{\mathbf{x}} \in \overline{\mathcal{M}}$, a contradiction to the assumption that $\mathbf{x} \notin \overline{\mathcal{M}}$. □

E20. **Theorem.** *Suppose $\mathcal{M}$ is a closed subspace of $\mathcal{H}$. Then, if $\mathbf{x}$ is any element of $\mathcal{H}$, there exists $\mathbf{x}_1 \in \mathcal{M}$, $\mathbf{x}_2 \in \mathcal{M}^\perp$ such that*

$$\mathbf{x} = \mathbf{x}_1 + \mathbf{x}_2.$$

Moreover, $\mathbf{x}_1$ and $\mathbf{x}_2$ are unique.

Proof. Define $\mathcal{K} = \mathcal{M} + \mathcal{M}^\perp \triangleq \{\mathbf{y}+\mathbf{z} \mid \mathbf{y} \in \mathcal{M}, \mathbf{z} \in \mathcal{M}^\perp\}$. Clearly, $\mathcal{K}$ is a subspace of $\mathcal{H}$. We must show that $\mathcal{H} \subset \mathcal{K}$. Since $\mathcal{M}$ is closed, we know that for any $\mathbf{x} \in \mathcal{H}$ there is an $\hat{\mathbf{x}} \in \mathcal{M}$ such that $(\mathbf{x}-\hat{\mathbf{x}}) \perp \mathcal{M}$, that is, $\mathbf{x}-\hat{\mathbf{x}} \in \mathcal{M}^\perp$. Hence,

$$\mathbf{x} = \hat{\mathbf{x}} + (\mathbf{x}-\hat{\mathbf{x}}) \in \mathcal{M} + \mathcal{M}^\perp = \mathcal{K}.$$

It remains to show uniqueness.

Suppose

$$\mathbf{x} = \mathbf{x}_1 + \mathbf{x}_2 = \mathbf{y}_1 + \mathbf{y}_2,$$

where $\mathbf{x}_1, \mathbf{y}_1 \in \mathcal{M}$ and $\mathbf{x}_2, \mathbf{y}_2 \in \mathcal{M}^\perp$. Then

$$\mathbf{x}_1 - \mathbf{y}_1 = \mathbf{y}_2 - \mathbf{x}_2.$$

However, $\mathbf{x}_1 - \mathbf{y}_1 \in \mathcal{M}$ and $\mathbf{y}_2 - \mathbf{x}_2 \in \mathcal{M}^\perp$, so since they are equal,

$$\mathbf{x}_i - \mathbf{y}_i \in \mathcal{M} \cap \mathcal{M}^\perp = \{\mathbf{0}\} \quad \text{for} \quad i = 1, 2.$$

It follows that $\mathbf{x}_1 = \mathbf{y}_1$ and $\mathbf{x}_2 = \mathbf{y}_2$, as required. □

E21. **Definition.** (a) Let $\mathcal{H}_1$ and $\mathcal{H}_2$ be Hilbert spaces. By an *operator* or *linear transformation* A from $\mathcal{H}_1$ to $\mathcal{H}_2$, we will mean a function

$$A : \mathcal{H}_1 \to \mathcal{H}_2$$

such that for all $\alpha, \beta \in \mathrm{F}$, all $\mathbf{x}, \mathbf{y} \in \mathcal{H}$.

$$A(\alpha\mathbf{x} + \beta\mathbf{y}) = \alpha A(\mathbf{x}) + \beta A(\mathbf{y}).$$

(b) $\mathcal{L}(\mathcal{H}_1, \mathcal{H}_2) \triangleq \{A \mid A : \mathcal{H}_1 \to \mathcal{H}_2 \text{ is linear}\}$.
(c) If $A, B \in \mathcal{L}(\mathcal{H}_1, \mathcal{H}_2)$, $\lambda \in \mathrm{F}$, define $A + B$ and λA by

$$(A + B)(\mathbf{x}) = A(\mathbf{x}) + B(\mathbf{x}) \quad \text{for all} \quad \mathbf{x} \in \mathcal{H}_1$$

$$(\lambda A)(\mathbf{x}) = \lambda \cdot A(\mathbf{x}) \quad \text{for all} \quad \mathbf{x} \in \mathcal{H}_1$$

(note, $\mathcal{H}_1$ and $\mathcal{H}_2$ have to be over the same field).

E22. **Theorem.** *$\mathcal{L}(\mathcal{H}_1, \mathcal{H}_2)$ with the above operations is a vector space.*

E23. **Definition.** (a) Let $\mathcal{H}_1$ and $\mathcal{H}_2$ be the Hilbert spaces, $A \in \mathcal{L}(\mathcal{H}_1, \mathcal{H}_2)$. Then A is *bounded* if and only if there exists a scalar α, called a *bound,* such that

$$||\Lambda(\mathbf{x})|| \leq \alpha ||\mathbf{x}||$$

for all $\mathbf{x} \in \mathcal{H}_1$. We denote by $\mathcal{B}(\mathcal{H}_1, \mathcal{H}_2)$ the set of all such bounded operators. If $A : \mathcal{H}_1 \to \mathcal{H}_1$ we simply write $\mathcal{B}(\mathcal{H}_1)$.
(b) Let $A \in \mathcal{B}(\mathcal{H}_1, \mathcal{H}_2)$. We define the *norm* of A, written $||A||$, by

$$||A|| = \inf\{\alpha \mid \alpha \text{ is a bound for } A\}.$$

It turns out that $\mathcal{B}(\mathcal{H}_1, \mathcal{H}_2)$ is a vector subspace of $\mathcal{L}(\mathcal{H}_1, \mathcal{H}_2)$. The proof is quite easy and has as a consequence, the fact that $||\cdot||$ does act like a norm on $\mathcal{B}(\mathcal{H}_1, \mathcal{H}_2)$, that is to say, the conclusions in Theorem E6 hold. The interested reader might try this to check his/her understanding of the definition of the norm of an operator.

E24. **Lemma.** *If $A \in \mathcal{B}(\mathcal{H}_1, \mathcal{H}_2)$, then $||A\mathbf{x}|| \leq ||A|| \, ||\mathbf{x}||$ for all $\mathbf{x} \in \mathcal{H}$.*

Proof. Suppose not. Then, for some $\mathbf{x}_0 \in \mathcal{H}_1$, we would have $\mathbf{x}_0 \neq \mathbf{0}$ and

$$||A\mathbf{x}_0|| > ||A|| \, ||\mathbf{x}_0||.$$

Hence,

$$\frac{||A\mathbf{x}_0||}{||\mathbf{x}_0||} > ||A||,$$

so we can find β_0 such that

$$\frac{||A\mathbf{x}_0||}{||\mathbf{x}_0||} > \beta_0 > ||A||.$$

It follows that

$$\|A\mathbf{x}_0\| > \beta_0\|\mathbf{x}_0\|.$$

But from the definition of $\|A\|$, $\beta_0 > \|A\|$ implies that $\|A\mathbf{x}_0\| \leq \beta_0\|\mathbf{x}\|$, a contradiction. □

E25. **Lemma.** *Let $A \in \mathcal{L}(\mathcal{H}_1, \mathcal{H}_2)$ and define*

$$\eta \triangleq \sup\{\|A\mathbf{x}\| \mid \|\mathbf{x}\| = 1\}.$$

Then

(a) $A \in \mathcal{B}(\mathcal{H}_1, \mathcal{H}_2) \Longleftrightarrow \eta < \infty$.

(b) *If* $A \in \mathcal{B}(\mathcal{H}_1, \mathcal{H}_2)$, *then* $\|A\| = \eta$.

E26. **Theorem.** *Let $A \in \mathcal{L}(\mathcal{H}_1, \mathcal{H}_2)$. Then the following are equivalent.*

(a) *A is uniformly continuous,*

(b) *A is continuous,*

(c) *A is continuous at some point,*

(d) *A is continuous at* $\mathbf{0}$, *and*

(e) $A \in \mathcal{B}(\mathcal{H}_1, \mathcal{H}_2)$.

Proof. Here, (a) ⇒ (b) ⇒ (c) is obvious. For (c) ⇒ (d) suppose A is continuous at $\mathbf{x}_0 \in \mathcal{H}_1$. Let $\mathbf{x}_n \to 0$. It is then easy to see that $\mathbf{x}_n + \mathbf{x}_0 \to \mathbf{x}_0$. Thus, by (c)

$$T(\mathbf{x}_n + \mathbf{x}_0) \to T(\mathbf{x}_0),$$

and since T is linear,

$$T(\mathbf{x}_n) + T(\mathbf{x}_0) \to T(\mathbf{x}_0),$$

whence

$$T(\mathbf{x}_n) \to \mathbf{0}.$$

For (d) ⇒ (e) we will show the contrapositive. If (e) fails, then by E25

$$\infty = \eta = \sup\{\|A\mathbf{x}\| \mid \|\mathbf{x}\| = 1\}.$$

This means that there is a sequence $\mathbf{x}_n \in \mathcal{H}_1$ with $\|\mathbf{x}_n\| = 1$ and $\|A\mathbf{x}_n\| \to \infty$. Let

$$\mathbf{y}_n = \frac{1}{\|A\mathbf{x}_n\|}\mathbf{x}_n.$$

Then $\|\mathbf{y}_n\| \to 0$ so that $\mathbf{y}_n \to 0$; but,

$$A(\mathbf{y}_n) = \frac{1}{\|A\mathbf{x}_n\|}A(\mathbf{x}_n),$$

so $||A(\mathbf{y}_n)|| = 1$ for all n. Thus, $A(\mathbf{y}_n) \not\to 0$ and so (d) fails.

For (e) $\Rightarrow$ (a): this follows easily from the inequality

$$||A\mathbf{x} - A\mathbf{y}|| = ||A(\mathbf{x} - \mathbf{y})|| \leq ||A|| \, ||\mathbf{x} - y||.$$

E27. **Definition.** $\ker(A) = \{\mathbf{x} \mid A\mathbf{x} = \mathbf{0}\}$. This set is called the *kernel* of A.

E28. **Theorem.** (a) $\ker(A)$ *is a subspace.*

(b) *If A is bounded, then* $\ker(A)$ *is closed.*

E29. **Definition.**

(a) If $f \in \mathcal{L}(\mathcal{H}, \mathrm{F})$, then f is called a *linear functional.*

(b) $\mathcal{H}^* \triangleq \mathcal{B}(\mathcal{H}, \mathrm{F})$. Thus, $\mathcal{H}^*$ is the set of all bounded linear functionals on $\mathcal{H}$; $\mathcal{H}^*$ is often called the *continuous dual* of $\mathcal{H}$.

There is a whole circle of ideas upon which one could embark at this point, namely, the subject of linear programming. Linear programming is the study of maximizing (or minimizing) linear functionals over closed, bounded, convex sets. The solution to the linear programming problem is intrinsically tied to the geometry of Hilbert spaces. We, unfortunately, do not have the time or space to pursue these ideas.

E30. **Example.** Let us look at the structure of linear functionals on R^3. Suppose $f \in (\mathrm{R}^3)^*$ and $\mathbf{e}_1$, $\mathbf{e}_2$, $\mathbf{e}_3$ are the standard bases for R^3. Define

$$\alpha_i = f(\mathbf{e}_i); \quad i = 1, 2, 3$$

and then define the vector

$$\mathbf{a} = (\alpha_1, \alpha_2, \alpha_3).$$

We claim that

$$f(\mathbf{x}) = \langle \mathbf{x}, \mathbf{a} \rangle,$$

so that $\mathbf{a}$ is a representation of f in R^3. For, if

$$\mathbf{x} = \sum_{i=1}^{3} \mathbf{x}_i \mathbf{e}_i,$$

then

$$\begin{aligned} f(x) &= f\left(\sum_{i=1}^{3} x_i \mathbf{e}_i\right) \\ &= \sum_{i=1}^{3} x_i f(\mathbf{e}_i) \\ &= \sum_{i=1}^{3} x_i \alpha_i \\ &= \langle \mathbf{x}, \mathbf{a} \rangle. \end{aligned}$$

Remarkably, this idea carries over to general Hilbert spaces.

E31. **Theorem** (Riesz Representation Theorem). *Let $\mathcal{H}$ be a Hilbert space and let $f \in \mathcal{H}^*$. Then there is a unique vector $\mathbf{f} \in \mathcal{H}$ such that*

(a) *For each $x \in \mathcal{H}$, $f(x) = \langle \mathbf{x}, \mathbf{f} \rangle$,*

(b) $\|f\| = \|\mathbf{f}\|$.

Proof. (a) If $\ker(f) = \mathcal{H}$, then $\mathbf{f} = \mathbf{0}$ will work since in this case $f = 0$. Thus, we suppose that $\ker(f) \neq \mathcal{H}$. By E28, $\ker(f)$ is closed, and so by E20,

$$\mathcal{H} = \ker(f) + \ker(f)^{\perp}.$$

Now $\ker(f)^{\perp} \neq \{\mathbf{0}\}$ or else $\mathcal{H} = \{0\}^{\perp} = \ker(f)$, a contradiction. Thus, we can choose $\mathbf{z} \in \ker(f)^{\perp}$, $\mathbf{z} \neq \mathbf{0}$, and scale $\mathbf{z}$ so that $f(\mathbf{z}) = 1$. We will show that there is a scalar $\alpha \in \mathrm{F}$ such that $\mathbf{f} \triangleq \alpha\mathbf{z}$ satisfies the conclusions (a) and (b) above. Given any $\mathbf{x} \in \mathcal{H}$, we have

$$\begin{aligned} f(\mathbf{x} - f(\mathbf{x})\mathbf{z}) &= f(\mathbf{x}) - f(f(\mathbf{x})\mathbf{z}) \\ &= f(\mathbf{x}) - f(\mathbf{x})f(\mathbf{z}) \\ &= f(\mathbf{x}) - f(\mathbf{x}) \\ &= 0, \end{aligned}$$

so $\mathbf{x} - f(\mathbf{x})\mathbf{z} \in \ker(f)$. Since $\mathbf{z} \in (\ker f)^{\perp}$, we have

$$(\mathbf{x} - f(\mathbf{x})\mathbf{z}) \perp \mathbf{z},$$

and so

$$\begin{aligned} 0 &= \langle \mathbf{x} - f(\mathbf{x})\mathbf{z}, \mathbf{z} \rangle \\ &= \langle \mathbf{x}, \mathbf{z} \rangle - f(\mathbf{x})\langle \mathbf{z}, \mathbf{z} \rangle \\ &= \langle \mathbf{x}, \mathbf{z} \rangle - f(\mathbf{x})\|\mathbf{z}\|^2. \end{aligned}$$

Hence, we have

$$f(\mathbf{x}) = \langle , \frac{1}{\|\mathbf{z}\|^2}\mathbf{z} \rangle,$$

and so

$$\alpha = \frac{1}{\|\mathbf{z}\|^2}$$

is the scalar we sought. Thus, for all $\mathbf{x} \in \mathcal{H}$,

$$f(\mathbf{x}) = \langle \mathbf{x}, \mathbf{f} \rangle.$$

Uniqueness follows easily from E11.

(b)

$$\|f(\mathbf{x})\| = |\langle \mathbf{x}, \mathbf{f} \rangle| \leq \|\mathbf{x}\|\,\|\mathbf{f}\| = \|\mathbf{f}\|\,\|\mathbf{x}\|,$$

so that

$$\|f\| \leq \|\mathbf{f}\|$$

from the definition in E23.

However,

$$|f(\mathbf{f})| = |\langle \mathbf{f}, \mathbf{f} \rangle| = \|\mathbf{f}\|^2,$$

and so

$$\|f\| = \sup_{\mathbf{x} \in \mathcal{H}} \frac{|f(\mathbf{x})|}{\|\mathbf{x}\|} > \frac{|f(\mathbf{f})|}{\|\mathbf{f}\|} = \frac{\|\mathbf{f}\|^2}{\|\mathbf{f}\|} = \|\mathbf{f}\|,$$

and we are done. □

Appendix F

The Uniform Boundedness Principle and Invertibility of Operators

F1. **Definition.** (1) Let $\mathcal{M} \subset \mathcal{H}$, $\mathcal{H}$ a Hilbert space. We say that $\mathcal{M}$ is a *weakly bounded* set providing there is a real valued (positive) function $\alpha : \mathcal{H} \to \mathrm{R}^+$ such that

$$|\langle \mathbf{x}, \mathbf{m} \rangle| \le \alpha(\mathbf{x})$$

for all $\mathbf{m} \in \mathcal{M}$.

(2) $\mathcal{M}$ is called *bounded* on set $\mathcal{A}$ provided there is a positive constant such that

$$|\langle \mathbf{x}, \mathbf{m} \rangle| \le \beta \|\mathbf{x}\|$$

for all $\mathbf{m} \in \mathcal{M}$, all $\mathbf{x} \in \mathcal{A}$. The name "bounded" is reinforced by the following lemma.

F2. **Lemma.** $|\langle \mathbf{x}, \mathbf{m} \rangle| \le \beta \|\mathbf{x}\|$ *for all* $\mathbf{x} \in \mathcal{H}$, $\mathbf{m} \in \mathcal{M} \Longleftrightarrow \|\mathbf{m}\| \le \beta$ *for all* $\mathbf{m} \in \mathcal{M}$.

Proof. If the first condition holds, simply let $\mathbf{x} = \mathbf{m}$ to obtain $\|\mathbf{m}\|^2 \le \beta \|\mathbf{m}\|$, whence $\|\mathbf{m}\| \le \beta$ (the conclusion is valid even if $\mathbf{m} = 0$).

Conversely, if $\|\mathbf{m}\| \le \beta$ for all $\mathbf{m} \in \mathcal{M}$, then from Cauchy–Schwarz

$$|\langle \mathbf{x}, \mathbf{m} \rangle| \le \|\mathbf{x}\| \, \|\mathbf{m}\| \le \beta \|\mathbf{x}\|. \quad \Box$$

F3. **Lemma.** *If $\mathcal{M}$ is weakly bounded on $\mathcal{H}$, $\mathcal{H}$ finite dimensional, then $\mathcal{M}$ is bounded on $\mathcal{H}$.*

Proof. If $\mathcal{M}$ is weakly bounded by $\alpha : \mathcal{H} \to \mathrm{R}^+$ and $\{\vec{\mathbf{e}}_1, \ldots, \vec{\mathbf{e}}_n\}$ is an orthonormal basis for $\mathcal{H}$, then for $\mathbf{m} \in \mathcal{M}$

$$\begin{aligned} |\langle \mathbf{x}, \mathbf{m} \rangle| &= \left| \sum_{i=1}^{n} \langle \mathbf{x}, \mathbf{e}_i \rangle \langle \mathbf{e}_i, \mathbf{m} \rangle \right| \\ &\le \sqrt{\sum_{i=1}^{n} |\langle \mathbf{x}, \mathbf{e}_i \rangle|^2} \cdot \sqrt{\sum_{i=1}^{n} |\langle \mathbf{e}_i, \mathbf{m} \rangle|^2} \\ &= \|x\| \sqrt{\sum_{i=1}^{n} |\langle \mathbf{e}_i, \mathbf{m} \rangle|^2} \end{aligned}$$

$$\leq \|x\|\sqrt{\sum_{i=1}^{n} \alpha(\mathbf{e}_i)^2}$$
$$\leq \|\mathbf{x}\| \cdot n \cdot \max\{\alpha(e_1), \ldots, \alpha(e_n)\}. \quad \square$$

F4. **Lemma.** *If $\mathcal{M}$ is bounded on M and $M^\perp$, M closed, then $\mathcal{M}$ is bounded on $\mathcal{H}$.*

Proof. Suppose β_1 is the bound for M, β_2 the bound for $M^\perp$. We claim $\beta \triangleq \sqrt{\beta_1^2 + \beta_2^2}$ will work for $\mathcal{H}$. Let $\mathbf{x} = \mathbf{x}_1 + \mathbf{x}_2$, $\mathbf{x}_1 \in M$, and $\mathbf{x}_2 \in M^\perp$. Then

$$\begin{aligned}
|\langle \mathbf{x}, \mathbf{m}\rangle| &= |\langle \mathbf{x}_1, \mathbf{m}\rangle + \langle \mathbf{x}_2, \mathbf{m}\rangle| \\
&\leq |\langle \mathbf{m}_1, \mathbf{m}\rangle| + |\langle \mathbf{x}_2, \mathbf{m}\rangle| \leq \|\mathbf{x}_1\|\beta_1 + \|\mathbf{x}_2\|\,\beta_2 \\
&\leq \sqrt{\|\mathbf{x}_1\|^2 + \|\mathbf{x}_2\|^2} \cdot \sqrt{\beta_1^2 + \beta_2^2} = \beta\|\mathbf{x}\|. \quad \square
\end{aligned}$$

F5. **Corollary.** *If $\mathcal{M}$ is weakly bounded on $\mathcal{H}$, M finite dimensional, $\mathcal{M}$ bounded on $M^\perp$, then $\mathcal{M}$ is bounded on $\mathcal{H}$.*

F6. **Theorem** (Uniform Boundedness Principle). *If $\mathcal{M}$ is weakly bounded on $\mathcal{H}$, then $\mathcal{M}$ is bounded on $\mathcal{H}$.*

Proof (Attributable to D.E. Sareson). If $\mathcal{H}$ is finite dimensional then the theorem is true (Lemma 3). Hence, we suppose $\mathcal{H}$ is infinite dimensional, $\mathcal{M}$ weakly bounded but not bounded on $\mathcal{H}$. This means that for every finite dimensional subspace M of $\mathcal{H}$, $\mathcal{M}$ cannot be bounded on $M^\perp$ (Corollary F5). We argue to a contradiction. Since $\mathcal{M}$ cannot be bounded on $M^\perp$ for any finite dimensional M, this means for any $\gamma > 0$ there is always a unit vector $e \in M^\perp$ and an $m \in \mathcal{M}$ such that $|\langle \mathbf{e}, \mathbf{m}\rangle| > \gamma$ (just take the conclusion $|\langle \mathbf{x}, \mathbf{m}\rangle| > \gamma\|\mathbf{x}\|$, $\mathbf{x} \in M^\perp$ and divide by $\|\mathbf{x}\|$).

Begin with $\gamma = 1$, $M = \{0\}$. Then there exists $\mathbf{e}_1$, $\mathbf{m}_1$ such that

$$|\langle \mathbf{e}_1, \mathbf{m}_1\rangle| > 1.$$

Let $M_1 = \operatorname{span}\{\mathbf{e}_1, \mathbf{m}_1\}$ so $\dim(M_1) \leq 2$; $\mathcal{M}$ cannot be bounded on $M_1^\perp$, so there exists unit vector $\mathbf{e}_2$ and $\mathbf{m}_2 \in \mathcal{M}$ such that

$$|\langle \mathbf{e}_2, \mathbf{m}_2\rangle| > 2(\alpha(\mathbf{e}_1) + 2).$$

[Here, $\gamma = 2(\alpha(\mathbf{e}_1) + 2)$ is our choice (which we are free to make), it is not imposed on us in any way!] Let

$$M_2 = \operatorname{span}\{\mathbf{e}_1, \mathbf{e}_2, \mathbf{m}_1, \mathbf{m}_2\},$$

$\dim(M_2) \leq 4$, so there exists $\mathbf{e}_3 \in M_2^\perp$ and $\mathbf{m}_3 \in \mathcal{M}$ such that

$$|\langle \mathbf{e}_3, \mathbf{m}_3\rangle| \geq 3\left(\alpha(\mathbf{e}_1) + \frac{1}{2}\alpha(\mathbf{e}_2) + 3\right).$$

Induction after n steps yields an element $m_{n+1} \in \mathcal{M}$ and $\mathbf{e}_{n+1} \perp \{\mathbf{e}_1, \ldots \mathbf{e}_n, \mathbf{m}_1, \ldots, \mathbf{m}_n\}$ such that

$$|\langle \mathbf{e}_{n+1}, \mathbf{m}_{n+1} \rangle| \geq (n+1) \left(\sum_{i=1}^{n} \frac{1}{i} \alpha(\mathbf{e}_i) + n + 1 \right). \qquad (*)$$

Now let

$$\mathbf{x} \triangleq \sum_{i=1}^{\infty} \frac{1}{i} \mathbf{e}_i,$$

which converges since the partial sums from a Cauchy sequence in $\mathcal{H}$.

Now since $\mathbf{e}_i \perp \mathbf{m}_{n+1}$ for $i > n+1$, we have

$$\begin{aligned} |\langle \mathbf{x}, \mathbf{m}_{n+1} \rangle| &= \left| \langle \sum_{i=1}^{\infty} \frac{1}{i} \mathbf{e}_i, \mathbf{m}_{n+1} \rangle \right| \\ &= \left| \langle \sum_{i=1}^{n+1} \frac{1}{i} \mathbf{e}_i, \mathbf{m}_{n+1} \rangle \right| \\ &= \left| \sum_{i=1}^{n} \frac{1}{i} \langle \mathbf{e}_i, \mathbf{m}_{n+1} \rangle + \frac{1}{n+1} \langle \mathbf{e}_{n+1}, \mathbf{m}_{n+1} \rangle \right|. \end{aligned}$$

Using the inequality $|\mathbf{a} + \mathbf{b}| \geq -|\mathbf{a}| + |\mathbf{b}|$, relation $(*)$, and $|\langle \mathbf{e}_i, \mathbf{m}_{n+1} \rangle| \geq \alpha(\mathbf{e}_i)$, we have

$$\begin{aligned} |\langle \mathbf{x}, \mathbf{m}_{n+1} \rangle| &\geq -\sum_{i=1}^{n} \frac{1}{i} \alpha(\mathbf{e}_i) + \frac{1}{n+1} (n+1) \left(\sum_{i=1}^{n} \frac{1}{i} \alpha(\mathbf{e}_i) + n + 1 \right) \\ &= n+1, \end{aligned}$$

so $\mathcal{M}$ is not weakly bounded, a contradiction. □

F7. **Theorem.** *If $A \in \mathcal{B}(\mathcal{H}_1, \mathcal{H}_2)$ and α is a positive real number such that $\|A\mathbf{x}\| \geq \alpha \|\mathbf{x}\|$ for all $\mathbf{x}$, then $\mathcal{R}(A) = \overline{\mathcal{R}(A)}$.*

Proof. If $\mathbf{y}_n = A\mathbf{x}_n$ and $\mathbf{y}_n \to \mathbf{y}$, then $\{\mathbf{y}_n\}$ is Cauchy and by hypothesis

$$\|\mathbf{y}_n - \mathbf{y}_m\| = \|A\mathbf{x}_n - A\mathbf{x}_m\| = \|A(\mathbf{x}_n - \mathbf{x}_m)\| \geq \alpha \|\mathbf{x}_n - \mathbf{x}_m\|,$$

so $\{\mathbf{x}_n\}$ is Cauchy. Therefore, $\mathbf{x}_n \to \mathbf{x}$ for some $\mathbf{x}$. Since A is bounded (continuous), we have $y = \lim_n A(\mathbf{x}_n) = A(\lim_n \mathbf{x}_n) = A\mathbf{x}$, so $\mathbf{y} \in \mathcal{R}(A)$. It follows that $\mathcal{R}(A)$ is closed. □

F8. **Theorem.** *An operator A is invertible (bounded) if and only if its range is dense in $\mathcal{H}_2$, that is, $\overline{\mathcal{H}(A)} = \mathcal{H}_2$, and there is a positive real number α such that $\|A\mathbf{x}\| \geq \alpha \|\mathbf{x}\|$ for all $\mathbf{x}$.*

Proof. If A is invertible and if $\mathbf{y} \in \mathcal{H}_2$ let $\mathbf{x} \triangleq A^{-1}\mathbf{y}$. Since $\mathbf{y} = A\mathbf{x}$, it follows that $\mathcal{R}(A) = \mathcal{H}_2$, so certainly $\overline{\mathcal{R}(A)} = \mathcal{H}_2$ trivially. Moreover, $\|\mathbf{x}\| = \|A^{-1}A\mathbf{x}\| \leq \|A^{-1}\| \cdot \|A\mathbf{x}\|$, so $\alpha = 1/\|A^{-1}\|$ will work.

Conversely, suppose $\overline{\mathcal{R}(A)} = \mathcal{R}_2$ and that $\|A\mathbf{x}\| \geq \alpha\|\mathbf{x}\|$ all $x \in \mathcal{H}_1$. Then, by Theorem F7, $\mathcal{R}(A) = \overline{\mathcal{R}(A)} = \mathcal{H}_2$. If $A\mathbf{x}_1 = A\mathbf{x}_2$, that is, $A\mathbf{x}_1 - A\mathbf{x}_2 = 0$, we have $0 = \|A(\mathbf{x}_1 - \mathbf{x}_2)\| \geq \alpha\|\mathbf{x}_1 - \mathbf{x}_2\|$, so $\mathbf{x}_1 = \mathbf{x}_2$, that is, A is one to one. Hence, there exists $B : \mathcal{H}_2 \to \mathcal{H}_1$ such that $\mathbf{x} = B\mathbf{y} \Longleftrightarrow \mathbf{y} = A\mathbf{x}$. It is easily verified that B is linear. Finally $\|\mathbf{y}\| = \|A\mathbf{x}\| \geq \alpha\|\mathbf{x}\| = \alpha\|B\mathbf{y}\| \Rightarrow B$ is bounded. Hence, $B = A^{-1}$. □

F9. **Theorem.** *If $A \in \mathcal{B}(\mathcal{H}_1, \mathcal{H}_2)$ A is one to one and onto, then A^{-1} exists, that is, $A^{-1} \in \mathcal{B}(\mathcal{H}_2, \mathcal{H}_1)$.*

Proof. By 4.1.7, it suffices to show A^* is (bounded) invertible. Since A is one to one, $\ker(A) = \{0\}$, so $\overline{\mathcal{R}(A^*)} = (\ker A)^\perp = \mathcal{H}_1$, that is, $\mathcal{R}(A^*)$ is dense in $\mathcal{H}_1$. By Theorem F8, it suffices to show there exists α such that $\|A^*\mathbf{y}\| \geq \alpha\mathbf{y}$ for all $\mathbf{y} \in \mathcal{H}_2$. Since $\mathcal{R}(A) = \mathcal{H}_2$, $\ker(A^*) = \mathcal{R}(A)^\perp = \{0\}$. From this it follows that it is sufficient to prove that if $\|A^*\mathbf{y}\| = 1$, then $\|\mathbf{y}\| \leq 1/\delta$ for some δ. [If $\mathbf{y} \neq 0$, $\mathbf{y} \in \ker A^*$, then $\|A^*\mathbf{y}\| \not\geq \|\mathbf{y}\|$ for any $\mathbf{y}$!] Let $\mathcal{M} = \{\mathbf{y} \mid \|A^*\mathbf{y}\| = 1\}$. We want to show $\mathcal{M}$ bounded on all of $\mathcal{H}$. By Theorem F6, it suffices to show $\mathcal{M}$ weakly bounded. But that is easy. If $\mathbf{y} \in \mathcal{M}$ and $\mathbf{z} \in \mathcal{H}_2$, find $\mathbf{x} \in \mathcal{H}_1$ such that $\mathbf{z} = A\mathbf{x}$ (possible since A is onto). Then

$$\begin{aligned} |\langle \mathbf{z}, \mathbf{y}\rangle| = |\langle A\mathbf{x}, \mathbf{y}\rangle| &= |\langle \mathbf{x}, A^*\mathbf{y}\rangle| \\ &\leq \|\mathbf{x}\|\,\|A^*\mathbf{y}\| = \|\mathbf{x}\| = \alpha(\mathbf{z}). \end{aligned}$$

[Since A is one to one and onto, A^{-1} exists as a function, so we are defining $\alpha(\mathbf{z}) = \|A^{-1}(\mathbf{z})\|$]. □

Appendix G

The Spectral Theorem for Self-Adjoint Operators on C^n or R^n

If $T : C^n \to C^n$ is any self-adjoint operator (see Chapter 4) and $\{\mathbf{f}_1, \ldots, \mathbf{f}_n\}$ is an orthonormal basis for C^n, then the matrix representative with respect to this basis, call it A, is a self-adjoint matrix. If one chooses a different orthonormal basis for C^n, for example, $\{\mathbf{g}_1, \mathbf{g}_2, \ldots, \mathbf{g}_n\}$, and if

$$\mathbf{g}_i = \sum_{j=1}^{n} \rho_{ji}\mathbf{e}_j; \qquad i = 1, 2, \ldots, n,$$

then if B is the matrix representative of T with respect to $\{\mathbf{g}_1, \ldots, \mathbf{g}_n\}$, A and B are related by

$$B = P^{-1}AP,$$

where

$$P = \begin{bmatrix} \rho_{11} & \cdots & \rho_{1n} \\ \vdots & & \vdots \\ \rho_{n1} & \cdots & \rho_{nn} \end{bmatrix}.$$

Moreover, one can easily show that P is orthogonal, so the above relation becomes

$$B = P^T AP.$$

In this appendix, we address the question, "does there exist an orthonormal basis $\mathbf{g}_1, \ldots, \mathbf{g}_n$ such that the matrix representative of T with respect to this basis is a diagonal matrix?" From the above discussion, this is equivalent to asking if one can always find an orthogonal matrix P such that $P^T AP$ is diagonal. The answer turns out to be yes, as we now demonstrate.

G1. **Definition.** Let $T : C^n \to C^n$ be linear.

(a) A number λ is called an *eigenvalue* of T if and only if there is an $\mathbf{x} \neq \mathbf{0}$ such that

$$T(\mathbf{x}) = \lambda\mathbf{x}.$$

Any such x is called an *eigenvector* with respect to λ.

(b) The set of all eigenvalues for T is called the *spectrum* of T and is denoted $\text{sp}(T)$.

(c) If $\lambda \in \text{sp}(T)$, we define

$$\mathcal{M}_\lambda = \{\mathbf{x} \,|\, T(\mathbf{x}) = \lambda\mathbf{x}\};$$

$\mathcal{M}_\lambda$ is called the *eigenspace* corresponding to λ.

G2. **Lemma.** *$\mathcal{M}_\lambda$ is a subspace of C^n.*

G3. **Theorem.** *Let A be the standard matrix representative of T. Then* $\lambda \in \mathrm{sp}(T) \iff \det(A - \lambda I) = 0$.

Proof. $\lambda \in \mathrm{sp}(T) \iff T(\mathbf{x}) = \lambda\mathbf{x}$ for some $\mathbf{x} \neq \mathbf{0} \iff A\mathbf{x} = \lambda\mathbf{x}$ for some $\mathbf{x} \neq \mathbf{0} \iff (A - \lambda I)\mathbf{x} = \mathbf{0}$ for some $\mathbf{x} \neq \mathbf{0} \iff A - \lambda I$ is singular $\iff \det(A - \lambda I) = 0$. □

G4. **Theorem.** *If A is the standard matrix for T, then A^* (transpose conjugate) is the standard matrix for T^*.*

Proof. Let $A = (\alpha_{ij})$ be the standard matrix for T and $B = (\beta_{ij})$ be the standard matrix for T^*. Then, by definition

$$T(\mathbf{e}_i) = \sum_{k=1}^{m} \alpha_{ki}\mathbf{e}_k$$

$$T^*(\mathbf{e}_j) = \sum_{s=1}^{n} \beta_{sj}\mathbf{e}_s.$$

Then

$$\begin{aligned} \langle T(\mathbf{e}_i), \mathbf{e}_j \rangle &= \langle \sum_{k=1}^{m} \alpha_{ki}\mathbf{e}_k, \mathbf{e}_j \rangle \\ &= \sum_{k=1}^{m} \alpha_{ki} \langle \mathbf{e}_k, \mathbf{e}_j \rangle \\ &= \alpha_{ji}. \end{aligned}$$

Also,

$$\begin{aligned} \langle \mathbf{e}_i, T^*(\mathbf{e}_j) \rangle &= \langle \mathbf{e}_i, \sum_{s=1}^{n} \beta_{sj}\mathbf{e}_s \rangle \\ &= \sum_{s=1}^{n} \beta_{sj}^* \langle \mathbf{e}_i, \mathbf{e}_s \rangle \\ &= \beta_{ij}^*. \end{aligned}$$

By the definition of adjoint, the left-hand expressions are equal, hence $\alpha_{ji} = \beta_{ij}^*$. □

G5. **Theorem.** *If $T = T^*$, then* $\mathrm{sp}(T) \subset \mathrm{R}$.

Proof. If $\lambda \in \mathrm{sp}(T)$, then there is some $\mathbf{x} \neq \mathbf{0}$ such that $T(\mathbf{x}) = \lambda\mathbf{x}$. By (possibly) normalizing, we can suppose that $||\mathbf{x}|| = 1$. Then

$$\lambda = \lambda||\mathbf{x}||^2$$

$$
\begin{aligned}
&= \lambda\langle \mathbf{x}, \mathbf{x}\rangle \\
&= \langle \lambda\mathbf{x}, \mathbf{x}\rangle \\
&= \langle T\mathbf{x}, \mathbf{x}\rangle \\
&= \langle x, T^*\mathbf{x}\rangle \\
&= \langle x, T\mathbf{x}\rangle \\
&= \langle \mathbf{x}, \lambda\mathbf{x}\rangle \\
&= \lambda^*\langle \mathbf{x}, \mathbf{x}\rangle \\
&= \lambda^*.
\end{aligned}
$$

Note that if $A = A^*$, then A is a real symmetric matrix. If we build $W : \mathbf{C}^n \to \mathbf{C}^n$ using matrix A, then $W = W^*$. By Theorem G5, we thus have that the eigenvalues of W, hence A, are real. □

G6. **Corollary** (to Theorem G5). *If $T : \mathbf{R}^n \to \mathbf{R}^n$ is self-adjoint, then* $\mathrm{Sp}(T) \neq \emptyset$.

Proof. Let A be the standard matrix for T. Using A, build $W : \mathbf{C}^n \to \mathbf{C}^n$ as in the above remark. Certainly by the fundamental theorem of algebra, $\mathrm{sp}(W) \neq \emptyset$, since $\det(W - \lambda I) = 0$ (the characteristic equation) has n complex roots. But by Theorem G5, $\mathrm{sp}(W) \subset \mathbf{R}$, and so since

$$\det(W - \lambda I) = \det(A - \lambda I) = \det(T - \lambda I),$$

we have $\mathrm{sp}(T) \neq \emptyset$. □

G7. **Convention.** In the following, $\mathbf{F}$ represents either $\mathbf{C}$ or $\mathbf{R}$.

G8. **Theorem.** *Let $T : \mathbf{F}^n \to \mathbf{F}^n$ be self-adjoint. If a subspace $\mathcal{M}$ is stable under T, that is, $T(M) \subset \mathcal{M}$, then $\mathcal{M}^{\perp}$ is also stable under T.*

Proof. Let $\mathbf{y} \in T(\mathcal{M}^{\perp})$. Then $\mathbf{y} = T(\mathbf{x})$ for some $\mathbf{x} \in \mathcal{M}^{\perp}$. Let $\mathbf{z} \in \mathcal{M}$ be arbitrary. Then

$$\langle \mathbf{y}, \mathbf{z}\rangle = \langle T(\mathbf{x}), \mathbf{z}\rangle = \langle \mathbf{x}, T(\mathbf{z})\rangle.$$

But by hypothesis, $T(\mathbf{z}) \in \mathcal{M}$ so $\langle \mathbf{x}, T(\mathbf{z})\rangle = 0$. Thus, $\langle \mathbf{y}, \mathbf{z}\rangle = 0$ for all $\mathbf{z} \in \mathcal{M}$; hence, $\mathbf{y} \in \mathcal{M}^{\perp}$. □

G9. **Theorem.** *Let $T : \mathbf{F}^n \to \mathbf{F}^n$ be self-ajoint. If $\lambda_1, \lambda_2 \in \mathrm{sp}(T)$, $\lambda_1 \neq \lambda_2$, then $\mathcal{M}_{\lambda_1} \perp \mathcal{M}_{\lambda_2}$.*

Proof. Let $\mathbf{x} \in \mathcal{M}_{\lambda_1}$, $\mathbf{y} \in \mathcal{M}_{\lambda_2}$. Then

$$T(\mathbf{x}) = \lambda_1\mathbf{x}, \quad T(\mathbf{y}) = \lambda_2\mathbf{y}.$$

It then follows that

$$
\begin{aligned}
\lambda_1\langle \mathbf{x}, \mathbf{y}\rangle &= \langle \lambda_1\mathbf{x}, \mathbf{y}\rangle \\
&= \langle T(\mathbf{x}), \mathbf{y}\rangle
\end{aligned}
$$

$$\begin{aligned} &= \langle \mathbf{x}, T^*(\mathbf{y}) \rangle \\ &= \mathbf{x}, T(\mathbf{y}) \\ &= \langle \mathbf{x}, \lambda_2 \mathbf{y} \rangle \\ &= \lambda_2^* \langle \mathbf{x}, \mathbf{y} \rangle \\ &= \lambda_2 \langle \mathbf{x}, \mathbf{y} \rangle. \end{aligned}$$

Hence, $(\lambda_1 - \lambda_2)\langle \mathbf{x}, \mathbf{y} \rangle = 0$, and since $\lambda_1 \neq \lambda_2$, it must be the case that $\langle \mathbf{x}, \mathbf{y} \rangle = 0$. □

G10. **Theorem.** *Let $\mathcal{E}$ denote the collection of all eigenvectors of T, $T : \mathrm{F}^n \to \mathrm{F}^n$ self-adjoint. Then if we denote*

$$\mathcal{M} \triangleq \mathrm{span}(\mathcal{E})$$

we have that $\mathcal{M} = \mathrm{F}^n$.

Proof. Suppose that $\mathcal{M} \neq \mathrm{F}^n$, then $\mathcal{M}$ is a proper subset of F^n and so $\mathcal{M}^\perp \neq \{0\}$. Define

$$W : \mathcal{M}^\perp \to \mathcal{M}^\perp$$

via

$$W(\mathbf{x}) = T(\mathbf{x}) \quad \text{for} \quad \mathbf{x} \in \mathcal{M}^\perp.$$

By Theorem G8, W is well defined. Moreover, if $\mathbf{x}, \mathbf{y} \in \mathcal{M}^\perp$,

$$\langle W\mathbf{x}, \mathbf{y} \rangle = \langle T\mathbf{x}, \mathbf{y} \rangle = \langle \mathbf{x}, T^*\mathbf{y} \rangle = \langle \mathbf{x}, T\mathbf{y} \rangle = \langle \mathbf{x}, W\mathbf{y} \rangle,$$

so W is self-adjoint. If $\mathrm{F} = \mathrm{C}$, then obviously $\mathrm{sp}(W) \neq \emptyset$. But if $\mathrm{F} = \mathrm{R}$, $\mathrm{sp}(W) \neq \emptyset$ by Theorem G6. Hence, either way, W has a nonzero eigenvector $\mathbf{z}$, that is, $W(\mathbf{z}) = \lambda \mathbf{z}$ for some $\lambda \in \mathrm{F}$. But this implies that $T(\mathbf{z}) = \lambda \mathbf{z}$, whence $\mathbf{z} \in \mathcal{M}$. Thus, $\mathbf{z} \in \mathcal{M} \cap \mathcal{M}^\perp$ so $\mathbf{z} = \mathbf{0}$, a contradiction. □

G11. **Corollary.** *If $T : \mathrm{F}^n \to \mathrm{F}^n$ is self-adjoint, then there is an orthonormal basis $\{\mathbf{f}_1, \ldots, \mathbf{f}_n\}$ for F^n such that each f_i is an eigenvector of T.*

Proof. Let $\mathrm{sp}(T) = \{\lambda_1, \lambda_2, \ldots, \lambda_p\}$. Choose an orthonormal basis for each $\mathcal{M}_{\lambda_i}$ and let $\{\mathbf{f}_1, \mathbf{f}_2, \ldots, \mathbf{f}_m\}$ be the collection of all of these vectors. If $\mathbf{f}_i$ and $\mathbf{f}_j$ are in the same eigenspace, they are orthogonal by construction, and if they are in different eigenspaces, they are orthogonal by Theorem G9. Hence, $m \leq n$ since the set is independent. However, $\{\mathbf{f}_1, \ldots, \mathbf{f}_m\}$ is also a spanning set. Because, if $\mathbf{x} \in \mathrm{F}^n$, then by Theorem G10, $\mathbf{x}$ can be written as

$$\mathbf{x} = \sum_{j=1}^{\infty} \alpha_j \mathbf{x}_j,$$

where $\mathbf{x}_j \in \mathcal{M}_{\lambda_j}$. But each $\mathbf{x}_j$ is a linear combination of the $\mathbf{f}_i$'s that span $\mathcal{M}_{\lambda_j}$. Hence, $m \geq n$ and we are done. □

Corollary G11 is the spectral theorem for self-adjoint operators on finite dimensional spaces. Combining this with the discussion at the beginning of this appendix, we have the following result for real symmetric matrices.

G12. **Corollary.** *Every real symmetric matrix A is orthogonally similar to a diagonal matrix D, that is, there exists an orthogonal matrix P and a diagonal matrix D such that*

$$D = P^T AP.$$

Note that the diagonal entries of D are the eigenvalues of A.

Bibliography

1. Box, G.E.P., and Jenkins, G.M., *Time Series Analysis, Forecasting and Control*, Holden-Day, San Francisco, California, 1970.

2. Catlin, D.E., The independence of foreward and backward estimation errors in the two-filter form of the fixed interval Kalman smoother, *IEEE Transactions on Automatic Control* **AC-25**, 6, December 1980, 1111–1115.

3. Chen, H.-F., *Recursive Estimation and Control for Stochastic Systems*, Wiley, New York, 1985, 254–257.

4. Ellis, R., *Entropy, Large Deviations, and Statistical Mechanics*, *Grundlehhren der mathematischen Wissenschaften*, Vol. 271, Springer-Verlag, Berlin and New York, 1985.

5. Feinstein, A., *Foundations of Information Theory*, McGraw-Hill, New York, 1958.

6. Feller, W., *An Introduction to Probability Theory and Its Applications*, Vol. I, 2nd ed., Wiley, New York, 1957, 115.

7. Foulis, D.J., Randall, C.H., Operational statistics I. Basic concepts, *J. Math. Phys.* **13**, 1667–1675, 1972.

8. Fraser, D.C., *A New Technique for the Optimal Smoothing of Data*, Ph.D. dissertation, Massachusetts Institute of Technology, Cambridge, Massachusetts, January 1974.

9. Fraser, D.C., and Potter, J.E., The optimum linear smoother as a combination of two optimum linear filters, *IEEE Transactions on Automatic Control* **AC-14**, 387–390, August 1969.

10. Gelb, A., *Applied Optimal Estimation*, MIT Press, Cambridge, Massachusetts, 1974.

11. Geldbaum, B.R., and Olmsted, J.M.H., *Counterexamples in Analysis*, Holden-Day, San Francisco, California, 1964, 98.

12. Jaynes, E.T., Information theory and statistical mechanics I., *Phys. Rev.* **106**, 4, May 1957, 620–630.

13. Jaynes, E.T., Information theory and statistical mechanics II., *Phys. Rev.* **108**, 2, October 1957, 171–190.

14. Kalman, R.E., A new approach to linear filtering and prediction problems, *Journal of Basic Engineering (ASME)*, **82D**, March 1960, 35–45.

15. Kalman, R.E., and Bucy, R., New results in linear filtering and prediction, *Journal of Basic Engineering (ASME)*, **83D**, 1961, 366–368.

16. Larimore, W.E., System identification, reduced-order filtering, and modeling via canonical variate analysis, *Proceedings of the 1983 American Control Conference*, San Francisco, California, June 22–24, 445–451.

17. Liebelt, P.B., *An Introduction to Optimal Estimation*, Addison-Wesley, Reading, Massachusetts, 1967.

18. Luenberger, D.G., *Optimization by Vector Space Methods*, Wiley, New York, 1969, 84–91.

19. Munroe, M.E., *Measure and Integration*, Addison-Wesley, Reading, Massachusetts, 1959, 85–98 and 121–124.

20. Randall, C.H., and Foulis, D.J., Operational Statistics II. Manuals of operations and their logics, *J. Math. Phys.* **14**, 1472–1480, 1973.

21. Rao, C.R., Inference from linear models with fixed effects: recent results and some problems, *Statistics, An Appraisal*, H.A. David and H.T. David, Eds., Iowa State Univ. Press, Ames, 1984, 345–369.

22. Rauch, H.E., Tung, F., and Streibel, C.T., Maximum likelihood estimators of linear dynamic systems, *AIAA Journal* **3**, 8, August 1965, 1445–1450.

23. Royden, H.L., *Real Analysis*, Macmillan, New York, 1968, 259.

24. Schweppe, F.C., *Uncertain Dynamical Systems*, Prentice-Hall, Englewood Cliffs, New Jersey, 1973, 100–104.

25. Shannon, C.E., A mathematical theory of communication, *Bell Systems Technical Journal* **27**, July 1948, 379–423.

26. White, J.V., Stochastic state-space models from empirical data, *Proceedings ICASSP 83* **1**, *IEEE Intern. Conf. Acoustics, Speech, and Signal Processing, Paper No. 6.3*, April 1983, 243–246.

27. Wilkinson, J.H., and Reinch, C., *Linear Algebra*, Springer-Verlag, Berlin and New York, 1971.

Index

Q

R

Applied Mathematical Sciences

cont. from page ii

55. *Yosida:* Operational Calculus: A Theory of Hyperfunctions.
56. *Chang/Howes:* Nonlinear Singular Perturbation Phenomena: Theory and Applications.
57. *Reinhardt:* Analysis of Approximation Methods for Differential and Integral Equations.
58. *Dwoyer/Hussaini/Voigt (eds.):* Theoretical Approaches to Turbulence.
59. *Sanders/Verhulst:* Averaging Methods in Nonlinear Dynamical Systems.
60. *Ghil/Childress:* Topics in Geophysical Dynamics: Atmospheric Dynamics, Dynamo Theory and Climate Dynamics.
61. *Sattinger/Weaver:* Lie Groups and Algebras with Applications to Physics, Geometry, and Mechanics.
62. *LaSalle:* The Stability and Control of Discrete Processes.
63. *Grasman:* Asymptotic Methods of Relaxation Oscillations and Applications.
64. *Hsu:* Cell-to-Cell Mapping: A Method of Global Analysis for Nonlinear Systems.
65. *Rand/Armbruster:* Perturbation Methods, Bifurcation Theory and Computer Algebra.
66. *Hlaváček/Haslinger/Nečas/Lovíšek:* Solution of Variational Inequalities in Mechanics.
67. *Cercignani:* The Boltzmann Equation and Its Applications.
68. *Temam:* Infinite Dimensional Dynamical System in Mechanics and Physics.
69. *Golubitsky/Stewart/Schaeffer:* Singularities and Groups in Bifurcation Theory, Vol. II.
70. *Constantin/Foias/Nicolaenko/Temam:* Integral Manifolds and Inertial Manifolds for Dissipative Partial Differential Equations.
71. *Catlin:* Estimation, Control, and the Discrete Kalman Filter.
72. *Lochak/Meunier:* Multiphase Averaging for Classical Systems.
73. *Wiggins:* Global Bifurcations and Chaos.
74. *Mawhin/Willem:* Critical Point Theory and Hamiltonian Systems.
75. *Abraham/Marsden/Ratiu:* Manifolds, Tensor Analysis, and Applications, 2nd ed.
76. *Lagerstrom:* Matched Asymptotic Expansion: Ideas and Techniques.